Master Handbook
of Acoustics

About the Authors

F. Alton Everest was a leading acoustics consultant. He was cofounder and director of the Science Film Production division of the Moody Institute of Science, and was also section chief of the Subsea Sound Research section of the University of California.

Ken C. Pohlmann is well known as an audio educator, consultant, and author. He is professor emeritus at the University of Miami in Coral Gables, consultant for many audio manufacturers and car makers, and author of numerous articles and books including *Principles of Digital Audio*.

Contributions are included from Peter D'Antonio, Geoff Goacher and Doug Plum.

Master Handbook of Acoustics

F. Alton Everest

Ken C. Pohlmann

Fifth Edition

New York Chicago San Francisco
Lisbon London Madrid Mexico City
Milan New Delhi San Juan
Seoul Singapore Sydney Toronto

The McGraw·Hill Companies

Library of Congress Cataloging-in-Publication Data

Everest, F. Alton (Frederick Alton), date.
   Master handbook of acoustics / F. Alton Everest.—5th ed. / Ken Pohlmann.
      p.   cm.
   ISBN 978-0-07-160332-4 (alk. paper)
   1. Sound.   2. Electronics.   I. Pohlmann, Ken C.   II. Title.
   QC225.15.E93   2009
   620.2—dc22   2009019102

McGraw-Hill books are available at special quantity discounts to use as premiums and sales promotions, or for use in corporate training programs. To contact a representative please e-mail us at bulksales@mcgraw-hill.com.

Master Handbook of Acoustics, Fifth Edition

1 2 3 4 5 6 7 8 9 0   DOC/DOC   0 1 4 3 2 1 0 9

ISBN 978-0-07-160332-4
MHID 0-07-160332-8

The pages within this book were printed on acid-free paper.

All trademarks are trademarks of their respective owners. Rather than put a trademark symbol after every occurrence of a trademarked name, we use names in an editorial fashion only, and to the benefit of the trademark owner, with no intention of infringement of the trademark. Where such designations appear in this book, they have been printed with initial caps. Following are listed trademark owners (and trademarks): RPG Diffusor Systems, Inc. (Abffusor™, Korner Killer™, Modffusor™, RFZ™, RPG™, and Triffusor™ and Acoustic Sciences Corporation (Snap Trap™ and Tube Trap™).

Sponsoring Editor Judy Bass	**Proofreader** Surendra N. Shivam
Acquisitions Coordinator Michael Mulcahy	**Indexer** Robert Swanson
Editorial Supervisor David E. Fogarty	**Production Supervisor** Richard C. Ruzycka
Project Manager Ekta Dixit	**Composition** International Typesetting and Composition
Copy Editor Megha Roy Choudhary	**Art Director, Cover** Jeff Weeks

Dedicated to the memory of F. Alton Everest

Introduction

What makes a book a classic? More than anything, a classic book is one that we know and trust, a well-used book with dog-eared covers and underlined passages. Without question, Mr. Everest's *The Master Handbook of Acoustics* qualifies as a classic. It was extremely well received when the first edition appeared in 1981, and strong reader demand prompted the preparation of new editions, through the fourth edition in 2001. In fact, the title was the best-selling book on acoustics for over 20 years. The acoustical engineering community grieved when Mr. Everest passed away in 2005 at the age of 95. He was a giant in the field of acoustical engineering, and a great example of the high caliber of engineers of his generation. He will be missed.

I was honored when McGraw-Hill asked me to prepare a fifth edition. I had used the *Handbook* for 25 years, and was well familiar with its value as a teaching text and reference book. Fully cognizant of the challenges of tackling such a project, I agreed to lend a hand. Some readers who are familiar with my book, *Principles of Digital Audio*, may be surprised to learn that my passion for digital technology is equaled by my enthusiasm for acoustics. I taught courses in architectural acoustics (in addition to classes in digital audio) for 30 years at the University of Miami, where I directed the Music Engineering Technology program. Throughout that time, I also consulted on many acoustics projects, ranging from recording studio to listening room design, from church acoustics to community noise intrusion. As with many practitioners in the field, it was important for me to understand the fundamentals of acoustical properties, and also to stay current with the practical applications and solutions to today's acoustical problems. This essential balance was the guiding principle of Mr. Everest's previous editions of this book, and I have continued to seek that same equilibrium.

A personal comment, particularly to some newbies to the field of acoustics. The question might arise, "Why is it important to study acoustics?" One reason, among many, is that you will be joining in, and hopefully contributing to, a noble and impressive scientific evolution. For literally thousands of years, acoustics and its elegant complexities have encouraged some of the world's greatest scientists and engineers to study its mysteries. The development of this art and science has profoundly influenced entire cultures and individual lives. But in today's binary world, is acoustics still important? Consider this: our eyes close when we sleep and we cannot see in the dark; someone can sneak up on us from behind. But from birth to death, awake or asleep, in light and in dark, our ears are always sensitive to our world all around us. Whether we are hearing sounds that give us pleasure, or sounds that alert us to danger, whether they are sounds of nature, or sounds of technology, the properties of acoustics and the way that architectural spaces affect those sounds is woven into every moment of our lives. Is acoustics important? You bet it is.

Ken C. Pohlmann
Durango, Colorado

CHAPTER 1

Fundamentals of Sound

Sound can be viewed as a wave motion in air or other elastic media. In this case, sound is a stimulus. Sound can also be viewed as an excitation of the hearing mechanism that results in the perception of sound. In this case, sound is a sensation. These two views of sound are familiar to those interested in audio and music. The type of problem at hand dictates our approach to sound. If we are interested in the disturbance in air created by a loudspeaker, it is a problem in physics. If we are interested in how that disturbance sounds to a person near the loudspeaker, psychoacoustical methods must be used. Because this book addresses acoustics in relation to people, both aspects of sound will be treated.

Sound is characterized by a number of basic phenomena. For example, frequency is an objective property of sound; it specifies the number of waveform repetitions per unit of time (usually one second). Frequency can be readily measured on an oscilloscope or a frequency counter. On the other hand, pitch is a subjective property of sound. Perceptually, the ear hears different pitches for soft and loud 100-Hz tones. As intensity increases, the pitch of a low-frequency tone goes down, while the pitch of a high-frequency tone goes up. Fletcher found that playing pure tones of 168 and 318 Hz at a modest level produces a very discordant sound. At a high intensity, however, the ear hears the pure tones in the 150- to 300-Hz octave relationship as a pleasant sound. We cannot equate frequency and pitch, but they are analogous.

A similar duality exists between intensity and loudness. The relationship between the two is not linear. Similarly, the relationship between waveform (or spectrum) and perceived quality (or timbre) is complicated by the function of the hearing mechanism. A complex waveform can be described in terms of a fundamental and a series of harmonics of various amplitudes and phases. But perception of timbre is complicated by the frequency-pitch interaction as well as other factors.

The interaction between the physical properties of sound, and our perception of them, poses delicate and complex issues. It is this complexity in audio and acoustics that creates such interesting problems. On one hand, the design of a loudspeaker or a concert hall should be a straightforward and objective engineering process. But in practice, that objective expertise must be carefully tempered with purely subjective wisdom. As has often been pointed out, loudspeakers are not designed to play sine waves into calibrated microphones placed in anechoic chambers. Instead, they are designed to play music in our listening rooms. In other words, the study of audio and acoustics involves both art and science.

The Sine Wave

The weight (mass) on the spring shown in Fig. 1-1 is a vibrating system. If the weight is pulled down to the –5 mark and released, the spring pulls the weight back toward 0. However, the weight will not stop at 0; its inertia will carry it beyond 0 almost to +5. The weight will continue to vibrate, or oscillate, at an amplitude that will slowly decrease due to frictional losses in the spring and the air.

In the arrangement of a mass and spring, vibration or oscillation is possible because of the elasticity of the spring and the inertia of the weight. Elasticity and inertia are two things all media must possess to be capable of conducting sound.

The weight in Fig. 1-1 moves in what is called simple harmonic motion. The piston in an automobile engine is connected to the crankshaft by a connecting rod. The rotation of the crankshaft and the up-and-down motion of the pistons beautifully illustrate the relationship between rotary motion and linear simple harmonic motion. The piston position plotted against time produces a sine wave. It is a basic type of mechanical motion, and it yields an equally basic waveshape in sound and electronics.

If a pen is fastened to the pointer of Fig. 1-2, and a strip of paper is moved past it at a uniform speed, the resulting trace is a sine wave. The sine wave is a pure waveform closely related to simple harmonic motion.

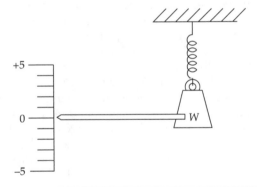

FIGURE 1-1   A weight on a spring vibrates at its natural frequency because of the elasticity of the spring and the inertia of the weight.

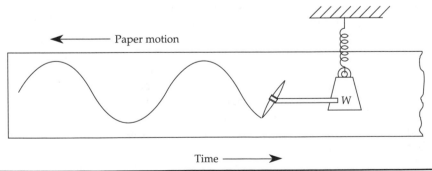

FIGURE 1-2   A pen fastened to the vibrating weight traces a sine wave on a paper strip moving at uniform speed. This shows the basic relationship between simple harmonic motion and the sine wave.

Sound in Media

If an air particle is displaced from its original position, elastic forces of the air tend to restore it to its original position. Because of the inertia of the particle, it overshoots the resting position, bringing into play elastic forces in the opposite direction, and so on.

Sound is readily conducted in gases, liquids, and solids such as air, water, steel, concrete, and so on, which are all elastic media. Imagine a friend stationed a distance away, who strikes a railroad rail with a rock. You will hear two sounds, one sound coming through the rail and one through the air. The sound through the rail arrives first because the speed of sound in the dense steel is faster than in tenuous air. Similarly, sounds can be detected after traveling thousands of miles through the ocean.

Without a medium, sound cannot be propagated. In the laboratory, an electric buzzer is suspended in a heavy glass bell jar. As the button is pushed, the sound of the buzzer is readily heard through the glass. As the air is pumped out of the bell jar, the sound becomes fainter and fainter until it is no longer audible. The sound-conducting medium, air, has been removed between the source and the ear. Because air is such a common agent for the conduction of sound, it is easy to forget that other gases as well as solids and liquids are also conductors of sound. Outer space is an almost perfect vacuum; no sound can be conducted except in the tiny island of atmosphere within a spaceship or a spacesuit.

Particle Motion

Waves created by the wind travel across a field of grain, yet the individual stalks remain firmly rooted as the wave travels on. In a similar manner, particles of air propagating a sound wave do not move far from their undisplaced positions, as shown in Fig. 1-3. The disturbance travels on, but the propagating particles move only in localized regions

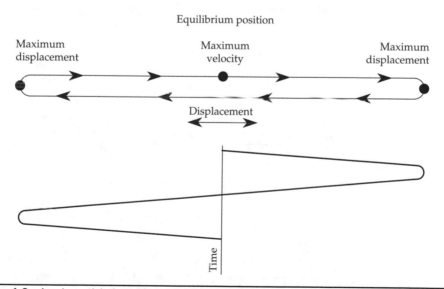

FIGURE 1-3   An air particle is made to vibrate about its equilibrium position by the energy of a passing sound wave because of the interaction of the elastic forces of the air and the inertia of the air particle.

(perhaps a maximum displacement of a few ten-thousandths of an inch). Note also that the velocity of a particle is maximum at its equilibrium position, and zero at the points of maximum displacement (a pendulum has the same property). The maximum velocity is called the velocity amplitude, and the maximum displacement is called the displacement amplitude. The maximum particle velocity is very small, less than 0.5 in/sec for even a loud sound. As we will see, to lower the level of a sound, we must reduce the particle velocity.

There are three distinct forms of particle motion. If a stone is dropped on a calm water surface, concentric waves travel out from the point of impact, and the water particles trace circular orbits (for deep water, at least), as shown in Fig. 1-4A. Another type of wave motion is illustrated by a violin string, as shown in Fig. 1-4B. The tiny elements of the string move transversely, or at right angles to the direction of travel of the waves along the string. For sound traveling in a gaseous medium such as air, the particles move in the direction the sound is traveling. These are called longitudinal waves, as shown in Fig. 1-4C.

Propagation of Sound

How are air particles, moving slightly back and forth, able to carry music from a loudspeaker to our ears, at the speed of a rifle bullet? The dots of Fig. 1-5 represent air molecules with different density variations. In reality, there are more than a million molecules in a cubic inch of air. The molecules crowded together represent areas of compression (crests) in which the air pressure is slightly greater than the prevailing atmospheric pressure (typically about 14.7 lb/in^2 at sea level). The sparse areas represent rarefactions (troughs) in which the pressure is slightly less than atmospheric pressure. The arrows (see Fig. 1-5) indicate that, on average, the molecules are moving to the right of the compression crests and to the left in the rarefaction troughs between the crests. Any given molecule, because of elasticity, after an initial displacement, will return toward its original position. It will move a certain distance to the right and then

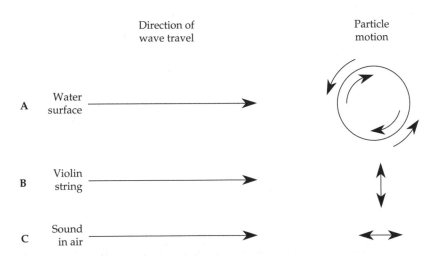

Figure 1-4   Particles involved in the propagation of sound waves can move with (A) circular motion, (B) transverse motion, or (C) longitudinal motion.

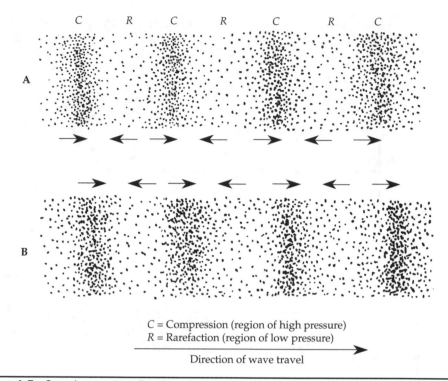

C = Compression (region of high pressure)
R = Rarefaction (region of low pressure)

Direction of wave travel

FIGURE 1-5 Sound waves traveling through a medium change localized air particle density. (A) The sound wave causes the air particles to be pressed together (compression) in some regions and spread out (rarefaction) in others. (B) An instant later the sound wave has moved slightly to the right.

the same distance to the left of its undisplaced position as the sound wave progresses uniformly to the right. Sound exists because of the transfer of momentum from one particle to another.

In this example, why does the sound wave move to the right? The answer is revealed by a closer look at the arrows (see Fig. 1-5). The molecules tend to bunch up where two arrows are pointing toward each other, and this occurs a bit to the right of each compression region. When the arrows point away from each other, the density of molecules decreases. Thus, the movement of the higher pressure crest and the lower pressure trough accounts for the progression of the sound wave to the right.

As mentioned previously, the pressure at the crests is higher than the prevailing atmospheric barometric pressure and the troughs lower than the atmospheric pressure, as shown in the sine wave of Fig. 1-6. These fluctuations of pressure are very small indeed. The faintest sound the ear can hear (20 µPa) exists at a pressure some 5,000 million times smaller than atmospheric pressure. Normal speech and music signals are represented by correspondingly small ripples superimposed on the atmospheric pressure.

Speed of Sound

The speed of sound in air is about 1,130 ft/sec (344 sec) at normal temperature and pressure. This is about 770 mi/hr (1,240 km/hr) and is not particularly fast in relation to

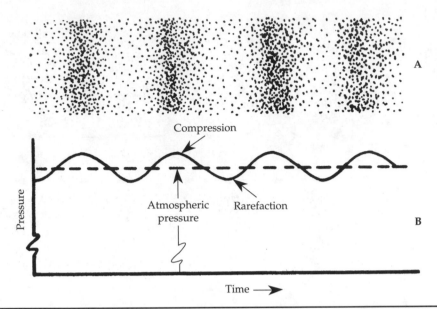

FIGURE 1-6   Pressure variations of sound waves are superimposed on prevailing barometric pressure. (A) An instantaneous view of the compressed and rarefied regions of a sound wave in air. (B) The compressed regions are very slightly above and the rarefied regions very slightly below atmospheric pressure.

familiar things. For example, the Boeing 787 jetliner has a cruising speed of 561 mi/hr (Mach 0.85). The speed of sound is dramatically slower than the speed of light (670,616,629 mi/hr). It takes sound about 5 seconds to travel 1 mile; you can gauge the distance of a thunderstorm by counting the time between the sight of the lightning strike and the sound of its thunder. The speed of sound in the audible range is not appreciably affected by intensity of the sound, its frequency, or by changes in atmospheric pressure.

Sound will propagate at a certain speed that depends on the medium, and other factors. The more dense the molecular structure, the easier it is for the molecules to transfer sound energy; compared to air, sound travels faster in denser media such as liquids and solids. For example, sound travels at about 4,900 ft/sec in freshwater, and about 16,700 ft/sec in steel. Sound also travels faster in air as temperature increases (an increase of about 1.1 ft/sec for every degree Fahrenheit). Finally, humidity affects the velocity of sound in air; the more humid the air, the faster the speed. It should be noted that the speed (velocity) of sound is different from the particle velocity. The velocity of sound determines how fast sound energy moves across a medium. Particle velocity is determined by the loudness of the sound.

Wavelength and Frequency

A sine wave is illustrated in Fig. 1-7. The wavelength λ is the distance a wave travels in the time it takes to complete one cycle. A wavelength can be measured between successive peaks or between any two corresponding points on the cycle. This also holds for

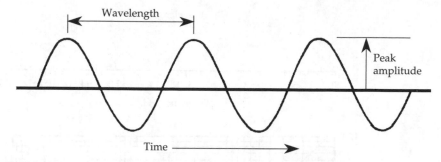

Figure 1-7 Wavelength is the distance a wave travels in the time it takes to complete one cycle. It can also be expressed as the distance from one point on a periodic wave to the corresponding point on the next cycle of the wave.

periodic waves other than the sine wave. The frequency *f* specifies the number of cycles per second, measured in hertz (Hz). Frequency and wavelength are related as follows:

$$\text{Wavelength (ft)} = \frac{\text{Speed of sound (ft/sec)}}{\text{Frequency (Hz)}} \qquad (1\text{-}1)$$

which can also be written as

$$\text{Frequency} = \frac{\text{Speed of sound}}{\text{Wavelength}} \qquad (1\text{-}2)$$

As noted, the speed of sound in air is about 1,130 ft/sec at normal conditions. For sound traveling in air, Eq. (1-2) becomes

$$\text{Wavelength} = \frac{1,130}{\text{Frequency}} \qquad (1\text{-}3)$$

This relationship is perhaps the most fundamentally important relationship in audio. Fig. 1-8 gives two graphical approaches for an easy solution to Eq. (1-3).

Complex Waves

Speech and music waveshapes depart radically from the simple sine wave, and are considered as complex waveforms. However, no matter how complex the waveform, as long as it is periodic, it can be reduced to sine components. The obverse of this states that any complex periodic waveform can be synthesized from sine waves of different frequencies, different amplitudes, and different time relationships (phase). Joseph Fourier was the first to prove these relationships. The idea is simple in concept, but often complicated in its application to specific speech or musical sounds. Let us see how a complex waveform can be reduced to simple sinusoidal components.

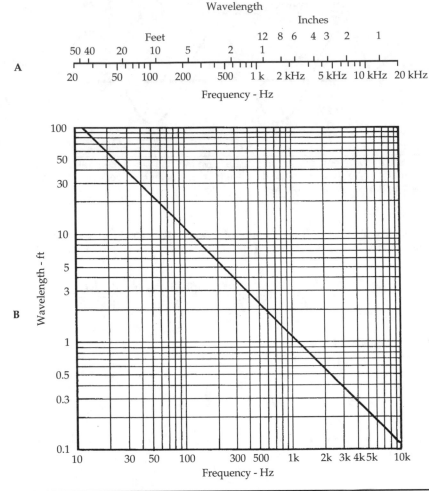

FIGURE 1-8 Wavelength and frequency are inversely related. (A) Scales for approximately determining wavelength of sound in air from a known frequency, or vice versa. (B) A chart for determining the wavelength in air of sound waves of different frequencies. (Both based on speed of sound of 1,130 ft/sec.)

Harmonics

A simple sine wave of a given amplitude and frequency, f_1, is shown in Fig. 1-9A. Figure 1-9B shows another sine wave f_2 that is half the amplitude and twice the frequency. Combining A and B at each point in time, the waveshape of Fig. 1-9C is obtained. In Fig. 1-9D, another sine wave f_3 that is half the amplitude of A and three times its frequency is shown. Adding this to the $f_1 + f_2$ waveshape of C, Fig. 1-9E is obtained. The simple sine wave of Fig. 1-9A has been progressively changed as other sine waves have been added to it. Whether these are acoustic waves or electronic signals, the process can be reversed. The complex waveform of Fig. 1-9E can be disassembled, as it were, to the simple $f_1, f_2,$ and f_3 sine components by either acoustical or electronic filters. For example, passing the waveform of Fig. 1-9E through a filter permitting only f_1 and rejecting f_2 and f_3, the original f_1 sine wave of Fig. 1-9A emerges in pristine condition.

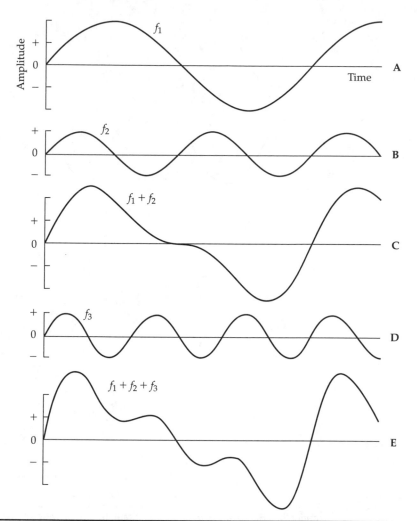

Figure 1-9 A study in the combination of sine waves. (A) The fundamental of frequency f_1. (B) A second harmonic of frequency $f_2 = 2f_1$ and half the amplitude of f_1. (C) The sum of f_1 and f_2 obtained by adding ordinates point by point. (D) A third harmonic of frequency $f_3 = 3f_1$ and half the amplitude of f_1. (E) The waveform resulting from the addition of f_1, f_2, and f_3. All three components are in phase, that is, they all start from zero at the same instant.

The sine wave with the lowest frequency (f_1) of Fig. 1-9A is called the fundamental, the sine wave that is twice the frequency (f_2) of Fig. 1-9B is called the second harmonic, and the sine wave that is three times the frequency (f_3) of Fig. 1-9D is the third harmonic. The fourth harmonic and the fifth harmonic are four and five times the frequency of the fundamental, and so on.

Phase

In Fig. 1-9, all three components, $f_1, f_2,$ and $f_3,$ start from zero together. This is called an in-phase condition. In some cases, the time relationships between harmonics or between harmonics and the fundamental are quite different from this. We observed that one

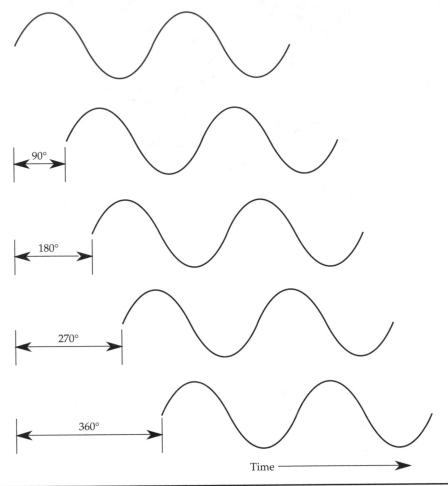

Figure 1-10 Illustration of the phase relationships between waveforms with the same amplitude and frequency. A rotation of 360° is analogous to one complete sine wave cycle.

revolution (360°) of the crankshaft of an automobile engine was equated with one cycle of simple harmonic motion of the piston. The up-and-down travel of the piston spread out in time traces a sine wave such as that in Fig. 1-10. One complete sine-wave cycle represents 360° of rotation. If another sine wave of identical frequency is delayed 90°, its time relationship to the first one is a quarter wave late (time increasing to the right). A half-wave delay would be 180°, and so on. For the 360° delay, the waveform at the bottom of Fig. 1-10 synchronizes with the top one, reaching positive peaks and negative peaks simultaneously and producing the in-phase condition.

In Fig. 1-9, all three components of the complex waveform of Fig. 1-9E are in phase. That is, the f_1 fundamental, the f_2 second harmonic, and the f_3 third harmonic all start at zero at the same time. What happens if the harmonics are out of phase with the fundamental? Figure 1-11 illustrates this case. The second harmonic f_2 is now advanced 90°, and the third harmonic f_3 is retarded 90°. By combining f_1, f_2, and f_3 for each instant of time,

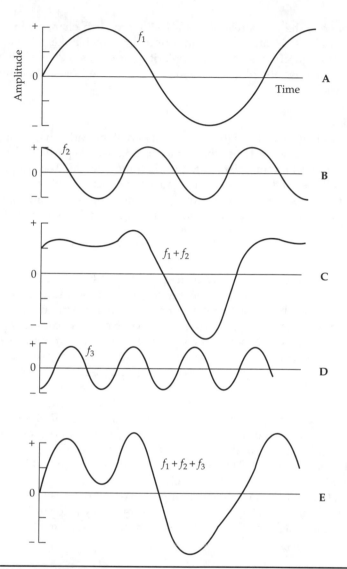

FIGURE 1-11   A study of the combination of sine waves that are not in phase. (A) The fundamental frequency f_1. (B) The second harmonic f_2 with twice the frequency and half the amplitude of f_1 advanced 90° with respect to f_1. (C) The combination of f_1 and f_2 obtained by adding ordinates point by point. (D) The third harmonic f_3 with phase 90° behind f_1, and with half the amplitude of f_1. (E) The sum of f_1, f_2, and f_3. Compare this resulting waveform with that of Fig. 1-9E. The difference in waveforms is due entirely to the shifting of the phase of the harmonics with respect to the fundamental.

with due regard to positive and negative signs, the contorted waveform of Fig. 1-11E is obtained.

The only difference between Figs. 1-9E and 1-11E is that a phase shift has been introduced between harmonics f_2 and f_3, and the fundamental f_1. That is all that is needed to produce drastic changes in the resulting waveshape. Curiously, even though the shape of the waveform is dramatically changed by shifting the time relationships of the components, the ear is relatively insensitive to such changes. In other words, wave-forms E of Figs. 1-9 and 1-11 would sound very much alike.

A common error is confusing polarity with phase. Phase is the time relationship between two signals, while polarity is the +/− or the −/+ relationship of a given pair of signal leads.

Partials

Musicians may be inclined to use the term partial instead of harmonic, but it is best that a distinction be made between the two terms because the partials of many musical instruments are not harmonically related to the fundamental. That is, partials might not be exact multiples of the fundamental frequency, yet richness of tone can still be imparted by such deviations from the true harmonic relationship. For example, the partials of bells, chimes, and piano tones are often in a nonharmonic relationship to the fundamental.

Octaves

Audio engineers and acousticians frequently use the integral multiple concept of harmonics, closely allied as it is to the physical aspect of sound. Musicians often refer to the octave, a logarithmic concept that is firmly embedded in musical scales and terminology because of its relationship to the ear's characteristics. Audio people are also involved with the human ear, hence their common use of logarithmic scales for frequency, logarithmic measuring units, and various devices based on octaves.

Harmonics and octaves are compared in Fig. 1-12. Harmonics are linearly related; each next harmonic is an integer multiple of the preceding one. An octave is defined as a 2:1 ratio of two frequencies. For example, middle C (C4) has a frequency close to 261 Hz. The next highest C (C5) has a frequency of about 522 Hz. Ratios of frequencies are very much a part of the musical scale. The frequency ratio 2:1 is the octave; the ratio 3:2 is the fifth; 4:3 is the fourth, and so on.

The interval from 100 to 200 Hz is an octave, as is the interval from 200 to 400 Hz. The interval from 100 to 200 Hz is perceived as being larger than the interval from 200 to 300 Hz; this demonstrates that the ear perceives intervals as ratios rather than arithmetic differences. In particular, our perception of frequency is logarithmic. Because the octave is important in acoustical work, it is useful to consider the mathematics of the octave.

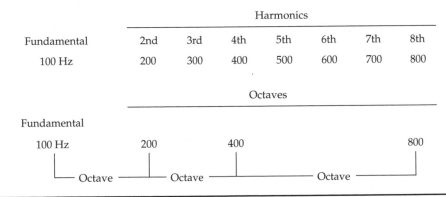

	Harmonics						
Fundamental	2nd	3rd	4th	5th	6th	7th	8th
100 Hz	200	300	400	500	600	700	800

Figure 1-12 Comparison of harmonics and octaves. Harmonics are linearly related; octaves are logarithmically related.

As the ratio of 2:1 is defined as the octave, its mathematical expression is

$$\frac{f_2}{f_1} = 2^n \qquad (1\text{-}4)$$

where f_2 = frequency of the upper edge of the octave interval, Hz
$\quad\;\; f_1$ = frequency of the lower edge of the octave interval, Hz
$\quad\;\; n$ = number of octaves

For one octave, $n = 1$ and Eq. (1-4) becomes $f_2/f_1 = 2$, which is the definition of the octave. Other applications of Eq. (1-4) are as follows:

Example 1

The low-frequency limit of a band is 20 Hz, what is the high-frequency limit of a band that is 10 octaves wide?

$$\frac{f_2}{20 \text{ Hz}} = 2^{10}$$

$$f_2 = (20)\,(2^{10})$$

$$f_2 = (20)\,(1{,}024)$$

$$f_2 = 20{,}480 \text{ Hz}$$

Example 2

If 446 Hz is the lower limit of a $\frac{1}{3}$-octave band, what is the frequency of the upper limit?

$$\frac{f_2}{446} = (2^{1/3})$$

$$f_2 = (446)(2^{1/3})$$

$$f_2 = (446)(1.2599)$$

$$f_2 = 561.9 \text{ Hz}$$

Example 3

What is the lower limit of a $\frac{1}{3}$-octave band centered on 1,000 Hz? The f_1 is 1,000 Hz but the lower limit would be $\frac{1}{6}$-octave lower than the $\frac{1}{3}$-octave, so $n = \frac{1}{6}$:

$$\frac{f_2}{f_1} = \frac{1{,}000}{f_1} = 2^{1/6}$$

$$f_1 = \frac{1{,}000}{2^{1/6}}$$

$$f_1 = \frac{1{,}000}{1.12246}$$

$$f_1 = 890.9 \text{ Hz}$$

Example 4

What is the frequency of the lower limit of an octave band centered on 2,500 Hz?

$$\frac{2,500}{f_1} = 2^{1/2}$$

$$f_1 = \frac{2,500}{2^{1/2}}$$

$$f_1 = \frac{2,500}{1.4142}$$

$$f_1 = 1,767.8 \text{ Hz}$$

What is the upper limit?

$$\frac{f_2}{2,500} = (2^{1/2})$$

$$f_2 = (2,500)(2^{1/2})$$

$$f_2 = (2,500)(1.4142)$$

$$f_2 = 3,535.5 \text{ Hz}$$

In many acoustical applications, sound is considered as falling in eight octave bands, with center frequencies of 63; 125; 250; 500; 1,000; 2,000; 4,000; and 8,000 Hz. In some cases, sound is considered in terms of $\frac{1}{3}$-octave bands, with center frequencies falling at 31.5; 50; 63; 80; 100; 125; 160; 200; 250; 315; 400; 500; 630; 800; 1,000; 1,250; 1,600; 2,000; 2,500; 3,150; 4,000; 5,000; 6,300; 8,000; and 10,000 Hz.

Spectrum

The commonly accepted scope of the audible spectrum is 20 Hz to 20 kHz; the range is one of the specific characteristics of the human ear. Here, in the context of sine waves and harmonics, we need to establish the concept of spectrum. The visible spectrum of light has its counterpart in sound in the audible spectrum, the range of frequencies that fall within the perceptual limits of the human ear. We cannot see far-ultraviolet light because the frequency of its electromagnetic energy is too high for the eye to perceive. Nor can we see far-infrared light because its frequency is too low. There are likewise sounds that are too low (infrasound) and sounds that are too high (ultrasound) in frequency for the ear to hear.

Figure 1-13 shows several waveforms that typify the infinite number of different waveforms commonly encountered in audio. These waveforms have been captured from the screen of an oscilloscope. To the right of each capture is the spectrum of that particular signal. The spectrum tells how the energy of the signal is distributed in frequency. In all but the signal of Fig. 1-13D, the audible range of the spectrum was processed with a waveform analyzer having a very sharp filter with a passband 5 Hz wide. In this way, concentrations of energy were located and measured.

For a sine wave, all the energy is concentrated at one frequency. The sine wave produced by this particular signal generator is not really a pure sine wave. No oscillator is perfect and all have some harmonic content, but in scanning the spectrum of this sine wave, the harmonics measured were too low to show on the graph scale of Fig. 1-13A.

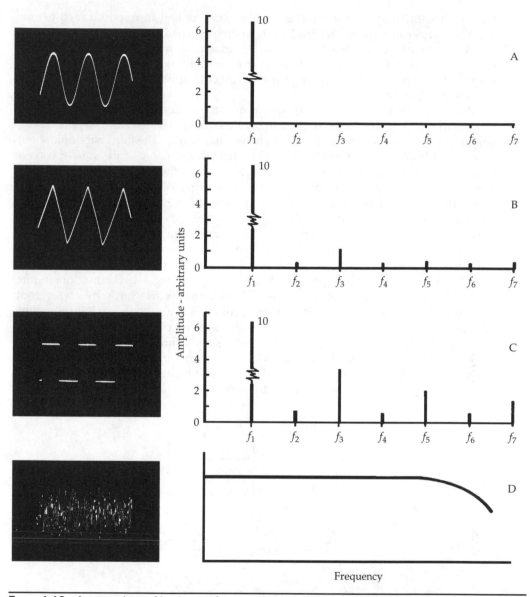

FIGURE 1-13 A comparison of basic waveforms and noise. (A) The spectral energy of a pure sinusoid is contained entirely at a single frequency. The triangular (B) and square (C) waveforms each have a prominent fundamental and numerous harmonics at integral multiples of the fundamental frequency. (D) Random noise (white noise) has energy distributed uniformly throughout the spectrum up to some point at which energy begins to fall off due to generator limitations. Random noise may be considered as statistically distributed signals with a continuous frequency spectrum.

The triangular waveform of this signal generator has a major fundamental component of 10 unit's magnitude, as shown in Fig. 1-13B. The waveform analyzer detected a significant second harmonic component at f_2, twice the frequency of the fundamental with a magnitude of 0.21 units. The third harmonic showed an amplitude of 1.13 units, the fourth of 0.13 units, and so on. The seventh harmonic had an amplitude of 0.19 units and the fourteenth harmonic (about 15 kHz in this case) had an amplitude of 0.03 units, but was still easily detectable. We see that this triangular waveform has both odd and even components of modest amplitude through the audible spectrum. If we know the amplitude and phases of each of these, the original triangular waveform can be synthesized by combining them.

A comparable analysis reveals the spectrum of the square wave shown in Fig. 1-13C. It has harmonics of far greater amplitude than the triangular waveform with a distinct tendency toward more prominent odd than even harmonics. The third harmonic shows an amplitude 34% of the fundamental. The fifteenth harmonic of the square wave is 0.52 units. Figure 1-14A shows a square wave; it can be synthesized by adding harmonics to a fundamental. However, many harmonics would be needed. For example, Fig. 1-14B shows the waveform that results from adding two nonzero harmonic components, and Fig. 1-14C shows the result from adding nine nonzero harmonic components. This demonstrates why a bandlimited "square wave" does not have a square appearance.

The spectra of sine, triangular, and square waves reveals energy concentrated at harmonic frequencies, but nothing between. These are all periodic waveforms, which repeat themselves cycle after cycle. The fourth example in Fig. 1-13D is random (white) noise. The spectrum of this signal cannot be measured satisfactorily by a waveform analyzer with a 5-Hz passband because the fluctuations are so great that it is impossible to get an accurate reading. Analyzed by a wider passband of fixed bandwidth and with the help of various integrating devices to get a steady indication, the spectral shape shown is obtained. This spectrum tells us that the energy of the random-noise signal is equally distributed throughout the spectrum; the roll-off at high frequencies indicates that the upper frequency limit of the random noise generator has been reached.

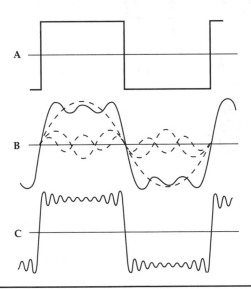

Figure 1-14   A square wave can be synthesizing by adding harmonics to the sine-wave fundamental. (A) A square wave (with an infinite number of harmonic components). (B) The summation of the fundamental and two nonzero harmonic components. (C) The summation of the fundamental and nine nonzero harmonic components. Clearly, many components would be needed to smooth the ripples and produce the square corners of (A).

There is little visual similarity between the sine and the random-noise signals as revealed by an oscilloscope, yet there is a hidden relationship. Even random noise can be considered as comprising sine-wave components constantly shifting in frequency, amplitude, and phase. If we pass random noise through a narrow filter and observe the filter output on an oscilloscope, we will see a restless, sinelike wave continually shifting in amplitude. Theoretically, an infinitely narrow filter would sift out a pure sine wave.

Electrical, Mechanical, and Acoustical Analogs

An acoustical system such as a loudspeaker can be represented in terms of an equivalent electrical or mechanical system. An engineer uses these equivalents to set up a mathematical approach for analyzing a given system. For example, the effect of a cabinet on the functioning of a loudspeaker is clarified by thinking of the air in the enclosed space as acting like a capacitor in an electrical circuit, absorbing and giving up the energy imparted by the cone movement.

Figure 1-15 shows the graphical representation of the three basic elements in electrical, mechanical, and acoustical systems. Inductance in an electrical circuit is equivalent to mass in a mechanical system and inertance in an acoustical system. Capacitance in an electrical circuit is analogous to compliance in a mechanical system and capacitance in an acoustical system. Resistance is resistance in all three systems, whether it is the frictional losses offered to air-particle movement in glass fiber, frictional losses in a wheel bearing, or resistance to the flow of current in an electrical circuit.

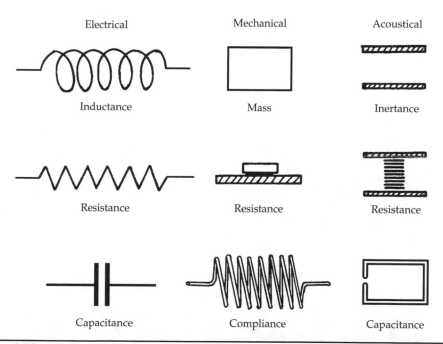

FIGURE 1-15 The three basic elements of electrical systems and their analogs in mechanical and acoustical systems.

CHAPTER 2

Sound Levels and the Decibel

The decibel is one of the most important units of measure in the audio field. The decibel is an extraordinarily efficient way to describe audio phenomena, and our perception of them. The goal of this chapter is to explain the decibel concept and show how decibels can be applied in many different ways. In particular, we will see how decibels are used to measure sound levels in various applications.

Sound levels expressed in decibels clearly demonstrate the wide range of sensitivity in human hearing. The threshold of hearing matches the ultimate lower limit of perceptible sound in air, the noise of air molecules against the eardrum. At the other end of the range, the ear can tolerate very high intensities of sound. A level expressed in decibels is a convenient way of handling the billion-fold range of sound pressures to which the ear is sensitive.

Ratios versus Differences

Imagine a sound source set up in a room completely protected from interfering noise. The sound source is adjusted for a weak sound with a sound pressure of 1 unit, and its loudness is carefully noted.

In observation A, when the sound pressure is increased until it sounds twice as loud, the level dial reads 10 units. For observation B, the source pressure is increased to 10,000 units. To double the loudness, we find that the sound pressure must be increased from 10,000 to 100,000 units. The results of this experiment can be summarized as follows:

Observations	Two Pressures	Ratio of Two Pressures
A	10–1	10:1
B	100,000–10,000	10:1

Observations A and B accomplish the same doubling of perceived loudness. In observation A, this was accomplished by an increase in sound pressure of only 9 units, where in observation B it took 90,000 units. Ratios of pressures seem to describe loudness changes better than differences in pressure. Ernst Weber, Gustaf Fechner, Hermann von Helmholtz, and other early researchers pointed out the importance of using ratios in such measurements. Ratios apply equally well to sensations of vision, hearing, vibration, or even electric shock. Ratios of stimuli come closer to matching human perception than do differences of stimuli. This matching is not perfect, but close enough to make a strong case for expressing levels in decibels.

Ratios of powers or ratios of intensities, or ratios of sound pressure, voltage, current, or anything else are dimensionless. For example, the ratio of 1 W to 100 W is 1 W/100 W, and the watt unit in the numerator and the watt unit in the denominator cancel, leaving, $\frac{1}{100} = 0.01$, a pure number without dimension. This is important because logarithms can only be taken of nondimensional numbers.

Expressing Numbers

Table 2-1 illustrates three different ways numbers can be expressed. The decimal and arithmetic forms are familiar in everyday activity. The exponential form, while not as commonly used, has an almost unique ability to simplify the expression of many relationships. When writing "one hundred thousand" watts, we can express the number as 100,000 W or 10^5 W. When writing a "millionth of a millionth" of a watt, the string of zeros behind the decimal point is clumsy, but writing 10^{-12} is easy. Engineering calculators display the exponential form in scientific notation, by which very large or very small numbers can be expressed. Moreover, the prefix pico means 10^{-12}, so the value can be expressed as 1 pW (shown later in Table 2-4).

Decimal Form	Arithmetic Form	Exponential Form
100,000	$10 \times 10 \times 10 \times 10 \times 10$	10^5
10,000	$10 \times 10 \times 10 \times 10$	10^4
1,000	$10 \times 10 \times 10$	10^3
100	10×10	10^2
10	10×1	10^1
1	$10/10$	10^0
0.1	$1/10$	10^{-1}
0.01	$1/(10 \times 10)$	10^{-2}
0.001	$1/(10 \times 10 \times 10)$	10^{-3}
0.0001	$1/(10 \times 10 \times 10 \times 10)$	10^{-4}
100,000	$(100)(1,000)$	$10^2 + 10^3 = 10^{2+3} = 10^5$
100	$10,000/100$	$10^4/10^2 = 10^{4-2} = 10^2$
10	$100,000/10,000$	$10^5/10^4 = 10^{5-4} = 10^{-1} = 10$
10	$\sqrt{100} = \sqrt[2]{100}$	$100^{1/2} = 100^{0.5}$
4.6416	$\sqrt[3]{100}$	$100^{1/3} = 100^{0.333}$
31.6228	$\sqrt[4]{100^3}$	$100^{3/4} = 100^{0.75}$

TABLE 2-1   Ways of Expressing Numbers

The softest sound intensity we can hear (the threshold of audibility) is about 10^{-12} W/m². A very loud sound (causing a sensation of pain) might be 10 W/m². (Acoustic intensity is acoustic power per unit area in a specified direction.) This range of intensities from the softest sound to a painfully loud sound is 10,000,000,000,000. Clearly, it is more convenient to express this range as an exponent, 10^{13}. Furthermore, it is useful to establish the intensity of 10^{-12} W/m² as a reference intensity I_{ref} and express other sound intensities I as a ratio I/I_{ref} to this reference. For example, the sound intensity of 10^{-9} W/m² would be written as 10^3 or 1,000 (the ratio is dimensionless). We see that 10^{-9} W/m² is 1,000 times the reference intensity.

Logarithms

Representing 100 as 10^2 simply means that $10 \times 10 = 100$. Similarly, 10^3 means $10 \times 10 \times 10 = 1,000$. But how about 267? That is where logarithms are useful. Logarithms are proportional numbers, and a logarithmic scale is one that is calibrated proportionally. It is agreed that 100 equals 10^2. By definition we can say that the logarithm of 100 to the base 10 equals 2, commonly written $\log_{10} 100 = 2$, or simply $\log 100 = 2$, because common logarithms are to the base 10. The number 267 can be expressed as 10 to some power between 2 and 3. Avoiding the mathematics, we can use a calculator to enter 267, push the "log" button, and 2.4265 appears. Thus, $267 = 10^{2.4265}$, and $\log 267 = 2.4265$. Logarithms are handy because, as Table 2-1 demonstrates, they reduce multiplication to addition, and division to subtraction.

Logarithms are particularly useful to audio engineers because they can correlate measurements to human hearing, and they also allow large ranges of numbers to be expressed efficiently. Logarithms are the foundation for expressing sound levels in decibels where the level is a logarithm of a ratio. In particular, a level in decibels is 10 times the logarithm to the base 10 of the ratio of two powerlike quantities.

Decibels

We observed that it is useful to express sound intensity in ratios. Furthermore, we can express the intensities as logarithms of the ratios. An intensity I can be expressed in terms of a reference I_{ref} as follows:

$$\log_{10} \frac{I}{I_{ref}} \text{ bels} \qquad (2\text{-}1)$$

The intensity measure is dimensionless, but to clarify the value, we assign it the unit of a bel (from Alexander Graham Bell). However, when expressed in bels, the range of values is somewhat small. To make the range easier to use, we usually express values in decibels. The decibel is $\frac{1}{10}$ bel. A decibel (dB) is 10 times the logarithm to base 10 of the ratio of two quantities of intensity (or power). Thus, the intensity ratio in decibels becomes:

$$\text{IL} = 10 \log_{10} \frac{I}{I_{ref}} \text{ decibels} \qquad (2\text{-}2)$$

This value is called the sound intensity level (IL in decibels) and differs from intensity (I in watts/m²). Using decibels is a convenience, and decibel values more closely follow the way we hear the loudness of sounds.

Questions sometimes arise when levels other than intensity need to be expressed in decibels. Equation (2-2) applies equally to acoustic intensity, as well as acoustic power, electric power, or any other kind of power. For example, we can write the sound-power level as:

$$PWL = 10 \log_{10} \frac{W}{W_{ref}} \text{ decibels} \qquad (2-3)$$

where PWL = sound-power level, dB
W = sound power, watts
W_{ref} = a reference power, 10^{-12} W

Sound intensity is difficult to measure. Sound pressure is usually the most accessible parameter to measure in acoustics (just as voltage is for electronic circuits). For this reason, the sound-pressure level (SPL) is often used. SPL is a logarithmic value of the sound pressure, in the same way that the sound intensity level (IL) corresponds to sound intensity. SPL is approximately equal to IL; both are often referred to as the sound level. Acoustic intensity (or power) is proportional to the square of the acoustic pressure p. This slightly alters the defining equation that we use. When the reference pressure is 20 μPa, a sound pressure p measured in micropascals has a sound-pressure level (SPL) of:

$$SPL = 10 \log_{10} \frac{p^2}{p_{ref}^2}$$

$$= 20 \log_{10} \frac{p}{20 \, \mu Pa} \text{ decibels} \qquad (2-4)$$

where SPL = sound-pressure level, dB
p = acoustic pressure, μPa or other
p_{ref} = acoustic reference pressure, μPa or other

The tabulation of Table 2-2 shows whether Eqs. (2-2) or (2-3), or Eq. (2-4) applies.

	Equations (2-2) or (2-3)	Equation (2-4)
Parameter	$10 \log_{10} \frac{a_1}{a_2}$	$20 \log_{10} \frac{b_1}{b_2}$
Acoustic		
Intensity	X	
Power	X	
Air particle velocity		X
Pressure		X
Electric		
Power	X	
Current		X
Voltage		X
Distance		X
(From source-SPL; inverse square)		

TABLE 2-2   Use of 10 log and 20 log

Reference Levels

As we have seen, reference levels are widely used to establish a baseline for measurements. For example, a sound-level meter is used to measure a certain sound-pressure level. If the corresponding sound pressure is expressed in normal pressure units, a great range of very large and very small numbers results. As we have seen, ratios are more closely related to human perception than linear numbers, and by expressing levels in decibels, we compress the large and small ratios into a more convenient and comprehensible range. Basically, a sound-level meter reading is a certain sound-pressure level (SPL), $20 \log (p/p_{ref})$, as in Eq. (2-4). The sound-pressure reference p_{ref} must be standardized, so that ready comparisons can be made. Several such reference pressures have been used over the years, but for sound in air the standard reference pressure is 20 μPa (micropascals). This might seem quite different from the reference pressure of 0.0002 microbar or 0.0002 dyne/cm², but it is the same standard merely written in different units. This is a very small sound pressure (0.0000000035 lb/in²) and corresponds closely to the threshold of human hearing at 1 kHz. The relationship between sound pressure in pascals, pounds/square inch, and sound-pressure level is shown in Fig. 2-1.

When a statement is encountered such as, "The sound-pressure level is 82 dB," the 82-dB sound-pressure level is normally used in direct comparison with other levels.

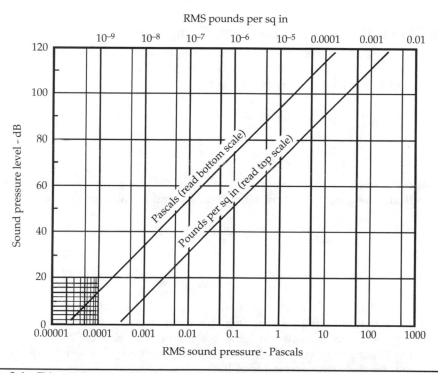

FIGURE 2-1 This graph shows the relationship between sound pressure in pascals or pounds/square inch and sound-pressure level (referred to 20 μPa). This is a graphical approach to the solution of Eq. (2-2).

However, if the sound pressure were needed, it can be computed readily by working backward from Eq. (2-4) as follows:

$$82 = 20 \ \log \ \frac{p}{20\,\mu Pa}$$

$$\log \frac{p}{20 \ \mu Pa} = \frac{82}{20}$$

$$\frac{p}{20 \ \mu Pa} = 10^{\frac{82}{20}}$$

The y^x button on a calculator helps us to evaluate $10^{4.1}$. Enter 10, enter 4.1, press the y^x power button, and the value 12,589 appears.

$$p = (20 \ \mu Pa)(12,589)$$

$$p = 251,785 \ \mu Pa$$

There is another lesson here. The 82 has two significant figures. The 251,785 has six significant figures and implies a precision that is not there. Just because a calculator says so doesn't make it so. A better answer is 252,000 µPa, or 0.252 Pa.

Logarithmic and Exponential Forms Compared

The logarithmic and exponential forms are equivalent as can be seen in Table 2-1. When working with decibels, it is important that this equivalence be understood.

Let's say we have a power ratio of 5:

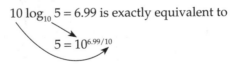

$$10 \ \log_{10} 5 = 6.99 \text{ is exactly equivalent to}$$

$$5 = 10^{6.99/10}$$

There are two 10s in the exponential statement but they come from different sources as indicated by the arrows. Now let us treat a sound pressure ratio of 5:

$$20 \ \log_{10} 5 = 13.98$$

$$5 = 10^{13.98/20}$$

Remember that sound-pressure level in air means that the reference pressure (p_{ref}) in the pressure ratio is 20 µPa. There are other reference quantities; some of the commonly used ones are listed in Table 2-3. The prefixes of Table 2-4 are often employed when dealing with very small and very large numbers. These prefixes are the Greek names for the power exponents of 10.

Level in Decibels	Reference Quantity
Acoustic	
Sound-pressure level in air (SPL, dB)	20 μPa
Power level (Lp, dB)	1 pW (10^{-12} W)
Electric	
Power level re 1 mW	10^{-3} W (1 mW)
Voltage level re 1 V	1 V
Volume level, VU	10^{-3} W

TABLE 2-3   Reference Quantities in Common Use

Prefix	Symbol	Multiple
tera	T	10^{12}
giga	G	10^{9}
mega	M	10^{6}
kilo	k	10^{3}
milli	m	10^{-3}
micro	μ	10^{-6}
nano	n	10^{-9}
pico	p	10^{-12}

TABLE 2-4   Prefixes, Symbols, and Exponents

Acoustic Power

It doesn't take many watts of acoustic power to produce very loud sounds. A 100-W amplifier may be driving a loudspeaker, but loudspeaker efficiency (output for a given input) is very low, perhaps on the order of 10%. A typical loudspeaker might radiate 1 W of acoustic power. Increasing amplifier power to achieve higher acoustic levels can be frustrating. Doubling amplifier power from 1 to 2 W is a 3-dB increase in power level (10 log 2 = 3.01), yielding a very small increase in loudness. Similarly, an increase in power from 100 to 200 W or 1,000 to 2,000 W yields the same 3-dB increase in level.

Sound Source	Sound Pressure (Pa)	Sound-Pressure Level* (dB, A-Weighted)
Saturn rocket	100,000	194
Ram jet	2,000	160
Propeller aircraft	200	140
Riveter	20	120
Heavy truck	2	100
Noisy office or heavy traffic	0.2	80
Conversational speech	0.02	60
Quiet residence	0.002	40
Leaves rustling	0.0002	20
Hearing threshold, excellent ears at frequency maximum response	0.00002	0

* Reference pressure (these are identical):
  20 μPa (micropascals)
  0.00002 Pa (pascals)
  2×10^{-5} N/m²
  0.0002 dyne/cm² or μbar

TABLE 2-5   Examples of Sound Pressures, and Sound-Pressure Levels

Table 2-5 lists sound pressure and sound-pressure levels of some common sounds. There is a vast difference in sound pressure from 0.00002 Pa (20 μPa) to 100,000 Pa, but this range is reduced to a convenient form when expressed in sound levels. The same information is present in graphical form in Fig. 2-2.

Another way to generate a 194-dB sound-pressure level, besides launching a Saturn rocket, is to detonate 50 lb of TNT 10 ft away. Sound waves are tiny ripples on the steady-state atmospheric pressure. A 194-dB sound-pressure level approaches the atmospheric pressure and hence is a ripple of the order of magnitude of atmospheric pressure. The 194-dB sound pressure is an RMS (root-mean-square) value. A peak sound pressure 1.4 times as great would modulate the atmospheric pressure completely.

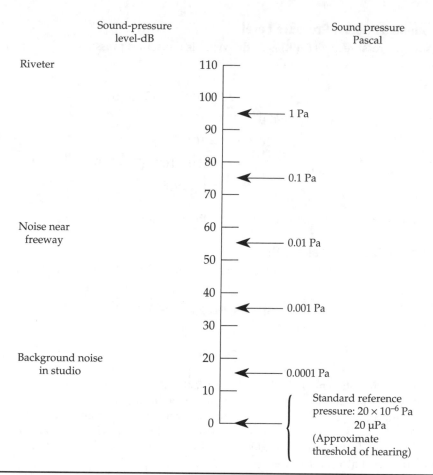

FIGURE 2-2 An appreciation of the relative magnitude of a sound pressure of 1 Pa can be gained by comparison to known sounds. The standard reference pressure for sound in air is 20 µPa, which corresponds closely to the minimum audible pressure.

Using Decibels

As we have seen, a level is a logarithm of a ratio of two powerlike quantities. When levels are computed from other than power ratios, certain conventions are observed. The convention for Eq. (2-4) is that sound power is proportional to sound pressure squared. The voltage-level gain of an amplifier in decibels is 20 log (output voltage/ input voltage), which holds true regardless of the input and output impedances. However, for power-level gain, the impedances must be considered if they are differ- ent. It is important to clearly indicate what type of level is intended, or else label the gain in level as "relative gain, dB." The following examples illustrate the use of the decibel.

Example 1: Sound-Pressure Level

A sound-pressure level (SPL) is 78 dB. What is the sound pressure?

$$78 \text{ dB} = 20 \log \ p/(20 \times 10^{-6})$$
$$\log p/(20 \times 10^{-6}) = 78/20$$
$$p/(20 \times 10^{-6}) = 10^{3.9}$$
$$p = (20 \times 10^{-6}) \ (7{,}943.3)$$
$$p = 0.159 \text{ Pa}$$

Remember that the reference level in SPL measurements is 20 μPa.

Example 2: Loudspeaker SPL

An input of 1 W produces a SPL of 115 dB at 1 m. What is the SPL at 6.1 m (20 ft)?

$$SPL = 115 - 20 \log \ (6.1/1)$$
$$= 115 - 15.7$$
$$= 99.3 \text{ dB}$$

The assumption made in the 20 log 6.1 factor is that the loudspeaker is operating in a free field and that the inverse square law is valid in this case. This is a reasonable assumption for a 20-ft distance if the loudspeaker is remote from reflecting surfaces.

A loudspeaker is rated at a sound-pressure level of 115 dB on axis at 1 m with 1 W into 8 Ω. If the input were decreased from 1 to 0.22 W, what would be the sound-pressure level at 1-m distance?

$$SPL = 115 - 10 \log \ (0.22/1)$$
$$= 115 - 6.6$$
$$= 108.4 \text{ dB}$$

Note that 10 log is used because two powers are being compared.

Example 3: Microphone Specifications

An omnidirectional dynamic microphone open-circuit voltage is specified as −80 dB for the 150-Ω case. It is also specified that 0 dB = 1 V/μbar. What would be the open-circuit voltage v in volts?

$$-80 \text{ dB} = 20 \log v/1$$
$$\log v/1 = -80/20$$
$$v = 0.0001 \text{ V}$$
$$= 0.1 \text{ mV}$$

Example 4: Line Amplifier

A line amplifier (600 Ω in, 600 Ω out) has a gain of 37 dB. With an input of 0.2 V, what is the output voltage?

$$37 \text{ dB} = 20 \log (v/0.2)$$
$$\log (v/02) = 37/20$$
$$= 1.85$$
$$v/0.2 = 10^{1.85}$$
$$v = (0.2)(70.79)$$
$$v = 14.16 \text{ V}$$

Example 5: General-Purpose Amplifier

An amplifier has a bridging input impedance of 10,000 Ω and an output impedance of 600 Ω. With a 50-mV input, an output of 1.5 V is observed. What is the gain of the amplifier? The voltage gain is:

$$\text{Voltage gain} = 20 \log (1.5/0.05)$$
$$= 29.5 \text{ dB}$$

It must be emphasized that this is not a power level gain because of the differences in impedance. However, voltage gain may serve a practical purpose in certain cases.

Example 6: Concert Hall

A seat in a concert hall is 84 ft from the tympani. The tympanist strikes a single note. The sound-pressure level of the direct sound of the note at the seat is measured to be 55 dB. The first reflection from the nearest sidewall arrives at the seat 105 msec after the arrival of the direct sound. (A) How far does the reflection travel to reach the seat? (B) What is the SPL of the reflection at the seat, assuming perfect reflection at the wall? (C) How long will the reflection be delayed after arrival of the direct sound at the seat?

(A) $$\text{Distance} = (1,130 \text{ ft/sec}) (0.105 \text{ sec})$$
$$= 118.7 \text{ ft}$$

(B) First, the level L, 1 ft from the tympani, must be estimated:

$$55 = L - 20 \log (84/1)$$
$$L = 55 + 38.5$$
$$L = 93.5 \text{ dB}$$

The SPL of the reflection at the seat is:

$$\text{dB} = 93.5 - 20 \log (118.7/1)$$
$$= 93.5 - 41.5$$
$$= 52 \text{ dB}$$

(C) The reflection will arrive after the direct sound at the seat after:

$$\text{Delay} = (118.7 - 84)/1{,}130 \text{ ft/sec}$$
$$= 30.7 \text{ msec}$$

A free field is also assumed here. The 30.7-msec reflection might be called an incipient echo.

Example 7: Combining Decibels

A studio has a heating, ventilating, and air-conditioning (HVAC) system, as well as a floor-standing fan. If both the HVAC and fan are turned off, a very low noise level prevails, low enough to be neglected in the calculation. If the HVAC alone is running, the sound-pressure level at a given position is 55 dB. If the fan alone is running, the sound-pressure level is 60 dB. What will be the sound-pressure level if both are running at the same time? The answer, most certainly, is not $55 + 60 = 115$ dB. Rather,

$$\text{Combined dB} = 10 \log \left(10^{\frac{55}{10}} + 10^{\frac{60}{10}}\right)$$
$$= 61.19 \text{ dB}$$

We see that the 55-dB sound only slightly increases the overall sound-pressure level established by the 60-dB level.

In another example, if the combined level of two noise sources is 80 dB and the level with one of the sources turned off is 75 dB, what is the level of the remaining source?

$$\text{Difference dB} = 10 \log \left(10^{\frac{80}{10}} - 10^{\frac{75}{10}}\right)$$
$$= 78.3 \text{ dB}$$

In other words, combining the 78.3-dB level with the 75-dB level gives the combined level of 80 dB. Both the HVAC and the fan are assumed to yield a wideband noise signal. It is important to remember that different sound levels can only be compared when the sounds have the same or similar spectral properties.

Measuring Sound-Pressure Level

A sound-level meter is designed to give readings of sound-pressure level. Sound pressure in decibels is referenced to the standard reference level, 20 μPa. The human hearing response is not flat across the audio band. For example, our hearing sensitivity particularly rolls off at low frequencies, and also at high frequencies. Moreover, this roll-off is more pronounced at softer listening levels. For this reason, to emulate human hearing, sound level meters usually offer a selection of weighting networks designated A, B, and C, having frequency responses shown in Fig. 2-3. The networks reduce the measured sound-pressure level at low and high frequencies. The A network is an inversion of the 40-phon hearing response, the B network is an inversion of 70-phon response, and the C network is an inversion of 100-phon response. Network selection is based on the

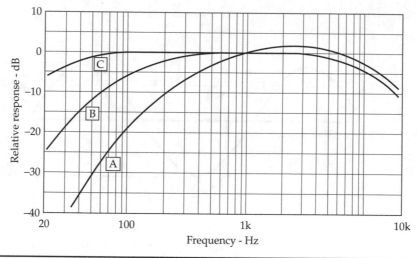

FIGURE 2-3   The A, B, and C weighting response characteristics for sound-level meters. (ANSI S1.4-1971.) Generally, only the A weighting is used.

general level of sounds to be measured (background noise, jet engines, and so on). For example:

- For sound-pressure levels of 20 to 55 dB, use network A.
- For sound-pressure levels of 55 to 85 dB, use network B.
- For sound-pressure levels of 85 to 140 dB, use network C.

These network response shapes were designed to bring the sound-level meter readings into closer conformance to the relative loudness of sounds. However, the B and C weightings often do not correspond to human perception. The A weighting is more commonly used. When a measurement is made with the A weighting, the value is designated as dBA. Such dBA readings are usually lower than unweighted dB readings. Since the A weighting is essentially flat above 1 kHz, any difference between a dBA and unweighted reading primarily shows differences in the low-frequency content of the signal. For example, a large difference in readings shows that the signal has significant low-frequency content.

Simple frequency weightings such as these cannot accurately represent loudness. Measurements made with simple weightings are not accepted as measurement loudness levels, but are only used for comparison of levels. Frequency analysis of sounds is recommended, using octave or $1/3$-octave bands.

Sine-Wave Measurements

The sine wave is a specific kind of alternating signal and is described by its own set of specific terms. Viewed on an oscilloscope, the easiest value to read is the peak-to-peak value (of voltage, current, sound pressure, or whatever the sine wave represents), as shown in Fig. 2-4. If the wave is symmetrical, the peak-to-peak value is twice the peak value.

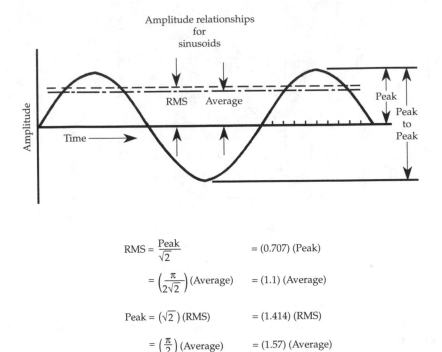

$$RMS = \frac{Peak}{\sqrt{2}} \qquad\qquad = (0.707)\,(Peak)$$

$$= \left(\frac{\pi}{2\sqrt{2}}\right)(Average) \qquad = (1.1)\,(Average)$$

$$Peak = (\sqrt{2}\,)(RMS) \qquad\qquad = (1.414)\,(RMS)$$

$$= \left(\frac{\pi}{2}\right)(Average) \qquad\quad = (1.57)\,(Average)$$

Figure 2-4   Amplitude relationships for sinusoids apply to sine waves of electrical voltage or current, as well as to acoustical parameters such as sound pressure. Another term used in the audio field is crest factor, or peak divided by RMS. These mathematical relationships apply to sine waves and cannot be applied to complex waveforms.

The common ac (alternating current) voltmeter is, in reality, a dc (direct current) instrument fitted with a rectifier that changes the alternating sine current to pulsating unidirectional current. The dc meter then responds to the average value, as indicated in Fig. 2-4. However, such meters are almost universally calibrated in terms of RMS (root-mean-square, described in the next paragraph). For pure sine waves, this is quite an acceptable fiction, but for nonsinusoidal waveforms, the reading will be in error.

An alternating current of one ampere RMS (or effective) is exactly equivalent in heating power to one ampere of direct current as it flows through a resistance of known value. After all, alternating current can heat a resistor or do work no matter which direction it flows; it is just a matter of evaluating it. In the right-hand positive curve of Fig. 2-4, the ordinates (height of lines to the curve) are read for each marked increment of time. Then (a) each of these ordinate values is squared, (b) the squared values are added together, (c) the average is found, and (d) the square root is taken of the average (or mean). Taking the square root of this average gives the RMS (root-mean-square) value of the positive curve of Fig. 2-4. The same can be done for the negative curve (squaring a negative ordinate gives a positive value), but simply doubling the positive curve of a symmetrical waveform is easier. In this way, the RMS or "heating-power" value of any alternating or periodic waves can be determined, whether the wave is for voltage, current, or sound pressure. Figure 2-4 also provides a summary of relationships pertaining only to the sine wave. It is important to remember that simple mathematical relationships properly describe sine waves, but they cannot be applied to complex sounds.

CHAPTER 3

Sound in the Free Field

M any practical acoustic problems are invariably associated with structures such as buildings and rooms, and vehicles such as airplanes and automobiles. These can generally be classified as problems in physics. These acoustical problems can be very complex in a physical sense; for example, a sound field might be comprised of thousands of reflected components, or temperature gradients might bend sound in an unpredictable manner. In contrast to practical problems, the simplest way to consider sound is in a free field, where its behavior is very predictable and analysis is straightforward. This analysis is useful because it allows us to understand the nature of sound waves in this undisturbed state. Then, these basic characteristics can be adapted to more complex problems.

The Free Field

Sound in a free field travels in straight lines and is unimpeded. Unimpeded sound is not subject to the many influences that we will consider in later chapters. Sound in a free field is unreflected, unabsorbed, undeflected, undiffracted, unrefracted, undiffused, and not subjected to resonance effects. In most practical applications, these are all factors that could (and do) affect sound leaving a source. An approximate free field can exist in anechoic chambers, special rooms where all the interior surfaces are covered by sound absorbers. But generally, a free field is a theoretical invention, a free space that allows sound to travel without interference.

A free space must not be confused with cosmological space. Sound cannot travel in a vacuum; it requires a medium such as air. Here, free space means any air space in which sound acts as though it is in a theoretical free field. In this unique environment, we must consider how sound diverges from a source, and how its intensity varies as a function of distance from the source.

Sound Divergence

Consider the point source of Fig. 3-1, radiating sound at a fixed power. The source can be considered as a point because its largest dimension is small (perhaps one-fifth or less) compared to the distances at which it is measured. For example, if the largest dimension of a source is 1 ft, it can be considered as a point source when measured at 5 ft or farther. Looked at in another way, the farther we are from a sound source, the more it behaves like a point source. In a free field, far from the influence of reflecting objects, sound from a point source is propagated spherically and uniformly in all directions. In addition, as described below, the intensity of sound decreases as the distance from the source increases.

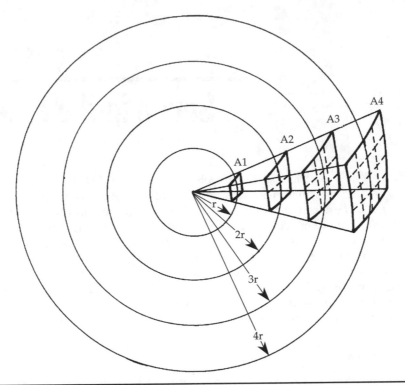

Figure 3-1   In the solid angle shown, the same sound energy is distributed over spherical surfaces of increasing area as radius *r* is increased. The intensity of sound is inversely proportional to the square of the distance from the point source.

This sound is of uniform intensity (power per unit area) in all directions. The circles represent spheres having radii in simple multiples. All of the sound power passing through the small square area *A*1 at radius *r* also passes through the areas *A*2, *A*3, and *A*4 at radii 2*r*, 3*r*, and 4*r*, respectively. The same sound power flows out through *A*1, *A*2, *A*3, and *A*4, but an increment of the total sound power traveling in this single direction is spread over increasingly greater areas as the radius is increased. Thus, intensity decreases with distance. This decrease is due to geometric spreading of the sound energy, and is not loss in the strict sense of the word.

Sound Intensity in the Free Field

Building on the discussion above (and referring again to Fig. 3-1), we observe that sound from a point source travels outward spherically. We also note that the area of a sphere is $4\pi r^2$. Therefore, the area of any small segment on the surface of the sphere also varies as the square of the radius. This means that the sound intensity (sound power per unit area) decreases as the square of the radius. This is an inverse square law. The intensity of a point-source sound in a free field is inversely proportional to the square of the distance from the source. In other words, intensity is proportional to $1/r^2$. More specifically:

$$I = \frac{W}{4\pi r^2}$$

(3-1)

where I = intensity of sound per unit area
       W = power of source
       r = distance from source (radius)

In this equation, since W and 4π are constants, we see that doubling the distance from r to $2r$ reduces the intensity I to $I/4$; this is because at twice the distance, the sound passes through an area that is four times the previous area. Likewise, tripling the distance reduces the intensity to $I/9$, and quadrupling the distance reduces intensity to $I/16$. Similarly, halving the distance from $2r$ to r increases the intensity to $4I$.

Sound Pressure in the Free Field

The intensity of sound (power per unit area) is a difficult parameter to measure. However, sound pressure is easily measured, for example, by using ordinary microphones. Because sound intensity is proportional to the square of sound pressure, the inverse square law (for sound intensity) becomes the inverse distance law (for sound pressure). In other words, sound pressure is inversely proportional to distance r. In particular:

$$P = \frac{k}{r} \tag{3-2}$$

where P = sound pressure
      k = a constant
      r = distance from source (radius)

   For every doubling of distance r from the sound source, sound pressure will be halved (not quartered). In Fig. 3-2, the sound-pressure level in decibels is plotted against

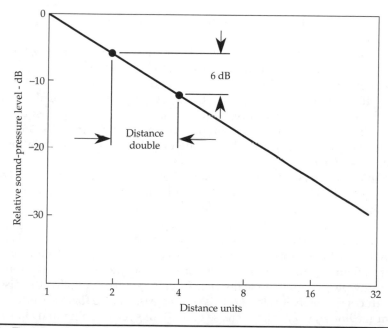

FIGURE 3-2 The inverse square law for sound intensity equates to the inverse distance law for sound pressure. This means that sound-pressure level is reduced 6 dB for each doubling of the distance.

distance. This illustrates the basis for the inverse distance law: when the distance from the source is doubled, the sound-pressure level decreases by 6 dB. This applies only for a free field. This law provides the basis for estimating the sound-pressure level in many practical circumstances.

Examples: Free-Field Sound Divergence

When the sound-pressure level L_1 at distance r_1 from a point source is known, the sound-pressure level L_2 at another distance r_2 can be calculated:

$$L_2 = L_1 - 20 \log \frac{r_2}{r_1} \text{ decibels} \qquad (3\text{-}3)$$

In other words, the difference in sound-pressure level between two points that are distances r_1 and r_2 from the source is:

$$L_2 - L_1 = 20 \log \frac{r_2}{r_1} \text{ decibels} \qquad (3\text{-}4)$$

For example, if a sound-pressure level of 80 dB is measured at 10 ft, what is the level at 15 ft?

$20 \log 10/15 = -3.5$ dB, and the level is $80 - 3.5 = 76.5$ dB.

What is the sound-pressure level at 7 ft?

$20 \log 10/7 = +3.1$ dB, and the level is $80 + 3.1 = 83.1$ dB.

This is only true for a free field in which sound diverges spherically, but this procedure may be helpful for rough estimates even under other conditions.

If a microphone is 5 ft from a singer and a VU meter in the control room peaks at +6 dB, moving the microphone to 10 ft would bring the reading down approximately 6 dB. The 6-dB figure is approximate because these distance relationships hold true only for free-field conditions. In practice, the effect of sound energy reflected from walls would change the doubling of distance relationship to something less than 6 dB.

An awareness of these relationships helps in estimating acoustical situations. For example, a doubling of the distance from 10 to 20 ft would, in free space, be accompanied by the same sound-pressure level decrease, 6 dB, as for a doubling from 100 to 200 ft. This accounts for the great carrying power of sound outdoors.

Even at a distance, not all sound sources behave as point sources. For example, traffic noise on a busy road can be better modeled as many point sources acting as a line source. Sound spreads in a cylinder around the line. In this case, sound intensity is inversely proportional to the distance from the source. Doubling the distance from r to $2r$ reduces the intensity I to $I/2$. This is a decrease of 3 dB for every doubling of distance from the line source, or an increase of 3 dB for every halving of distance from the line source.

Sound Fields in Enclosed Spaces

Free fields exist in enclosed spaces only in anechoic circumstances. In most rooms, in addition to the direct sound, reflections from the enclosing surfaces affect the way sound level decreases with distance. No longer does the inverse square law or the inverse distance law describe the entire sound field. In a free field, we can calculate sound level in terms of distance. In contrast, in a perfectly reverberant sound field, the sound level is equal everywhere in the sound field. In practice, rooms yield a combination of these two extremes, with both direct sound and reflected sound.

For example, assume that there is a loudspeaker in an enclosed space that is capable of producing a sound-pressure level of 100 dB at a distance of 4 ft. This is shown in the graph of Fig. 3-3. In the region very close to the loudspeaker, the sound field is in considerable disarray. The loudspeaker, at such close distances, cannot be considered a point source. This region is called the near field. In the near field, sound level decreases about 12 dB for every doubling of distance. This near-field region (not to be confused with "near-field" or "close-field" studio monitoring) is of limited practical interest.

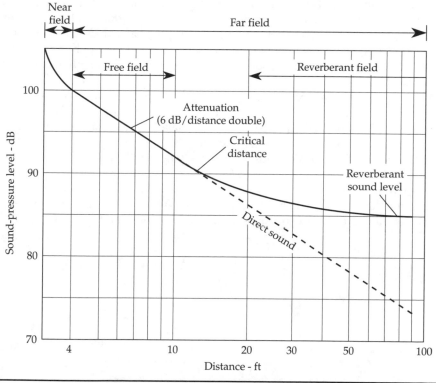

FIGURE 3-3 Even in an enclosed space, an approximate free field can exist close to the source, where sound-pressure level is attenuated 6 dB for every doubling of distance. By definition, the critical distance is that distance at which the direct sound-pressure level is equal to the reverberant sound-pressure level. Farther from the source, sound-pressure level becomes constant and depends on the amount of absorption in the room.

After moving several loudspeaker dimensions away from the loudspeaker, significant measurements can be made in the far field. This far field consists of the free field and the reverberant field, and a transition region between them. Free-field conditions exist near the loudspeaker where it operates as a point source. Direct sound is predominant, spherical divergence prevails in this limited space, and reflections from the surfaces are of negligible comparative level. In this region, sound-pressure level decreases 6 dB for every doubling of distance.

Moving away from the loudspeaker, sound reflected from the surfaces of the room affects the result. We define the critical distance, the location in a room where the direct and the reflected sounds are equal. The critical distance is useful as a rough single-figure description of the acoustics of the environment. Still farther away, the reverberant sound field dominates. Its level remains constant, even at greater distances from the source. This level depends on the amount of absorption in the room. For example, in a very absorptive room, the reverberant level will be lower.

Hemispherical Field and Propagation

True spherical divergence requires no reflecting surfaces at all. Because this is difficult to achieve, it is sometimes approximated by placing a sound source such as a loudspeaker face-up, on a hard, reflective surface. This creates a hemispherical sound field, radiating upward. This can be used, for example, to measure the response of the loudspeaker. In this case, the sound field radiates over a surface that is $2\pi r^2$ in area. Therefore the intensity varies as $I = W/2\pi r^2$. As in a spherical sound field, the sound-pressure level in a hemispherical field attenuates by 6 dB for every doubling of distance. However, in a hemispherical sound field, the level begins 3 dB higher than in a spherical sound field.

How do we characterize hemispherical sound propagation over the earth's surface? Estimates made by the "6 dB per doubling distance" rule are only rough approximations. Reflections from the earth outdoors usually tend to make the sound level with distance something less than that indicated by the 6 dB approximation. The reflective efficiency of the earth's surface varies from place to place. Consider the sound level of a sound at 10 ft and again at 20 ft from the source. In practice, the difference between the two will probably be closer to 4 dB than 6 dB. For such outdoor measurements, the distance law must be taken at "4 or 5 dB per doubling distance." General environmental noise can also influence the measurement of specific sound sources.

CHAPTER 4

The Perception of Sound

The study of the physical structure of the ear is a study in physiology. The study of human perception of sound is one of psychology and psychoacoustics. Psychoacoustics is an inclusive science embracing the physical structure of the ear, the sound pathways and their function, the human perception of sound, and their interrelationships. In many ways, psychoacoustics is the human basis for the entire field of audio engineering. Its role in the design of perceptual codecs such as MP3 and WMA is obvious, but psychoacoustics is also vital in architectural acoustics, telling us, for example, how a room's sound field is interpreted by the listener.

A stimulus wave striking the ear sets in motion mechanical movements that result in electrical discharges that are sent to the brain and create the sensation that we call sound. How are sounds recognized and interpreted? That basic question may seem simple, but despite vigorous research activities on all aspects of human hearing, our knowledge is still incomplete. The human ear is, by far, the most complex device in all of audio engineering.

Listening to a symphony orchestra, concentrate first on the violins, then the cellos and the double basses. Now focus your attention on the clarinets, then the oboes, bassoons, and flutes. This is a remarkable power of the human auditory system. In the ear canal, all these sounds are mixed together, and the ear-brain succeeds in separating them from a complex sound wave. By rigorous training, a keen observer can listen to the sound of a violin and pick out the various overtones apart from the fundamental.

Sensitivity of the Ear

The sensitive nature of our hearing can be underscored by a thought experiment. The bulky door of an anechoic chamber is opened, revealing extremely thick walls, and 3-ft wedges of glass fiber, pointing inward, lining all walls, ceiling, and what could be called the floor, except that you walk on an open steel grillwork.

You sit on a chair. This experiment takes time, and you lean back, patiently waiting. It is very eerie in here. The sea of sound and noises of life and activity in which we are normally immersed and of which we are ordinarily scarcely conscious is now conspicuous by its absence.

The silence presses down on you in the tomblike silence as the minutes pass. New sounds are discovered, sounds that come from within your body. The pounding of your heart and the blood coursing through the vessels become audible. If your ears are keen, your patience is rewarded by a hissing sound between the thumps of the heart and the slushing of blood—the sound of air particles striking your eardrums.

The human ear cannot detect sounds softer than the motion of air particles on the eardrum. This is the threshold of hearing. There would be no reason to have ears more sensitive, because any lower-level sound would be drowned by the air-particle noise. This means that the ultimate sensitivity of our hearing just matches the softest sounds possible in an air medium.

Leaving the anechoic chamber, imagine being thrust into the loudest imaginable acoustical environments. At this extreme, our ears can respond to the roar of a cannon, the noise of a rocket blastoff, or a jet aircraft under full power. Physiological features of the ear help protect the sensitive mechanism from damage at the threshold of feeling (where a tingling sensation is felt), but sudden or intense noises can easily cause temporary shifts in the hearing response. At the extreme threshold of pain, some permanent damage is probably inevitable.

Ear Anatomy

The three principal parts of the human auditory system, the outer ear, middle ear, and inner ear, are shown in Fig. 4-1. The outer ear is composed of the pinna and the auditory canal or auditory meatus. The auditory canal is terminated by the tympanic membrane or the eardrum. The middle ear is an air-filled cavity spanned by three tiny bones called the ossicles: the malleus, incus, and stapes. These three bones are also sometimes called the hammer, anvil, and stirrup, respectively, because of their shapes. The malleus is attached to the eardrum and the stapes is attached to the oval window of the inner ear. Together these three bones form a mechanical, lever-action connection between the air-actuated eardrum and the fluid-filled cochlea of the inner ear. The inner ear is terminated in the auditory nerve, which sends electrical impulses to the brain.

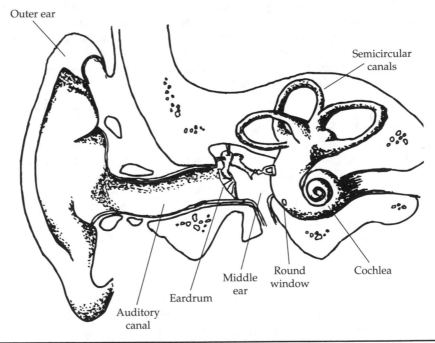

FIGURE 4-1   The human ear receives sound from the outer ear and it passes through the auditory canal. The middle ear connects the eardrum in air to the cochlea in fluid. The cochlea in the inner ear converts sound energy to electrical impulses that are transmitted to the brain.

The Outer Ear

Among its other functions, the outer ear, known as the pinna, operates as a sound-gathering device. Cupping your hand behind the ear increases the effective size of the pinna and thus the apparent loudness by an amount varying with frequency. For the important speech frequencies (2,000 to 3,000 Hz), sound pressure at the eardrum is increased about 5 dB. Some animals are able to move their ears to point at the source, to help amplify the sound; humans lack this ability, but it is common (and often unconscious) to slightly move ones head to better hear a sound and determine its location.

The pinna also performs a crucial function in imprinting directional information on all sounds picked up by the ear. This means that information concerning the direction of the source is superimposed on the sound content itself so that the resultant sound pressure on the eardrum enables the brain to interpret both the content of the sound and the direction from which it comes. The pinna provides a differentiation of sound from the front as compared to sound from the rear. It also allows differentiation of sound from all around the listener. At least in a free field, it is easy to close your eyes, and accurately point to sound sources.

Directional Cues: An Experiment

A simple psychoacoustical experiment can demonstrate how simple changes in sounds falling on the ear can yield subjective directional impressions. Listen with a headphone on one ear to an octave bandwidth of random noise arranged with an adjustable notch filter. Adjusting the filter to 7.2 kHz will cause the noise to seem to come from a source on the level of the observer. With the notch adjusted to 8 kHz, the sound seems to come from above. With the notch at 6.3 kHz, the sound seems to come from below. This experiment demonstrates that the human hearing system extracts directional information from the shape of the sound spectra at the eardrum.

The Ear Canal

The ear canal also increases the loudness of the sounds traversing it. In Fig. 4-2, the ear canal, with an average diameter of about 0.7 cm and length of about 2.5 cm, is idealized by straightening and giving it a uniform diameter throughout its length. Acoustically, this is a reasonable approximation. It is a pipelike duct, closed at the inner end by the eardrum.

The acoustical similarity of this ear canal to an organ pipe is apparent. As with a pipe closed at one end, the resonance effect of the ear canal increases sound pressure at the eardrum at certain frequencies. The maximum is near the frequency at which the 2.5-cm pipe is one-quarter wavelength—about 3,000 Hz.

Figure 4-3 shows the increase in sound pressure at the eardrum over that at the opening of the ear canal. A primary peak is noted around 3,000 Hz caused by the quarter-wave pipe resonance effect. The primary pipe resonance amplifies the sound pressure at the eardrum by approximately 12 dB at the major resonance at about 4,000 Hz. There is a secondary resonance nearer 9,000 Hz of lower peak pressure. In addition, a plane wave striking the front of the head will be diffracted; this diffraction also increases sound pressure at the ears at midrange frequencies. These effects combine to make the ear most sensitive to midrange frequencies, the same frequencies occupied by speech. Unfortunately, these same resonances also make the ear susceptible to hearing loss at these important midrange frequencies.

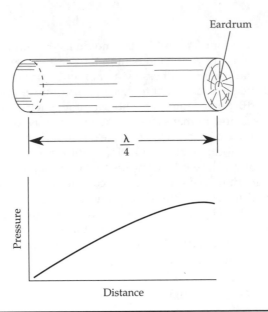

Figure 4-2   The auditory canal, closed at one end by the eardrum, acts as a quarter-wavelength pipe. Resonance provides acoustic amplification for voice frequencies.

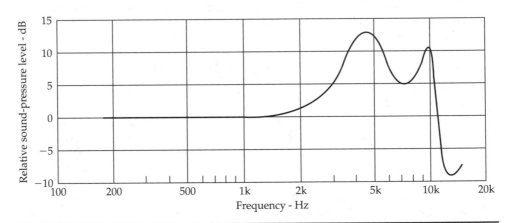

Figure 4-3   The transfer function (frequency response) of the ear canal shows the effect of resonance. This is a fixed component that is combined with every directionally encoded sound reaching the eardrum. See also Figs. 4-15 and 4-16. (*Mehrgardt and Mellart*)

The Middle Ear

Transmitting sound energy from a tenuous medium such as air into a dense medium like water poses challenges. Without some transfer mechanism, sound originating in air bounces off water like light off a mirror. For efficient energy transfer, the two imped-ances must be matched; in this case the impedance ratio is about 4,000:1. It would be very unsatisfactory to drive the 1-Ω voice coil of a loudspeaker with an amplifier having an output impedance of 4,000 Ω; not much power would be transferred. Similarly, the ear must provide a way for energy in air to enter the ear's fluid interior.

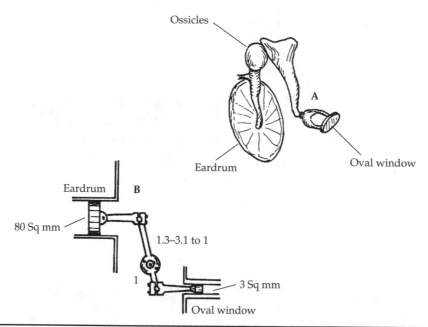

FIGURE 4-4   The middle ear provides impedance matching. (A) The ossicles (hammer, anvil, and stirrup) of the middle ear transmit mechanical vibrations of the eardrum to the oval window of the cochlea. (B) A mechanical analog of the impedance-matching function of the middle ear. The difference in area between the eardrum and the oval window, coupled with the step-down mechanical linkage, match the motion of the air-actuated eardrum to the fluid-loaded oval window.

The object is to get the energy represented by the vibratory motion of an eardrum diaphragm, transferred with maximum efficiency to the fluid of the inner ear. The two-fold solution is suggested in Fig. 4-4. The three ossicles (hammer, anvil, and stirrup, as shown in Fig. 4-4A) form a mechanical linkage between the eardrum and the oval window, which is in intimate contact with the fluid of the inner ear. The first of the three bones, the malleus, is fastened to the eardrum. The third, the stapes, is actually a part of the oval window. There is a lever action in this linkage with a ratio leverage ranging from 1.3:1 to 3.1:1. That is, the eardrum motion is reduced by this amount at the oval window of the inner ear.

This is only part of the mechanical-impedance-matching device. The area of the eardrum is about 80 mm², and the area of the oval window is only 3 mm². Hence, a given force on the eardrum is reduced by the ratio of 80/3, or about 27-fold.

In Fig. 4-4B, the action of the middle ear is likened to two pistons with area ratios of 27:1 connected by an articulated connecting rod having a lever arm ranging from 1.3:1 to 3.1:1, making a total mechanical force increase of between 35 and 80 times. The acoustical impedance ratio between air and water being on the order of 4,000:1, the pressure ratio required to match two media would be $\sqrt{4000}$, or about 63.2. We note that this falls within the 35 to 80 range obtained from the mechanics of the middle ear illustrated in Fig. 4-4B. It is interesting that the ossicles in infants are fully formed and do not grow significantly larger over time; any change in size would only decrease the efficiency of the energy transfer.

The problem of matching sound in air to sound in the fluid of the inner ear is solved by the mechanics of the middle ear. The evidence that the impedance matching plus the

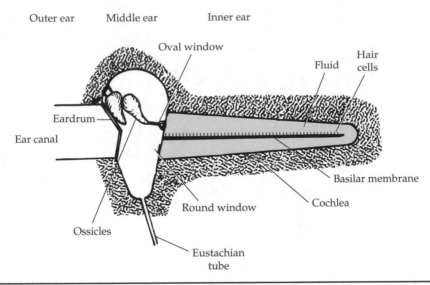

Outer ear   Middle ear   Inner ear

Oval window   Hair
                    Fluid   cells

Eardrum—

Ear canal

                                    Basilar membrane

                                    Cochlea

Ossicles   Round window

          Eustachian
          tube

Figure 4-5   Idealized sketch of the human ear showing the uncoiled fluid-filled cochlea. Sound entering the ear canal causes the eardrum to vibrate. This vibration is transmitted to the cochlea through the mechanical linkage of the middle ear. The sound is analyzed through standing waves set up on the basilar membrane.

resonance amplification of Fig. 4-3 really work is that a diaphragm motion comparable to molecular dimensions gives a threshold perception.

A schematic of the ear is shown in Fig. 4-5. The conical eardrum at the inner end of the auditory canal forms one side of the air-filled middle ear. The middle ear is vented to the upper throat behind the nasal cavity by the Eustachian tube. The eardrum operates as an "acoustic suspension" system, acting against the compliance of the trapped air in the middle ear. The Eustachian tube is suitably small and constricted so as not to destroy this compliance. The round window separates the air-filled middle ear from the practically incompressible fluid of the inner ear.

The Eustachian tube fulfills a second function by equalizing the static air pressure of the middle ear with the outside atmospheric pressure so that the eardrum and the delicate membranes of the inner ear can function properly. Whenever we swallow, the Eustachian tubes open, equalizing the middle ear pressure. Changes in external air pressure (such as when an aircraft without a pressurized cabin undergoes rapid changes in altitude) may cause momentary deafness or pain until the middle ear pressure is equalized by swallowing. Finally, the Eustachian tube has a third emergency function of drainage if the middle ear becomes infected.

The Inner Ear

The acoustical amplifiers and mechanical impedance matching features of the middle ear, discussed so far, are relatively well understood. The intricate operation of the cochlea is not as explicit.

The cochlea, the sound-analyzing organ, is in close proximity to the three mutually perpendicular; semicircular canals of the vestibular mechanism, the balancing organ (see Fig. 4-1). The same fluid permeates all, but their functions are independent. The cochlea is about the size of a pea and is encased in solid bone. It is coiled up like a cockleshell from which it gets its name. For the purposes of illustration, this 2¾-turn coil has been stretched out its full length, about 1 in, as shown in Fig. 4-5. The fluid-filled inner ear is divided lengthwise by two membranes, Reissner's membrane and the basilar membrane. Of immediate interest is the basilar membrane and its response to sound vibrations in the fluid.

Vibration of the eardrum activates the ossicles. The motion of the stapes, attached to the oval window, causes the fluid of the inner ear to vibrate. An inward movement of the oval window results in a flow of fluid around the distant end of the basilar membrane, causing an outward movement of the membrane of the round window; the round window thus provides pressure release. Sound actuating the oval window results in standing waves being set up on the basilar membrane. The position of the amplitude peak of the standing wave on the basilar membrane changes as the frequency of the exciting sound is changed.

Low-frequency sound results in maximum amplitude near the distant end of the basilar membrane; high-frequency sound produces peaks near the oval window. For a complex signal such as music or speech, many momentary peaks are produced, constantly shifting in amplitude and position along the basilar membrane. These resonant peaks on the basilar membrane were originally thought to be so broad as to be unable to explain the sharpness of frequency discrimination displayed by the human ear. More recent research shows that at low sound intensities, the basilar membrane tuning curves are very sharp, broadening only for intense sound. It appears that the sharpness of the basilar membrane's mechanical tuning curves is comparable to the sharpness of single auditory nerve fibers, which innervate it.

Stereocilia

Waves set up on the basilar membrane in the fluid-filled duct of the inner ear stimulate hairlike nerve terminals that convey signals to the brain in the form of neuron discharges. There is one row of inner hair cells and three to five rows of outer hair cells. Each hair cell contains a bundle of tiny hairs called stereocilia. As sound causes the cochlear fluid and the basilar membrane to move, the stereocilia vibrate according to the vibrations around them. Stereocilia at various locations along the basilar membrane are stimulated by characteristic frequencies corresponding to that location. The inner hair cells act like microphones, transducers that convert mechanical vibration to electrical signals that initiate neural discharges to the auditory nerve and the brain. The outer hair cells provide additional gain or attenuation to more sharply tune the output of the inner hair cells and to make the hearing system more sensitive.

When sound excites the fluid of the inner ear, the basilar membrane and hair cells are stimulated, sending an electrical wave through the surrounding tissue. These so-called microphonic potentials are picked up and amplified, reproducing the sound falling on the ear. These potentials are proportional to the sound pressure and linear in their response over an 80-dB range. This microphonic potential is different from the action potentials of the auditory nerve, which conveys information to the brain.

Bending of the stereocilia triggers the nerve impulses that are carried by the auditory nerve to the brain. The microphonic signals are analog, but the impulses sent to the brain are generated by neuron discharges. A single nerve fiber is either firing or not firing, in binary fashion. When a nerve fires, it causes an adjoining one to fire, and so on. Physiologists liken the process to a burning gunpowder fuse. The rate of travel bears no relationship to how the fuse was lighted. Presumably the loudness of the sound is related to the number of nerve fibers excited and the repetition rates of such excitation. When all the nerve fibers are excited, this is the maximum loudness that can be perceived. The threshold sensitivity would be represented by a single fiber firing. The sensitivity of the system is remarkable; at the threshold of hearing, the faintest sound we can hear, tiny filaments associated with the stereocilia move about 0.04 nm—about the radius of a hydrogen atom.

A well-accepted theory of how the inner ear and the brain function has not yet been formulated. The presentation here is a highly simplified explanation of a very complex mechanism. Some of the theories discussed here are not universally accepted.

Loudness versus Frequency

The seminal work on loudness was done by Fletcher and Munson at Bell Laboratories, and reported in 1933. Since that time, refinements have been added by others. The family of equal-loudness contours shown in Fig. 4-6, more recent work by Robinson and Dadson has been adopted as an international standard (ISO 226).

Each equal-loudness contour is identified by its value at the reference frequency of 1 kHz, and the term, loudness level in phons, is thus defined. For example, the equal-loudness contour passing through 40-dB sound-pressure level at 1 kHz is called the

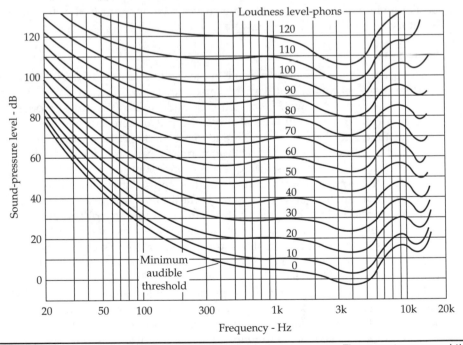

Figure 4-6 Equal-loudness contours of the human ear for pure tones. These contours reveal the relative lack of sensitivity of the ear to bass tones, especially at lower sound levels. Inverting these curves give the frequency response of the ear in terms of loudness level. These data are taken for a sound source directly in front of the listener, binaural listening, and subjects aged 18 to 25. (*Robinson and Dadson*)

40-phon contour. Similarly, the 100-phon contour passes through 100 dB at 1 kHz. The line of each contour shows how sound-pressure level must be varied at different frequencies to sound equally loud as the 1-kHz reference loudness of 40 phons. Each contour was obtained experimentally by asking subjects to state when levels at different frequencies sounded equally loud as the reference level at 1 kHz, at each of the 13 different reference levels. These data are for pure tones and do not apply directly to music and other audio signals. Loudness is a subjective term; sound-pressure level is strictly a physical term. Loudness level is also a physical term that is useful in estimating the loudness of a sound (in units of sones) from sound-level measurements. However, a loudness level is not the same as a SPL reading. The shapes of the equal-loudness contours contain subjective information because they were obtained by a subjective comparison of the loudness of a tone to its loudness at 1 kHz.

The curves of Fig. 4-6 reveal that perceived loudness varies greatly with frequency and sound-pressure level. For example, a sound-pressure level of 30 dB yields a loudness level of 30 phons at 1 kHz, but it requires a sound-pressure level of 58 dB more to sound equally loud at 20 Hz as shown in Fig. 4-7. The curves tend to flatten at greater loudness, showing that the ear's response is more uniform at high levels. The 90-phon curve rises only 32 dB between 1,000 and 20 Hz. Note that inverting the curves of Fig. 4-7 gives the frequency response of the ear in terms of loudness level. The ear is less sensitive to bass notes than midband notes at low levels. This bass deficiency of the ear means that the quality of reproduced music depends on the volume-control setting. Listening to background music at low levels requires a different frequency response

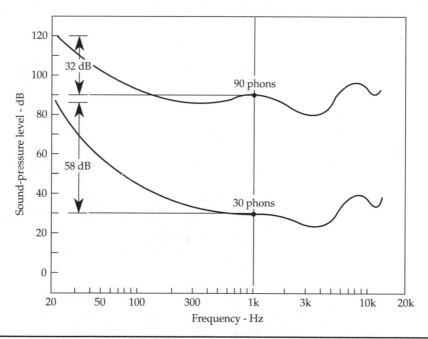

Figure 4-7   A comparison of the ear's response at 20 Hz, to the response at 1 kHz. At a loudness level of 30 phons, the sound-pressure level of a 20-Hz tone must be 58 dB higher than that at 1 kHz to have the same loudness. At a 90-phon loudness level, an increase of only 32 dB is required. The ear's response is somewhat flatter at high loudness levels. Loudness level is only an intermediate step to true subjective loudness.

than listening at higher levels. There are also deviations in the ear's high-frequency response, but these are relatively less noticeable.

Loudness Control

Suppose that you want to play back a recording at a quiet level (assume 50 phons). If the music was originally played and recorded at a higher level (assume 80 phons), you would need to increase both bass and treble for proper balance. The loudness control found on many audio devices compensates for the change in frequency response of the ear for different loudness levels. But the curve corresponding to a given setting of the loudness control might apply only to a specific loudness level of reproduced sound. Most loudness controls offer an incomplete solution to the problem. Think of all the things that affect the volume-control setting. The loudspeakers vary in acoustic output for a given input power. The gain of power amplifiers differs. Listening-room conditions vary from dead to highly reverberant. For a loudness control to function properly, the system must be calibrated and the loudness control must adaptively vary the frequency response relative to level at the listener.

Area of Audibility

Curves A and B shown in Fig. 4-8 were obtained from groups of trained listeners. The listeners faced the sound source and judged whether a tone at a given frequency was barely audible (curve A) or beginning to be painful (curve B). These two curves therefore represent the extremes of our perception of loudness.

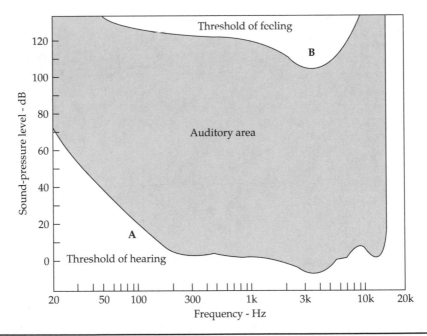

FIGURE 4-8   The auditory area of the human ear is bounded by two threshold curves. (A) The threshold of hearing delineates the lowest level sounds the ear can detect. (B) The threshold of feeling defines the upper extreme. All of our auditory experiences occur within this area.

Curve A, the threshold of hearing, represents the level at each frequency where sounds are just barely audible. The curve also shows that human ears are most sensitive around 3 kHz. Another way to state this is that around 3 kHz a lower-level sound elicits a greater threshold response than higher or lower frequencies. At this most sensitive region, a sound-pressure level defined as 0 dB can just barely be heard by a person of average hearing acuity. The reference level of pressure of 20 μPa was selected to establish this 0-dB level.

Curve B, the threshold of feeling, represents the level at each frequency at which a tickling sensation is felt in the ears. At 3 kHz, this occurs at a sound-pressure level of about 110 dB. Further increase in level results in an increase in feeling until a sensation of pain is produced. The threshold tickling is a warning that the sound is becoming dangerously loud and that ear damage is either imminent or has already taken place.

Between the threshold of hearing and the threshold of feeling is the area of audibility. This area has two dimensions: the vertical range of sound-pressure level and the horizontal range of frequencies that the ear can perceive. All the sounds that humans experience must be of such a frequency and level as to fall within this auditory area.

The area of audibility for humans is quite different from that of many animals. The bat specializes in sonar cries that are far above the upper frequency limit of our ears. The hearing of dogs extends higher than ours, hence the usefulness of ultrasonic dog whistles. Sound in the infrasonic and ultrasonic regions, as related to the hearing of humans, is no less true sound in the physical sense, but it does not result in human perception.

Loudness versus Sound-Pressure Level

The phon is a unit of physical loudness level that is referenced to a sound-pressure level at 1 kHz. This is useful, but it tells us little about human reaction to loudness of sound. Some sort of subjective unit of loudness is needed. Many experiments conducted with hundreds of subjects and many types of sound have yielded a consensus that for a 10-dB increase in sound-pressure level, the average person reports that loudness is doubled. For a 10-dB decrease in sound level, subjective loudness is cut in half. The sone is a unit of subjective loudness. One sone is defined as the loudness (not loudness level) experienced by a person listening to a 1-kHz tone of 40-phon loudness level. A sound of 2 sones is twice as loud and 0.5 sone half as loud.

Figure 4-9 shows a graph for translating sound-pressure levels to loudness in sones. One point on the graph is the definition of the sone, the loudness experienced by a person hearing a 1-kHz tone at 40-dB sound-pressure level, or 40 phons. A loudness of 2 sones is then 10 dB higher; a loudness of 0.5 sones is 10 dB lower. A straight line can be drawn through these three points, which can then be extrapolated for sounds of higher and lower loudness. This graph only applies to 1-kHz tones.

The idea of subjective loudness has great practical value. For example, a consultant might be required by a court to give an opinion on the loudness of an industrial noise that bothers neighbors. The consultant can make a $\frac{1}{3}$-octave analysis of the noise, translate the sound-pressure levels of each band to sones (using graphs such as Fig. 4-9), add together the sones of each band, and arrive at an estimate of the loudness of the noise. It is convenient to be able to add component sones; adding decibels of sound-pressure levels can be confusing.

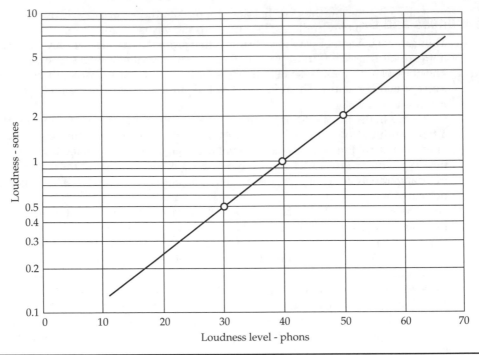

FIGURE 4-9   The graphical relationship between subjective loudness in sones and physical loudness level in phons. This graph only applies to 1-kHz tones.

Table 4-1 shows the relationship between loudness level in phons and the subjective loudness in sones. Although most audio engineers have little use for phons or sones, it is good to realize that a true subjective unit of loudness (sone) is related to loudness level (phon), which in turn is related by definition to what we measure with a sound-level meter. There are empirical methods for calculating the loudness of sound as perceived by humans from purely physical measurements of sound spectra, such as those measured with a sound-level meter and an octave or $\frac{1}{3}$-octave filter.

Loudness Level (Phons)	Subjective Loudness (Sones)	Typical Examples
100	64	Heavy truck passing
80	16	Talking loudly
60	4	Talking softly
40	1	Quiet room
20	0.25	Very quiet studio

TABLE 4-1   Loudness Level in Phons versus Loudness in Sones

Loudness and Bandwidth

Thus far, we have discussed loudness in terms of single-frequency tones, but tones do not give all the information we need to relate subjective loudness to meter readings. The noise of a jet aircraft taking off, for example, sounds much louder than a tone of the same sound-pressure level. The bandwidth of the noise affects the loudness of the sound, at least within certain limits.

Figure 4-10A represents three sounds having the same sound-pressure level of 60 dB. Their bandwidths are 100, 160, and 200 Hz, but their heights (representing sound intensity per hertz) vary so that their areas are equal. In other words, the three sounds have equal intensities. (Sound intensity has a specific meaning in acoustics and is not to be equated to sound pressure. Sound intensity is proportional to the square of sound pressure for a plane progressive wave.) However, the three sounds of Fig. 4-10A do not have the same loudness. Figure 4-10B shows how a bandwidth of noise having a constant 60-dB sound-pressure level and centered on 1 kHz is related to loudness as experimentally determined. The 100-Hz bandwidth noise has a loudness level of 60 phons and a loudness of 4 sones. The 160-Hz bandwidth noise has the same loudness. But something unexpected happens as the bandwidth is increased beyond 160 Hz. From 160 Hz upward, increasing bandwidth increases loudness. For example, the loudness of the 200-Hz bandwidth noise is louder. Why the sharp change at 160 Hz?

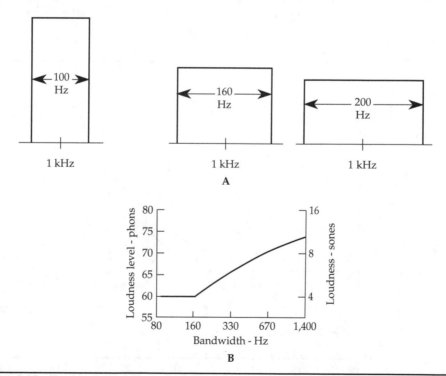

Figure 4-10   The bandwidth affects the loudness of sounds. (A) Three noises of different bandwidths, but all having the same sound-pressure level of 60 dB. (B) The subjective loudness of the 100- and 160-Hz noise is the same, but the 200-Hz band sounds louder because it exceeds the 160-Hz critical bandwidth of the ear at 1 kHz.

The reason is that 160 Hz is the width of the ear's critical band at 1 kHz. If a 1-kHz tone is presented to a listener along with random noise, only the noise in a band 160-Hz wide is effective in masking the tone. In other words, the ear acts like an analyzer composed of a set of bandpass filters stretching throughout the audible spectrum. This filter set is not like that found in an electronics laboratory. The common ⅓-octave filter set may have 28 adjacent filters fixed and overlapping at the −3 dB points. The ear's critical band filters are continuous; no matter what frequency we choose, there is a critical band centered on that frequency.

Research shows how the width of the critical-band filters varies with frequency. This bandwidth function is shown in Fig. 4-11. This shows that critical bands become much wider at higher frequencies. There are other methods for measuring critical bandwidth; they provide different estimates particularly below 500 Hz. For example, the equivalent rectangular bandwidth (ERB) (that applies to young listeners at moderate sound levels) is based on mathematical methods. The ERB offers the convenience of being calculable from the equation:

$$ERB = 6.23f^2 + 93.3f + 28.52 \ \text{Hz} \tag{4-1}$$

where f = frequency, kHz.

One-third-octave filter sets have been justified in certain measurements because the filter bandwidths approach those of the critical bands of the ear. For comparison, a plot of ⅓-octave bandwidths is included in Fig. 4-11. One-third-octave bands are 23.2% of

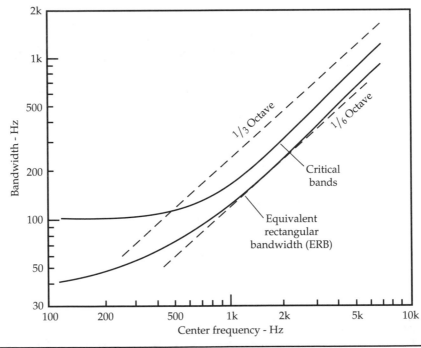

Figure 4-11   A comparison of bandwidths of ⅓- and ⅙-octave bands, critical bands of the ear, and equivalent rectangular bandwidth (ERB).

the center frequency. The critical-band function is about 17% of the center frequency. The ERB function is about 12%; this is close to that of $\frac{1}{6}$-octave bands (11.6%). This suggests that $\frac{1}{6}$-octave filter sets may be at least as relevant as $\frac{1}{3}$-octave filters.

Critical bands are important in many audio disciplines. For example, codecs such as AAC, MP3, and WMA are based on the principle of masking. A tone (a music signal) will mask quantization noise that lies within a critical band centered at the tone's frequency. However, if the noise extends outside the critical band, it will not be masked by the tone. One critical band is defined as having a width of 1 bark (named after German physicist Heinrich Barkhausen).

Loudness of Impulses

The examples discussed so far have been concerned with steady-state tones and noise. How does the ear respond to transients of short duration? This is important because music and speech are essentially made up of transients. To focus attention on this aspect of speech and music, play some audio tracks backward. The initial transients now appear at the ends of syllables and musical notes and stand out prominently.

As a 1-sec tone burst, a 1-kHz tone sounds like 1 kHz. But an extremely short burst of the same tone sounds like a click. The duration of such a burst also influences the perceived loudness. Short bursts do not sound as loud as longer ones. Figure 4-12 shows how much the level of shorter pulses must be increased to have the same loudness as a long pulse or steady tone. A 3-msec pulse must have a level about 15 dB higher to sound as loud as a 0.5-sec (500-msec) pulse. Tones and random noise follow roughly the same relationship in loudness versus pulse length.

The region less than 100 msec in Fig. 4-12 is significant. Only when the tones or noise bursts are shorter than this amount must the sound-pressure level be increased to produce a loudness equal to that of long pulses or steady tones or noise. This 100 msec

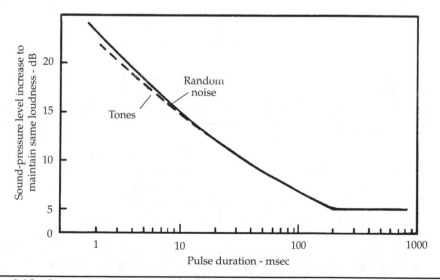

FIGURE 4-12 Short pulses of tones or noise are less audible than longer pulses. The discontinuity of the 100- to 200-msec region is related to the integrating time of the ear.

appears to be the maximum integrating time or the time constant of the human ear. In particular, events occurring within 35 msec, such as reflections from walls, are integrated by the ear with regard to level. This shows that the ear responds to sound energy averaged over time.

Figure 4-12 shows that our ears are less sensitive to short transients such as peaks in sound level. This has a direct bearing on understanding speech. The consonants of speech determine the meaning of many words. For instance, the only difference between bat, bad, back, bass, ban, and bath are the consonants at the end. The words led, red, shed, bed, fed, and wed have the all-important consonants at the beginning. No matter where they occur, consonants are transients having durations on the order of 5 to 15 msec. Figure 4-12 shows that transients this short must be louder to be comparable to longer sounds. In the above words, each consonant is not only much shorter than the rest of the word, it is also at a lower level. Thus we need good listening conditions to distinguish between such sets of words. Too much background noise or too much reverberation can seriously impair the understandability of speech because they can mask important, lower-level consonants.

Audibility of Loudness Changes

As we have seen, the ear is sensitive to a wide dynamic range of sounds from the softest to loudest. Within that range, how sensitive is the ear to small changes in loudness? Steps of 5 dB are definitely audible, while steps of 0.5 dB may be inaudible, depending on circumstances. Detecting differences in intensity varies somewhat with frequency and also with sound level.

At 1 kHz, for very low levels, a 3-dB change is the least detectable by the ear, but at high levels the ear can detect a 0.25-dB change. A very low-level, 35-Hz tone requires a 9-dB level change to be detectable. For the important midfrequency range and for commonly used levels, the minimum detectable change in level that the ear can detect is about 2 dB. In most cases, making level changes in increments less than these is usually unnecessary.

Pitch versus Frequency

Pitch is a subjective term. It is chiefly a function of frequency, but it is not linearly related to it. Because pitch is somewhat different from frequency, it requires another subjective unit, the mel. Frequency is a physical term measured in hertz. Although a soft 1-kHz signal is still 1 kHz if you increase its level, the pitch of a sound may depend on sound-pressure level. A reference pitch of 1,000 mels is defined as the pitch of a 1-kHz tone with a sound-pressure level of 60 dB. The relationship between pitch and frequency, determined by experiments with juries of listeners, is shown in Fig. 4-13. On the experimental curve, 1,000 mels coincides with 1 kHz; thus the sound-pressure level for this curve is 60 dB. The shape of the curve of Fig. 4-13 is similar to a plot of position along the basilar membrane as a function of frequency. This suggests that pitch is related to action on this membrane.

Researchers tell us that the human ear can detect about 280 discernible steps in intensity and some 1,400 discernible steps in pitch. As changes in intensity and pitch are vital to audio communication, it would be interesting to know how many combinations are possible. Offhand, it might seem that there would be $280 \times 1,400 = 392,000$ combinations

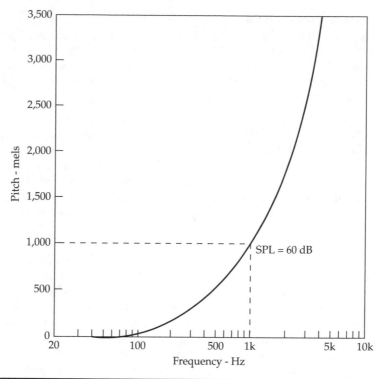

FIGURE 4-13 Pitch (in mels, a subjective unit) is related to frequency (in hertz, a physical unit) according to this curve obtained by juries of listeners. (*Stevens and Volkman*)

detectable by the ear. This is overly optimistic because the tests were conducted by comparing two simple, single-frequency sounds in rapid succession and bears little resemblance to the complexities of commonly heard sounds. Other experiments show that the ear can detect only about 7° of loudness and 7° of pitch or only 49 pitch-loudness combinations. This is not too far from the number of phonemes (the smallest unit in a language that distinguishes one utterance from another) which can be detected in a language.

An Experiment in Pitch

The level of sound affects the perception of pitch. For low frequencies, the pitch goes down as the level of sound is increased. At high frequencies, the reverse takes place; the pitch increases with sound level.

The following is an experiment suggested by Fletcher. Two audio oscillators are required, as well as a frequency counter. One oscillator is applied to the input of one channel of a playback system, the other oscillator to the other channel. Adjust the frequency of one oscillator to 168 Hz and the other to 318 Hz. At low level these two tones are quite discordant. Increase the level until the pitches of the 168-Hz and 318-Hz tones decrease to the 150- to 300-Hz octave relationship, which gives a pleasant sound. This illustrates the decrease of pitch at lower frequencies. A similar test would show that the pitch of higher frequency tones increases with sound level.

The Missing Fundamental

If tones such as 1,000; 1,200; and 1,400 Hz are reproduced together, a pitch of 200 Hz is heard. This can be interpreted as the fundamental with 1,000 Hz as the fifth harmonic; 1,200 Hz as the sixth harmonic; and so on. The auditory system recognizes that the upper tones are harmonics of the 200-Hz tone and perceptually supplies the missing fundamental that would have generated them.

Timbre versus Spectrum

Timbre describes our perception of the tonal quality of complex sounds. The term is applied chiefly to the sound of musical instruments. A flute and oboe sound different even though they are both playing the same pitch. The tone of each instrument has its own timbre. Timbre is determined by the number and relative strengths of the instrument's partials.

Timbre is a subjective term. The analogous physical term is spectrum. A musical instrument produces a fundamental and a set of partials (or harmonics) that can be analyzed with a wave analyzer. Suppose the fundamental is 200 Hz, the second harmonic 400 Hz, the third harmonic 600 Hz, and so on. The subjective pitch that the ear associates with our measured 200 Hz, for example, varies slightly with the level of the sound. The ear also has its own subjective interpretation of the harmonics. Thus, the ear's perception of the overall timbre of the instrument's note might be considerably different from the measured spectrum in an intricate way. In other words, timbre (a subjective description) and spectrum (an objective measurement) are not the same.

Localization of Sound Sources

The perception of the location of a sound source begins at the external ear, the pinna. Sound reflected from the ridges, convolutions, and surfaces of the pinna combines with the unreflected direct sound at the entrance to the auditory canal. This combination, now encoded with directional information, passes down the auditory canal to the eardrum and then to the middle and inner ear and finally to the brain for interpretation.

This directional encoding process of the sound signal is shown in Fig. 4-14. The sound wavefront can be considered as a multiplicity of sound rays coming from a specific source at a specific horizontal and vertical angle. As these rays strike the pinna, they are reflected from the surfaces, some of the reflections going toward the entrance to the auditory canal. At that point these reflected components combine with the unreflected component.

For a sound coming directly from the front of the listener (azimuth and vertical angle = 0°), the "frequency response" of the combination sound at the opening of the ear canal is shown in Fig. 4-15. A curve of this type is called a transfer function; it represents a vector combination involving phase angles.

For the sound at the entrance of the ear canal to reach the eardrum, the auditory canal must be traversed. As the transfer function at the entrance to the ear canal (Fig. 4-15) and that of the ear canal (Fig. 4-3) are combined, the shape of the resulting transfer function impinging on the eardrum is radically changed. Figure 4-3 showed a typical transfer function of the ear canal alone. It is a static, fixed function that does not change with

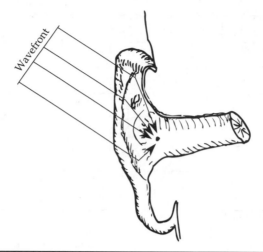

FIGURE 4-14 A wavefront of a sound can be considered as numerous rays perpendicular to that wavefront. Such rays, striking a pinna, are reflected from the various ridges and convolutions. Those reflections directed to the opening of the ear canal combine vectorially (according to relative amplitudes and phases). In this way the pinna encodes all sound falling on the ear with directional information, which the brain decodes as a directional perception.

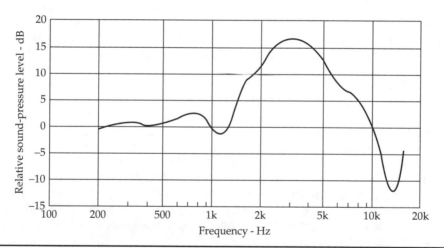

FIGURE 4-15 A measured example of the sound pressure (transfer function) at the opening of the ear canal corresponding to sound arriving from a point immediately in front of the subject. The shapes of such transfer functions vary with the horizontal and vertical angles at which the sound arrives at the pinna. (*Mehrgardt and Mellert*)

direction of arrival of the sound. The ear canal acts like a quarter-wave pipe closed at one end by the eardrum and exhibiting two prominent resonances.

The transfer function representing the specific direction to the source (Fig. 4-15) combining with the fixed transfer function of the ear canal (Fig. 4-3) gives the transfer function at the eardrum of Fig. 4-16. The brain translates this to a perception of sound coming from directly in front of the observer.

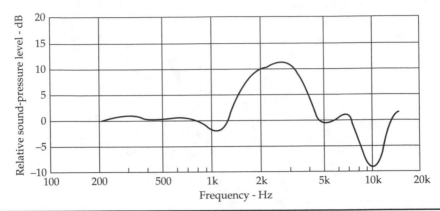

FIGURE 4-16   The transfer function of Fig. 4-15 at the opening of the ear canal is altered to this shape at the eardrum after being combined with the transfer function of the ear canal. In other words, a sound arriving at the opening of the ear canal from a source directly in front of the observer (Fig. 4-15) looks like Fig. 4-16 at the eardrum because it has been combined with the characteristics of the ear canal itself (Fig. 4-3). The brain discounts the fixed influence of the ear canal from every changing arriving sound.

The transfer function at the entrance to the ear canal is shaped differently for each horizontal and vertical direction. This is how the pinna encodes all arriving sound enabling the brain to yield different perceptions of direction. The sound arriving at the eardrum is the raw material for all directional perceptions. The brain neglects the fixed component of the ear canal and translates the differently shaped transfer functions to directional perceptions.

Another more obvious directional function of the pinna is that of forward-backward discrimination, which does not depend on encoding and decoding. At higher frequencies (shorter wavelengths), the pinna is an effective barrier. The brain uses this front-back differentiation to convey a general perception of direction.

The ear can also weakly perceive vertical localization. The median plane is a vertical plane passing symmetrically through the center of the head and nose. Sources of sound in this plane present identical transfer functions to the two ears. The auditory mechanism uses another system for such localization, that of giving a certain place identity to different frequencies. For example, signal components near 500 and 8,000 Hz are perceived as coming from directly overhead, components near 1,000 and 10,000 Hz as coming from the rear.

Sound arriving from directly in front of an observer results in a peak in the transfer function at the eardrum in the 2- to 3-kHz region. This is partly the basis of the technique of imparting "presence" to a recorded voice by adding an equalization boost in this frequency region. A voice can also be made to stand out from a musical background by adding such a peak to the voice response.

Binaural Localization

Two ears function together in binaural hearing to allow localization of sound sources in the horizontal plane. Signals from both ears are combined in the brain; thus localization largely takes place in the brain, and not in the individual ears. Two factors are involved,

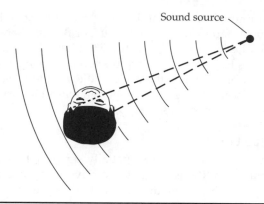

Sound source

FIGURE 4-17   Our binaural directional sense depends in part on the difference in intensity and phase of the sound falling on two ears.

the difference in intensity and the difference in time of arrival (phase) of the sound falling on the two ears. In Fig. 4-17, the ear nearest the source receives a greater intensity than the far ear because it is closer, and because the head casts an acoustic shadow. Because of diffraction, acoustic shadows are much weaker at low frequencies. However, at higher frequencies, the acoustic shadow, combined with the path-length difference, results in a higher intensity at the nearest ear.

Because of the difference in distance from the source, the near ear receives sound somewhat earlier than the far ear. Below 1 kHz, the phase (time) effect dominates while above 1 kHz the intensity effect dominates. There is one localization blind spot. A listener cannot tell whether sounds are coming from directly in front or from directly behind because the intensity of sound arriving at each ear is the same and in the same phase. Using these cues, the ear can localize sound sources in the horizontal plane to within 1° or 2°.

Law of the First Wavefront

The sound that arrives first creates in the listener the perception of direction; this is sometimes called the law of the first wavefront. Imagine two people in a small room, one person speaking and the other listening. The first sound to reach the listener is that traveling a direct path because it travels the shortest distance. This direct sound establishes the perception of the direction from which the sound came. Even though it is immediately followed by a multitude of reflections from the various surfaces of the room, this directional perception persists, and tends to diminish the effects of later reflections insofar as direction is concerned. This is the law of the first wavefront. This identification of the direction to the source of sound is accomplished within a small fraction of a millisecond.

The Franssen Effect

The ear is relatively adept at identifying the locations of sound sources. However, it also uses an auditory memory that can sometimes confuse direction. The Franssen effect demonstrates this. Two loudspeakers are placed to the left and right of a listener in a live room. The loudspeakers are about 3 ft from the listener at about 45° angles. A sine wave is played through the left loudspeaker, and the signal is immediately faded out and simultaneously faded in at the right loudspeaker, so there is no appreciable change in overall level. Most listeners will continue to locate the signal in the left loudspeaker, even though it is silent and the sound location has changed to the right loudspeaker. They are often surprised when the cable to the left loudspeaker is disconnected, and they continue to "hear" the signal coming from the left loudspeaker. This demonstrates the role of auditory memory in sound localization.

The Precedence Effect

Our hearing mechanism integrates spatially separated sounds over short intervals and under certain conditions, tends to perceive them as coming from one location. For example, in an auditorium, the ear and brain have the ability to gather all reflections arriving within about 35 msec (millisecond) after the direct sound, and combine (integrate) them to give the impression that all this sound is from the direction of the original source, even though reflections from other directions are involved. The sound that arrives first establishes the perceptual source location of later sounds. This is called the precedence effect, Haas effect, or law of the first wavefront. The sound energy integrated over this period also gives an impression of added loudness.

It is not too surprising that the human ear fuses sounds arriving during a certain time window. After all, at the cinema, our eyes fuse a series of still pictures, giving the impression of continuous movement. The rate of presentation of the still pictures is important; there must be at least 16 pictures per second (62-msec interval) to avoid seeing a series of still pictures or a flicker. Auditory fusion similarly is a process of temporal fusion. Auditory fusion works best during the first 35 msec; beyond 50 to 80 msec the integration breaks down, and with long delays, discrete echoes are heard.

Haas placed his subjects 3 m from two loudspeakers arranged so that they subtended an angle of 45°, the observer's line of symmetry splitting this angle (there is some ambiguity in the literature about the angle). The rooftop conditions were approximately anechoic. Both speakers played the same speech content and at the same level, but one speaker was delayed relative to the other. Clearly, sound from the undelayed speaker arrived at the listeners slightly before the sound from the delayed speaker. Haas studied the effects of varying the delay on speech signals. As shown in Fig. 4-18, Haas found that in the 5- to 35-msec delay range, the sound from the delayed loudspeaker was perceived as coming from the undelayed speaker. In other words, the listeners localized both sources to the location of the undelayed source.

Moreover, the level of the delayed sound had to be increased more than 10 dB over the undelayed sound before its location was heard as being separate. In a room, reflected energy arriving at the ear within 35 msec is spatially integrated with the direct sound and is perceived as being spatially part of the direct sound as opposed to reverberant sound. This is sometimes called the fusion zone or Haas zone. These integrated early reflections increase the loudness of the direct sound and can change its timbre. As Haas said, they result in ". . . a pleasant modification of the sound impression in the sense of broadening of the primary sound source while the echo source is not perceived acoustically."

The transition zone between the integrating effect for delays less than 35 msec and the perception of delayed sound as spatially discrete is gradual, and therefore, somewhat indefinite. Some researchers set the dividing line at 62 msec ($\frac{1}{16}$ sec), some at 80 msec, and some at 100 msec beyond which there is no question about the discrete location of the delayed sound. If the delayed sound is attenuated, the fusion zone is extended.

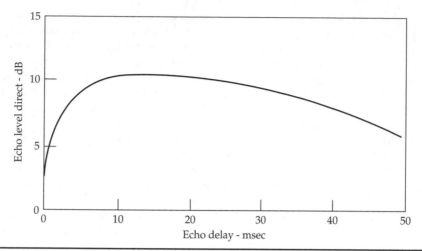

FIGURE 4-18 The precedence effect (Haas effect) in the human auditory system describes temporal fusion. In the 5- to 35-msec region, the echo level must be about 10 dB higher than the direct sound to be discernible as an echo. In this region, reflected components arriving from many directions are gathered by the ear. The resulting sound seems louder, because of the reflections and appears to come from the direct source. For delays 50 to 100 msec and longer, reflections are perceived as discrete echoes. (*Haas*)

For example, if the delayed sound is –3 dB relative to the first sound, integration extends to about 80 msec. Room reflections are lower in level than direct sound, so we expect that integration will extend over a longer time. However, with very long delays of perhaps 250 msec or more, the delayed sound is clearly heard as a discrete echo.

Other researchers had previously found that very short delays (less than 1 msec) were involved in our discerning the direction of a source by slightly different times of arrival at our two ears. Delays greater than this do not affect our directional sense.

The precedence effect is easily demonstrated. Stand 100 ft from a concrete wall and clap your hands—a distinct echo will be heard (177 msec). As you move closer and continue to clap your hands, the echo will arrive sooner and will be louder. But as you enter the fusion zone, your ears will spatially integrate the echo into the direct sound.

Perception of Reflected Sound

In the preceding section, "reflected" sound was considered in a rather limited way. A more general approach is taken in this section. The loudspeaker arrangement used by Haas was also used by other researchers, and this is the familiar stereo setup: two separated loudspeakers with the listener located symmetrically between the two loudspeakers. The sound from one loudspeaker is designated as the direct sound, that from the other loudspeaker, the delayed sound (the reflection). The delay injected between the two signals and their relative levels is adjustable.

With the sound of the direct loudspeaker set at a comfortable level, and with a delay of, say 10 msec, the level of the reflected, or delayed, loudspeaker sound is slowly increased from a very low level. The sound level of the reflection at which the observer first detects a difference in the sound is the threshold of reflection detection. For levels less than this, the reflection is inaudible; for levels greater than this, the reflection is clearly audible.

As the reflection level is gradually increased above the threshold value, a sense of spaciousness is imparted to the combined sound. This sense of spaciousness prevails even though the experiment is conducted in an anechoic space. As the level of the reflection is increased about 10 dB above the threshold value, another change is noticed in the sound; a broadening of the sound image and possibly a shifting of the image toward the direct loudspeaker is now added to the increasing spaciousness. As the reflection level is increased another 10 dB or so above the image broadening threshold, another change is noted; discrete echoes are heard.

What practical value does this have? Consider the example of a listening room used for playback of recorded music. Figure 4-19 shows the effect of lateral reflections being added to the direct sound from the loudspeakers. Speech is used as the signal. Reflections below the threshold of perception are unusable; reflections perceived as discrete echoes are also unusable. The usable area is the unshaded area between those two threshold curves, A and C. Calculations can give estimates of the level and delay of any specific reflection, knowing the speed of sound, the distance traveled, and applying the inverse square law. Figure 4-19 also shows the subjective reactions listeners will probably have to the combination of reflected and direct sound.

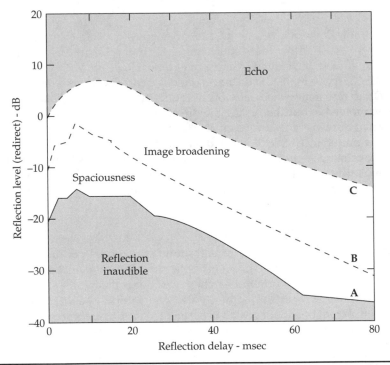

Figure 4-19   The effects of lateral reflections on the perception of the direct sound in a simulated stereo arrangement. These measurements were made in anechoic conditions, lateral angles 45° to 90°, with speech as the signal. (A) Absolute threshold of audibility of the reflection. (B) Image shift/broadening threshold. (C) Lateral reflection perceived as a discrete echo. (*A and B, Olive and Toole, and Toole*) (*C, Meyer and Schodder, and Lochner and Burger*)

To assist in the calculations mentioned previously, the following equations can be applied:

$$\text{Reflection delay} = \frac{(\text{Reflected path}) - (\text{Direct path})}{1{,}130} \qquad (4\text{-}2)$$

This assumes 100% reflection at the reflecting surface. Both path lengths are measured in feet; the speed of sound is measured in feet per second.

$$\text{Reflection level at listening position} = 20 \log \frac{\text{Direct distance}}{\text{Reflection distance}} \qquad (4\text{-}3)$$

This assumes inverse square propagation. Both path lengths are measured in feet.

In an auditorium, for example, the room geometry can be designed so that time delays are less than 50 msec and thus fall within the fusion zone. Consider a direct sound with a path length of 50 ft, and an early reflection with a path length of 75 ft, both reaching a listener. The resulting time delay of 22 msec is well within the fusion zone. Similarly, by limiting the time delay between direct sound and early reflections to less than 50 msec (a path length difference of about 55 ft), the listener will not hear the reflections as discrete echoes. If the typical attenuation of reflected sound is accounted for, differences slightly greater than 50 msec are permissible. Very generally, maximum differences of 50 msec are allowable for speech, while a maximum of 80 msec is allowable for music. Shorter differences are preferred. As we will see later, the precedence effect can also be used in the design of live-end–dead-end control rooms.

The Cocktail-Party Effect

The human auditory system possesses a powerful ability to distinguish between, and pay attention to, one sound amid many sounds. This is sometimes called "the cocktail-party effect" or "auditory scene analysis." Imagine yourself at a busy party with many talkers and music playing. You are able to listen to one talker while excluding many other conversations and sounds. But if someone from across the room calls your name, you will be alert to that. There is evidence that musicians and conductors are highly skilled at this auditory segregation; they can independently follow the sounds of multiple musical instruments simultaneously.

This ability to distinguish particular sounds is greatly assisted by our localization abilities. If the sounds of two talkers are played over one loudspeaker, it can be difficult to differentiate them. However, if the sounds of two talkers are each played over two physically separated loudspeakers, it can be quite easy to follow both (factors such as relative language, gender, and pitch of the talkers also play a role). While humans function well at differentiating sources at cocktail parties, electronic signal processing systems have a more difficult time. This field of signal processing is referred to as source separation, or blind source separation.

Aural Nonlinearity

When multiple frequencies are input to a linear system, the same frequencies are output. The ear is a nonlinear system. When multiple frequencies are input, the output can contain additional frequencies. This is a form of distortion that is introduced by the auditory system and it cannot be measured by ordinary instruments. It is a subjective effect requiring a different approach. This experiment demonstrates the nonlinearity of the ear and the output of aural harmonics. It can be performed with a playback system and two audio oscillators. Plug one oscillator into the left channel and the other into the right channel, and adjust both channels for an equal and comfortable volume level at some midband frequency. Set one oscillator to 23 kHz and the other to 24 kHz without changing the level settings. With either oscillator alone, nothing is heard because the signal is outside the range of the ear. However, if the tweeters are good enough, you might hear a distinct 1-kHz tone.

The 1-kHz tone is the difference between 23 and 24 kHz. The sum, 47 kHz, is another sideband. Such sum and difference sidebands are generated whenever two pure tones are mixed in a nonlinear element. The nonlinear element in this case is the middle and inner ear. In addition to intermodulation products, the nonlinearity of the ear generates new harmonics that are not present in the sound falling on the eardrum.

Another demonstration of auditory nonlinearity can be performed with the same equipment used above, with the addition of headphones. First, a 150-Hz tone is applied to the left earphone channel. If the hearing mechanism were perfectly linear, no aural harmonics would be heard as the exploratory tone is swept near the frequencies of the second, third, and other harmonics. Since it is nonlinear, the presence of aural harmonics is indicated by the generation of beats. When 150 Hz is applied to the left ear, and the exploratory tone of the right ear is slowly varied about 300 Hz, the second harmonic is indicated by the presence of beats between the two. If we change the exploratory oscillator to a frequency around 450 Hz, the presence of a third harmonic will be revealed by beats. Researchers have estimated the magnitude of the harmonics by the strength of such beats. Conducting the above experiment with tones of a higher level will make the presence of aural harmonics even more obvious.

Subjective versus Objective

There still remains a great divide between subjective assessments of sound quality and objective measurements. Consider the following descriptive words which are often applied to concert-hall acoustics: warmth, bassiness, definition, reverberance, fullness of tone, liveness, sonority, clarity, brilliance, resonance, blend, and intimacy. There is no instrument to directly measure qualities such as warmth or brilliance. However, there are some ways to relate subjective terms to objective measurements. For example, consider the term "definition." German researchers have adopted the term "deutlichkeit" which literally means clearness or distinctness. It is measured by taking the energy in an echogram during the first 50 to 80 msec and comparing it to the energy of the entire echogram. This compares the direct sound and early reflections, which are integrated by the ear, to the entire reverberant sound. This is a straightforward measurement of an impulsive sound from a pistol or other source.

Measurements are vitally important, but the ear is the final arbiter. Observations by human subjects provide valuable input to any acoustical evaluation. For example, in a loudness investigation, panels of listeners are presented with various sounds, and each observer is asked to compare the loudness of sound A with the loudness of B. The data submitted by the jury of listeners is then subjected to statistical analysis, and the dependence of human sensory factors, such as loudness, upon physical measurements of sound level is assessed. If the test is conducted properly and sufficient observers are involved, the results are trustworthy. In this way, for example, we discover that there is no linear relationship between sound level and loudness, pitch and frequency, or timbre and sound quality.

Occupational and Recreational Deafness

Nerve deafness resulting from occupational noise is recognized as a serious health hazard. The hearing of workers in industry is protected by law. The higher the environmental noise, the less exposure allowed, as shown in Table 4-2. Audiologists determine what noise exposure workers are subjected to in different environments. This is not easy because noise levels fluctuate and workers move about; wearable dosimeters are often used to integrate the exposure over the work day. Companies often install noise shields around offending equipment and mandate ear plugs or ear muffs for workers. Professional audio engineers operating with high monitoring levels are risking irreparable injury to their hearing; in most cases, their work is not subject to occupational noise-protection laws.

Dangerous noise exposure is more than an occupational problem; it is also recreational. An individual might work all day in a high-noise environment, then enjoy motorcycle or automobile racing, listen to music playback at high level, or spend hours in a night club. As high-frequency loss creeps in, volume controls are turned up to compensate, and the rate of deterioration is accelerated.

The audiogram is an important tool in the conservation of hearing. Comparing today's audiogram with earlier ones establishes the trend; if downward, steps can be taken to check it. The audiogram of Fig. 4-20 shows serious hearing loss of a mixing engineer in a recording studio. This hearing loss, centered on 4 kHz, may be the accumulation of many years of listening to high-level sounds in the control room.

Sound-Pressure Level, dB, A-Weighting, Slow Response	Maximum Daily Exposure Hours
85	16
90	8
92	6
95	4
97	3
100	2
102	1.5
105	1
110	0.5
115	0.25 or less

* Reference: OSHA 2206 (1978).

TABLE 4-2   OSHA Permissible Noise Exposure Times*

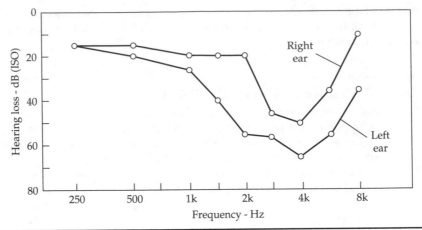

FIGURE 4-20 Audiograms showing serious hearing loss centered on 4 kHz, perhaps resulting from years of exposure to high-level sound in the control room of a recording studio.

Summary

- The auditory canal, acting as a quarter-wave pipe closed at one end by the eardrum, contributes an acoustical amplification of about 10 dB, and the head diffraction effect produces another 10 dB near 3 kHz. These are vital speech frequencies.

- The leverage of the ossicles bones of the middle ear and the ratio of areas of the eardrum and oval window efficiently match the impedance of air to the fluid of the inner ear.

- Waves set up in the inner ear by vibration of the oval window excite the sensory hair cells, which direct signals to the brain. There is a place effect, the peak of hair cell agitation for higher frequencies being nearer the oval window, and low frequencies at the distal end.

- The area of audibility is bounded by two threshold curves, the threshold of audibility at the lower extreme and the threshold of feeling or pain at the loud extreme. Our auditory experience occurs within these two extremes.

- The loudness of tone bursts decreases as the length of the burst is decreased. Bursts greater than 200 msec have full loudness, indicating a time constant of the ear at about 100 msec.

- Our ears are capable of accurately locating the direction of a source in the horizontal plane. In a vertical median plane, however, localization ability is less accurate.

- Pitch is a subjective term. Frequency is the associated physical term, and the two have only a general relationship.

- Subjective timbre or quality of sound and the physical spectrum of the sound are related, but not equal.

- The nonlinearity of the ear generates intermodulation products and spurious harmonics.

- The Haas, or precedence, effect describes the ability of the ear to integrate all sound arriving within the first 50 msec, giving the perception that the sounds originate from the earlier source, and making the sounds seem louder.

- Although the ear is not effective as a measuring instrument yielding absolute values, it is very keen in comparing frequencies, levels, and sound quality.

- Occupational and recreational noises can cause permanent hearing loss. Precautionary steps to minimize this type of environmentally caused deafness are recommended.

CHAPTER 5

Signals, Speech, Music, and Noise

Signals such as speech, music, and noise are within the common experience of everyone. We are profoundly familiar with the sound of speech; we hear speech virtually every day of our lives. Speech is one of the keys to human communication, and poor speech intelligibility is extremely frustrating. If we are lucky, we also hear music every day. Music can be one of the most pleasurable, and needful of human experiences. It would be hard to imagine a world without music. Noise is usually unwanted sound, often chaotic and disruptive to speech, music, or silence. The close relationship of speech, music, and noise is made evident in this chapter.

Sound Spectrograph

A consideration of speech sounds is necessary to understand how the sounds are produced. Speech is highly variable and transient in nature, comprising energy moving up and down the three-dimensional scales of frequency, sound level, and time. A sound spectrograph can show all three variables. Each common noise has its spectrographic signature that reveals the very stuff that characterizes it. The spectrographs of several commonly experienced sounds are shown in Fig. 5-1. In these spectrographs, time progresses horizontally to the right, frequency increases from the origin upward, and the sound level is indicated roughly by the density of the trace—the blacker the trace, the more intense the sound at that frequency and at that moment of time. Random noise on such a plot shows up as a gray, slightly mottled rectangle as all frequencies in the audible range and all intensities are represented as time progresses. The snare drum approaches random noise at certain points, but it is intermittent. The "wolf whistle" opens on a rising note followed by a gap, and then a similar rising note that then falls in frequency. The police whistle is a tone, slightly frequency modulated.

The human voice mechanism is capable of producing many sounds other than speech. Figure 5-2 shows a number of these sound spectrographs. Harmonic series appear on a spectrograph as more or less horizontal lines spaced vertically in frequency. These are particularly noticeable in the trained soprano's voice and the baby's cry, but traces are also evident in other spectrographs.

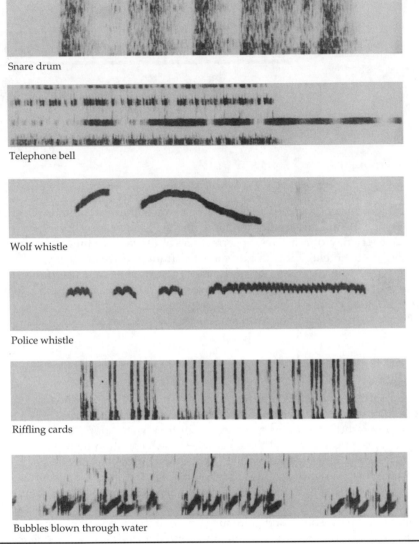

Snare drum

Telephone bell

Wolf whistle

Police whistle

Riffling cards

Bubbles blown through water

FIGURE 5-1   Sound spectrographs of common sounds. Time progresses to the right, the vertical scale is frequency, and the intensity of components is represented by the intensity of the trace. (*AT&T Bell Laboratories*)

Speech

There are two quasi-independent functions in the generation of speech sounds: the sound source and the vocal system. In general, speech is a series flow, as pictured in Fig. 5-3A, in which the raw sound is produced by a source and subsequently shaped in the vocal tract. To be more exact, three different sources of sound are shaped by the vocal tract, as shown in Fig. 5-3B. First, there is the sound we naturally think of—the sounds emitted by the vocal cords. These are formed into the voiced sounds. They are produced by air from the

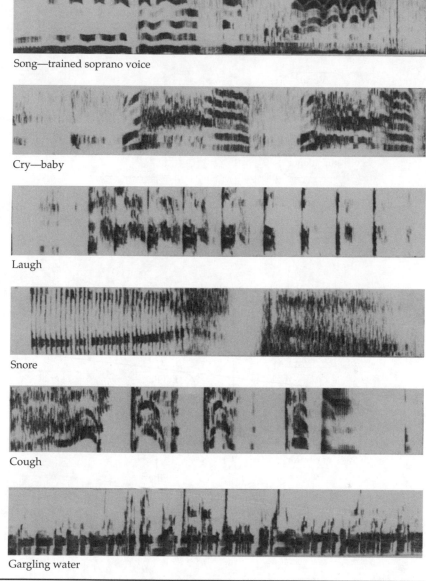

Song—trained soprano voice

Cry—baby

Laugh

Snore

Cough

Gargling water

Figure 5-2 Sound spectrographs of human sounds other than speech. (*AT&T Bell Laboratories*)

lungs flowing past the slit between the vocal cords (the glottis), which causes the cords to vibrate. This air stream, broken into pulses of air, produces a sound that can almost be called periodic, that is, repetitive in the sense that one cycle follows another.

The second source of speech sound is that made by forming a constriction at some point in the vocal tract with the teeth, tongue, or lips and forcing air through it under high enough pressure to produce significant turbulence. Turbulent air creates noise. This noise is shaped by the vocal tract to form the fricative sounds of speech such as the consonants *f, s, v,* and *z.* Try making these sounds, and you will see that high-velocity air is very much involved.

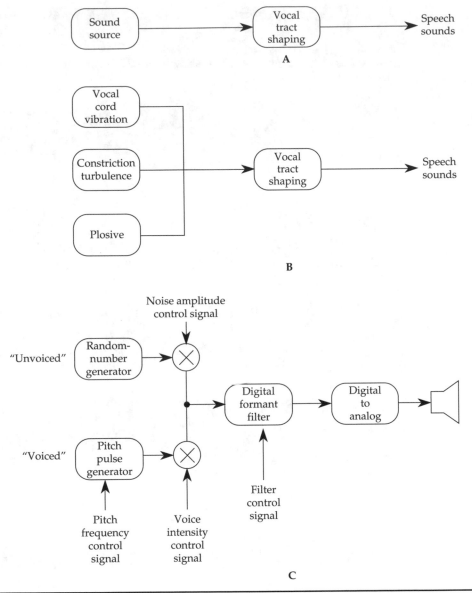

Figure 5-3 (A) The human voice is produced through the interaction of two essentially independent functions, a sound source and a time-varying-filter action of the vocal tract. (B) The sound source is comprised of vocal-cord vibration for voiced sounds, the fricative sounds resulting from air turbulence, and plosive sounds. (C) A digital system used to synthesize human speech.

The third source of speech sound is produced by the complete stoppage of the breath, usually toward the front, a building up of the pressure, and then the sudden release of the breath. Try speaking the consonants k, p, and t, and you will sense the force of such plosive sounds. They are usually followed by a burst of fricative or turbulent sound. These three types of sounds—voiced, fricative, and plosive—are the raw sources that are shaped into the words we speak.

These sound sources and signal processing can be implemented in digital hardware or software. A simple speech synthesis system is shown in Fig. 5-3C. A random-number generator produces the digital equivalent of the *s*-like sounds for the unvoiced components. A counter produces pulses simulating the pulses of sound of the vocal cords for the voiced components. These are shaped by time-varying digital filters simulating the ever-changing resonances of the vocal tract. Signals control each of these to form digitized speech, which is then converted to analog form.

Vocal Tract Molding of Speech

The vocal tract can be considered as an acoustically resonant system. This tract, from the lips to the vocal cords, is about 6.7 in (17 cm) long. Its cross-sectional area is determined by the placement of the lips, jaw, tongue, and velum (a sort of trapdoor that can open or close off the nasal cavity) and varies from 0 to about 3 in^2 (20 cm^2). The nasal cavity is about 4.7 in (12 cm) long and has a volume of about 3.7 in^3 (60 cm^3). These dimensions have a bearing on the resonances of the vocal tract and their effect on speech sounds.

Formation of Voiced Sounds

If the components of Fig. 5-3 are elaborated into source spectra and modulating functions, we arrive at something of great importance in audio—the spectral distribution of energy in the voice. We also gain a better understanding of the aspects of voice sounds that contribute to the intelligibility of speech in the presence of reverberation and noise. Figure 5-4 shows the steps in producing voiced sounds. First, sound is produced by the vibration of the vocal cords; these are pulses of sound with a fine spectrum that falls off at about 10 dB/octave as frequency is increased, as shown in Fig. 5-4A. The sounds of the vocal cords pass through the vocal tract, which acts as a filter varying with time. The humps of Fig. 5-4B are due to the acoustical resonances, called formants of the vocal pipe, which is open at the mouth end and essentially closed at the vocal cord end. Such an acoustical pipe 6.7-in long has resonances at odd quarter wavelengths, and these peaks occur at approximately 500; 1,500; and 2,500 Hz. The output sound, shaped by the resonances of the vocal tract, is shown in Fig. 5-4C. This applies to the voiced sounds of speech.

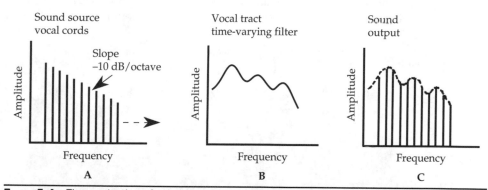

FIGURE 5-4   The production of voiced sounds can be considered as several steps. (A) Sound is first produced by the vibration of the vocal cords; these are pulses of sound with a spectrum that falls off at about 10 dB/octave. (B) The sounds of the vocal cords pass through the vocal tract, which acts as a filter varying with time. Acoustical resonances, called formants, are characteristic of the vocal pipe. (C) The output voiced sounds of speech are shaped by the resonances of the vocal tract.

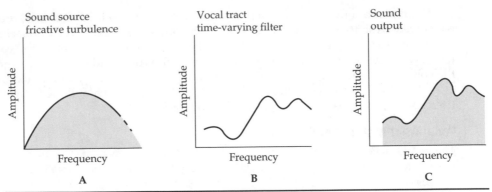

FIGURE 5-5   A diagram of the production of unvoiced fricative sounds such as *f*, *s*, *v*, and *z*. (A) The distributed spectrum of noise due to air turbulence resulting from constrictions in the vocal tract. (B) The time-varying filter action of the vocal tract. (C) The output sound resulting from the filter action of the distributed sound of (A).

Formation of Unvoiced Sounds

Unvoiced sounds are shaped in a similar manner, as shown in Fig. 5-5. Unvoiced sounds start with the distributed, almost random-noise like spectrum of the turbulent air as fricative sounds are produced. The distributed spectrum of Fig. 5-5A is generated near the mouth end of the vocal tract, rather than the vocal cord end; hence, the resonances of Fig. 5-5B are of a somewhat different shape. Figure 5-5C shows the sound output shaped by the time-varying filter action of Fig. 5-5B.

Frequency Response of Speech

The voiced sounds originating in vocal cord vibrations, the unvoiced sounds originating in turbulences, and plosives which originate near the lips, together form our speech sounds. As we speak, the formant resonances shift in frequency as the lips, jaw, tongue, and velum change position to shape the desired words. The result is the complexity of human speech evident in the spectrograph of Fig. 5-6. Information communicated via speech is a pattern of frequency and intensity shifting rapidly with time. Notice that there is little speech energy above 4 kHz in Fig. 5-6. Although it is not shown by the

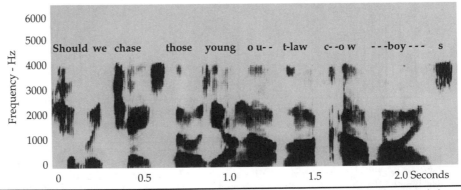

FIGURE 5-6   Sound spectrograph of a sentence spoken by a male voice. (*AT&T Bell Laboratories*)

spectrograph, there is relatively little speech energy below 100 Hz. It is understandable why presence filters peak in the 2- to 3-kHz region; that is where human pipes resonate.

Directionality of Speech

Speech sounds do not have the same strength in all directions. This is due primarily to the directionality of the mouth and the sound shadow cast by the head and torso. Two measurements of speech-sound directionality are shown in Fig. 5-7. Because speech sounds are variable and complex, averaging is necessary to give an accurate measure of directional effects.

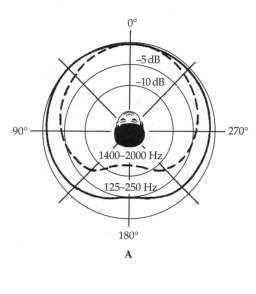

A

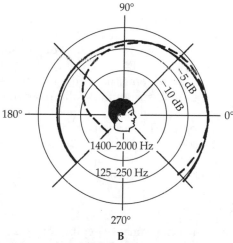

B

FIGURE 5-7   The human voice is highly directional. (A) Front-to-back directional effects of about 12 dB are found for critical speech frequencies. (B) In the vertical plane, the front-to-back directional effects for the 1,400- to 2,000-Hz band are about the same as for the horizontal plane. (*Heinrich Kuttruff and Applied Science Publishers Ltd., London*)

The horizontal directional effects shown in Fig. 5-7A demonstrate only a modest directional effect of about 5 dB in the 125- to 250-Hz band. This is expected because the head is small compared to wavelengths of 4.5 to 9 ft associated with this frequency band. However, there are significant directional effects for the 1,400- to 2,000-Hz band. For this band, which contains important speech frequencies, the front-to-back difference is about 12 dB.

In the vertical plane shown in Fig. 5-7B, the 125- to 250-Hz band shows about 5 dB front-to-back difference again. For the 1,400- to 2,000-Hz band, the front-to-back difference is also about the same as the horizontal plane, except for the torso effect. The discrimination against high speech frequencies picked up on a lapel microphone is obvious (see Fig. 5-7B), although the measurements were not carried to angles closer to 270°.

Music

Musical sounds are extremely variable in their complexity and can range from a near sine-wave form of a single instrument or voice to the highly complex mixed sound of a symphony orchestra. Each instrument and each voice has a different tonal texture for each note.

String Instruments

Musical instruments such as the violin, viola, cello, or bass, produce their tones by vibration of strings. On a stretched string, the overtones are all exact multiples of the fundamental, the lowest tone produced. These overtones may thus be called harmonics. If the string is bowed in the middle, odd harmonics are emphasized because the fundamental and odd harmonics have maximum amplitude there. Because the even harmonics have nodes in the center of the string, they will be subdued if bowed there. The usual place for bowing is near one end of the strings, which gives a better blend of even and odd harmonics. The "unmusical" seventh harmonic is anathema in most music (musically, it is a very flat minor seventh). By bowing (or striking or plucking) $\frac{1}{7}$ (or $\frac{2}{7}$) of the distance from one end, this harmonic is decreased. For this reason, the hammers in a piano are located near a node of the seventh harmonic.

The harmonic content of the E and G notes of a violin are displayed in Fig. 5-8. Harmonic multiples of the higher E tone are spaced wider and hence have a thinner timbre. The lower G tone, on the other hand, has a closely spaced spectral distribution and a richer timbre. The small size of the violin relative to the low frequency of the G string means that the resonating body cannot produce a fundamental at as high a level as the higher harmonics. The harmonic content and spectral shape depend on the shape and size of the resonating violin body, the type and condition of the wood, and even the varnish. Why there are so few superb violins among the many good ones is a problem that has not yet been solved completely.

Wind Instruments

In many musical instruments, resonance in pipes or tubes can be considered primarily one dimensional. (Resonances in three-dimensional rooms are discussed in later chapters.) Standing-wave effects are dominant in pipes. If air is enclosed in a narrow pipe closed at both ends, the fundamental (twice the length of the pipe) and all its harmonics will be formed. Resonances are formed in a pipe open at only one end at the frequency at

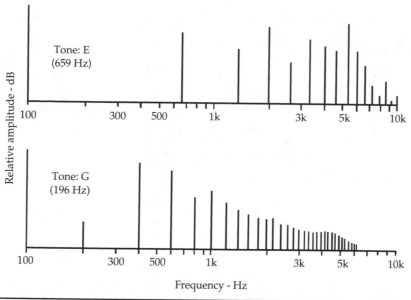

FIGURE 5-8 Harmonic content of open strings of the violin. The lower-frequency tones sound richer because of the closely packed harmonics.

which the pipe length is four times the wavelength, and results in odd harmonics. Wind instruments form their sounds this way; the length of the column of air is continuously varied, as in the slide trombone, or in jumps as in the trumpet or French horn, or by opening or closing holes along its length as in the saxophone, flute, clarinet, and oboe.

The harmonic content of several wind instruments is compared to that of the violin in the spectrographs of Fig. 5-9. Each instrument has its characteristic timbre as determined by the number and strength of its harmonics and by the formant shaping of the train of harmonics by the structural resonances of the instrument.

Nonharmonic Overtones

Some musical instruments generate a complex type of nonharmonic overtones. Bells produce a mixture of overtones. The overtones of drums are not harmonically related, although they give a richness to the drum sound. Triangles and cymbals give a mixture of overtones that blend reasonably well with other instruments. Piano strings are stiff strings and vibrate like a combination of solid rods and stretched strings. Thus piano overtones are not strictly harmonic. Nonharmonic overtones produce the difference between organ and piano sounds and give variety to musical sounds in general.

Dynamic Range of Speech and Music

The dynamic range of speech is relatively limited. From the softest to the loudest speech sounds, a voice spoken with normal effort might have a dynamic range of 30 to 40 dB. With more effort, the range of loud speech might be 60 to 70 dB. Even this range can be easily accommodated by many audio technologies. Music, however, has historically posed a more difficult challenge for recording and transmission.

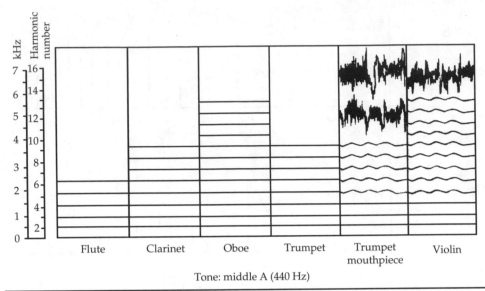

FIGURE 5-9 Spectrograph comparison of the harmonic content of woodwind instruments and the violin playing middle A (440 Hz). The differences displayed account for the difference in timbre of the various instruments. (*AT&T Bell Laboratories*)

In the concert hall, a full symphony orchestra is capable of producing very loud sounds, but also soft, delicate passages. Seated in the audience, one can fully appreciate this sweep of sound due to the great dynamic range of the human ear. The dynamic range between the loudest and the softest passage may be 100 dB (the ear's dynamic range is about 120 dB). To be effective, the soft passages must still be audible above the ambient background noise in the hall; hence the emphasis on adequate structural isolation to protect against traffic and other outside noises, and precautions to ensure that air-handling equipment noise is low.

For those not present in the music hall, live radio or television broadcast, or recordings must suffice. Conventional analog radio broadcasts are unable to handle the full dynamic range of the orchestra. Noise at the lower extreme and distortion at the upper extreme introduce limitations. In addition, regulatory restrictions prohibit interference with adjacent channels.

Digital audio, ideally, has the dynamic range and signal-to-noise ratio needed to fully capture music. The dynamic range in a digital system is directly related to the word length of binary digits (bits) as shown in Table 5-1. A Compact Disc, for example, stores 16-bit words and thus is capable of storing music with a 96 dB dynamic range; this can be extended if the signal is properly dithered. Consumer formats such as Super Audio CD and DVD-Audio as well as professional audio recorders provide 24-bit resolution and avoid the audibility of digital artifacts resulting from lower-resolution processing. Digital techniques, when taking full advantage of the technology, largely transfer dynamic range limitations from the recording medium to the playback environment. On the other hand, digital formats such as AAC, MP3, and WMA provide dynamic range and fidelity that can be quite variable; their quality depends on factors such as the bit rate of the recorded or streaming file.

Number of Binary Digits	Dynamic Range (dB)
4	24
8	48
12	72
16	96
24	144

TABLE 5-1   Theoretical Dynamic Ranges for Digital Word Lengths

Power in Speech and Music

In many applications, one must consider the power of sound sources. For conversational speech, the average power is perhaps 20 μW, but peaks might reach 200 μW. Most of the power of speech is at mid-to-low frequencies, with 80% below 500 Hz, yet there is very little power below 100 Hz. On the other hand, the small amount of power at high frequencies is where consonants are and determines the intelligibility of speech. Higher and lower frequencies outside this range add a natural quality to speech, but do not contribute to intelligibility.

Musical instruments produce more power than the human voice. For example, a trombone might generate a peak power of 6 W, and a full symphony orchestra's peak power might be 70 W. The peak power levels of various musical instruments are listed in Table 5-2.

Instrument	Peak Power (W)
Full orchestra	70
Large bass drum	25
Pipe organ	13
Snare drum	12
Cymbals	10
Trombone	6
Piano	0.4
Trumpet	0.3
Bass saxophone	0.3
Bass tuba	0.2
Double bass	0.16
Piccolo	0.08
Flute	0.06
Clarinet	0.05
French horn	0.05
Triangle	0.05

(from Sivian et al.)

TABLE 5-2   Peak Power of Musical Sources

Frequency Range of Speech and Music

It is instructive to compare the frequency range of various musical instruments with that of speech. This is best done graphically. Figure 5-10 shows the ranges of various musical instruments and voices. It is important to note that the figure only shows the fundamental pitches, and not the harmonic overtones. Very low organ notes are perceived mainly by their harmonics. Certain high-frequency noise accompanying musical instruments is not included, such as reed noise in woodwinds, bowing noise of strings, and key clicks and thumps of piano and percussion instruments.

Auditory Area of Speech and Music

The frequency range and the dynamic range of speech, music, and other sounds place varying demands on the human ear. Speech uses only a small portion of the ear's auditory capability. The portion of the auditory area used in speech is shown by the shaded area of Fig. 5-11. This area is located centrally in the auditory range; neither extremely soft or extremely loud sounds, nor sounds of very low or very high frequency, are used in common speech sounds. The speech area of Fig. 5-11 is derived from long time averages, and its boundaries should be fuzzy to represent the transient excursions in level and frequency. The speech area, as represented, shows an average dynamic range of about 42 dB. The 170- to 4,000-Hz frequency range covers about 4.5 octaves.

The music area shown in Fig. 5-12 is much greater than the speech area of Fig. 5-11. Music uses a greater proportion of the auditory range of the ear. Its excursions in both level and frequency are correspondingly greater than speech, as would be expected. Here again, long time averages are used in establishing the boundaries of the music area, and the boundaries should be fuzzy to show extremes. The music area shown is very conservative; it has a dynamic range of about 75 dB and a frequency range of about 50 to 8,500 Hz. This frequency span is about 7.5 octaves, compared to the 10-octave range of the human ear. High-fidelity standards demand a much wider frequency range than this. Without the averaging process involved in establishing the speech and music areas, both the dynamic range and the frequency range would be greater to accommodate the short-term transients that contribute little to the overall average, but are still of great importance.

Noise

The word "signal" implies that information is being conveyed. Noise can also be considered an information carrier. Interrupting noise to form dots and dashes is one way to shape noise into communication. We will also see how a decaying band of noise can give information on the acoustical quality of a room. On the other hand, there are types of noise that are undesirable, such as dripping faucets and traffic noise. Sometimes it is difficult to distinguish between objectionable noise and a legitimate carrier of information. For example, the noise of an automobile conveys considerable information on how well it is running. An audio playback system can produce sounds deemed very desirable by the owner, but to a neighbor they might be considered intrusive. A loud ambulance or fire truck siren is specifically designed to be both objectionable, and to carry important information. Society establishes limits to keep objectionable noise to a minimum while ensuring that information-carrying sounds can be heard by those who need to hear them.

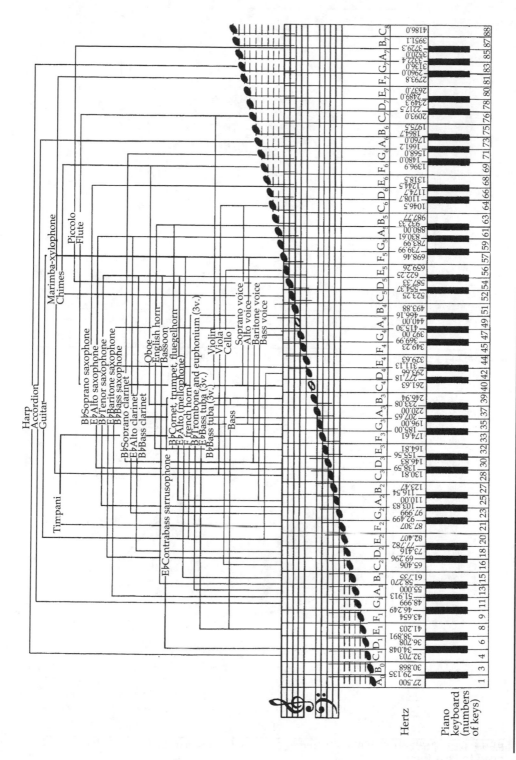

Figure 5-10 The audible frequency range of various musical instruments and voices. Only the fundamental tones are included; the partials (not shown) extend much higher. Also not shown are the many high-frequency incidental noises produced. (*C. G. Conn, Ltd.*)

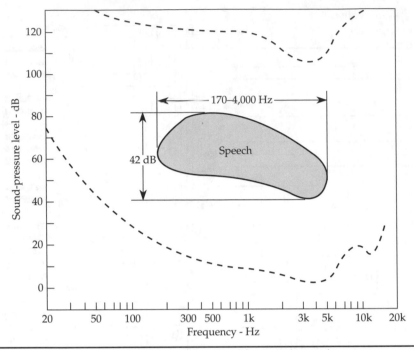

Figure 5-11 The portion of the auditory region utilized for speech sounds.

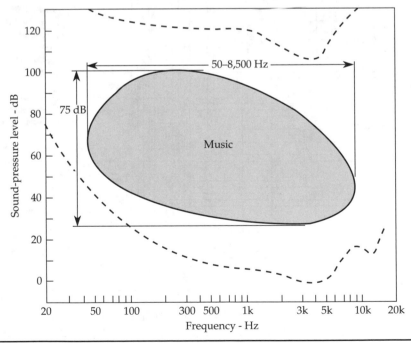

Figure 5-12 The portion of the auditory region utilized for music sounds.

Our evaluation of noise is very much a subjective response. Generally, high-frequency noise is more annoying than low-frequency noise. Intermittent noise is more annoying than steady or continuous noise. Moving and nonlocalized noise is more annoying than fixed and localized noise. No matter how it is evaluated, noise intrusion can be a minor annoyance, or it may cause hearing damage and must be taken seriously.

Noise Measurements

Defining noise as unwanted sound fits some kinds of noise, but noise is also an important tool for measurements in acoustics. This noise is not necessarily different from unwanted noise, it is just that the noise is put to a specific use.

In acoustical measurements, pure tones are often difficult to use, while a narrow band of noise centered on the same frequency can make satisfactory measurements possible. For example, a studio microphone picking up a pure tone signal of 1 kHz from a loudspeaker will have an output that varies greatly from position to position due to room resonances. If, however, a band of noise one octave wide centered at 1 kHz is radiated from the same loudspeaker, the level from position to position would tend to be more uniform, yet the measurement would contain information on what is occurring in the region of 1 kHz. Such measuring techniques make sense because we are usually interested in how a studio or listening room reacts to the complex sounds being recorded or reproduced, rather than to steady, pure tones.

Random Noise

Random noise is generated in any analog electrical circuit and minimizing its effect often poses a difficult problem. Figure 5-13 shows a sine wave and a random noise signal as viewed on an oscilloscope. The regularity of the one is in stark contrast to the

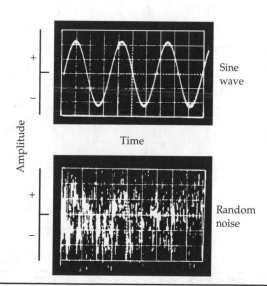

FIGURE 5-13 Oscillograms of a sine wave and of random noise. Random noise is continually shifting in amplitude, phase, and frequency.

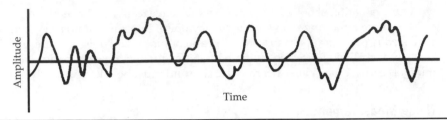

FIGURE 5-14   A section of the random noise signal of Fig. 5-13 spread out in time. The nonperiodic nature of a noise signal is evident, the fluctuations are random.

randomness of the other. If the horizontal sweep of the oscilloscope is expanded sufficiently and a capture is taken of the random noise signal, it would appear as in Fig. 5-14.

Noise is said to be purely random in character if it has a normal or Gaussian distribution of amplitudes. This means that if we sampled the instantaneous voltage at many equally spaced times, some readings would be positive, some negative, some greater, some smaller, and a plot of these samples would approach the familiar Gaussian distribution curve of Fig. 5-15.

White and Pink Noise

Both white noise and pink noise are often used as test signals. White noise is analogous to white light in that the energy of both is distributed uniformly throughout the spectrum. In other words, white-weighted random noise has the same average power in each 1-Hz frequency band. It is sometimes said that white noise has equal energy

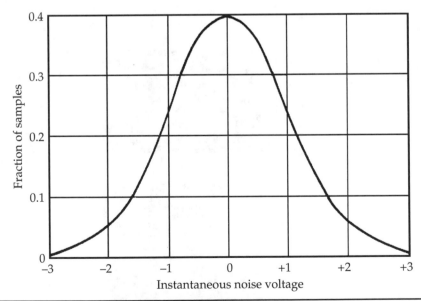

FIGURE 5-15   The proof of randomness of a noise signal lies in the sampling of instantaneous voltage, for example, at 1,000 points equally spaced in time, and plotting the results. If the noise is random, the familiar bell-shaped Gaussian distribution curve results.

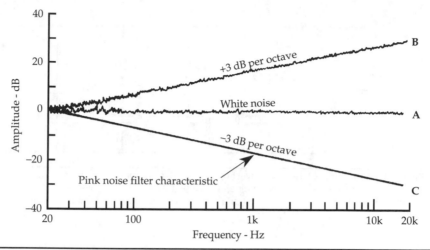

FIGURE 5-16   (A) Random white noise has constant energy per hertz. If the spectrum of random white noise is measured with an analyzer of fixed bandwidth, the resulting spectrum will be flat with frequency. (B) If measured with an analyzer whose passband width is a given percentage of the frequency to which it is tuned, the spectrum will slope upward at 3 dB/octave. (C) Pink noise is obtained by low-pass filtering white noise with a characteristic that slopes downward at 3 dB/octave. When using pink noise, a flat response results when constant percentage bandwidth filters are used such as octave or ⅓-octave filters. In measuring a system, pink noise is applied to the input and, if the system is flat, the output response will be flat if ⅓-octave filters, for example, are used.

per hertz. Therefore, when plotted on a logarithmic frequency scale, white noise exhibits a flat distribution of energy with frequency, as shown in Fig. 5-16A. Since each higher octave contains twice as many 1-Hz bands as the previous octave, white noise energy doubles in each higher octave. Audibly, white noise has a high-frequency hissing sound.

White light sent through a prism is broken down into a range of colors. The red color is associated with longer wavelengths of light, that is, light in the lower frequency region. Pink-weighted random noise has the same average power in each octave (or ⅓-octave) band. Since successive octaves encompass progressively larger frequency ranges, pink noise has relatively more energy in low frequencies. Audibly, pink noise has a stronger low-frequency sound than white noise. Pink noise is identified specifically as noise exhibiting more energy in the low-frequency region with a specific downward slope of 3 dB/octave, as shown in Fig. 5-16C. Very generally, pink noise is often used for acoustical measurements, while white noise is used for electrical measurements. The energy distribution of pink noise more closely matches the way the ear subjectively hears sound.

These white and pink terms arose because two types of spectrum analyzers are used. One is a constant-bandwidth analyzer, which has a passband of fixed width as it is tuned throughout the spectrum. For example, a bandwidth of 5 Hz may be used. If white noise with its flat spectrum is measured with a constant-bandwidth analyzer, another flat spectrum would result because the fixed bandwidth would measure a constant energy throughout the band (see Fig. 5-16A).

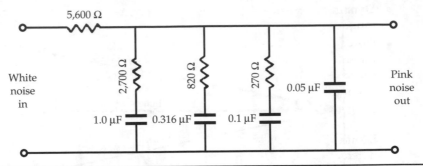

FIGURE 5-17   A simple filter for converting white noise to pink noise. It changes white noise of constant energy per hertz to pink noise of constant energy per octave. Pink noise is most useful in acoustical measurements utilizing analyzers having passbands with bandwidth of a constant percentage of the center frequency.

In contrast, in a constant-percentage bandwidth analyzer, the bandwidth changes with frequency. An example of this is the $\frac{1}{3}$-octave analyzer, commonly used because its bandwidth follows reasonably well with the critical bandwidth of the human ear throughout the audible frequency range. At 100 Hz the bandwidth of the $\frac{1}{3}$-octave analyzer is 23 Hz but at 10 kHz the bandwidth is 2,300 Hz. Obviously, it intercepts much greater noise energy in a $\frac{1}{3}$-octave band centered at 10 kHz than one centered at 100 Hz. Measuring white noise with a constant-percentage analyzer would give an upward-sloping result with a slope of 3 dB/octave (see Fig. 5-16B).

In many audio-frequency measurements, a flat response throughout the frequency range is the desired characteristic of many instruments and rooms. Assume that the system to be measured has a frequency characteristic that is almost flat. If this system is driven with white noise and measured with a constant-percentage analyzer, the result would have an upward slope of 3 dB/octave. It would be more desirable if the measured result was nominally flat so that deviations from flatness would be very apparent. This can be accomplished by using a noise with a downward slope of 3 dB/octave, that is, pink noise. By passing white noise through a low-pass filter, such as that of Fig. 5-17, such a downward sloping noise can be obtained. A close-to-flat system (such as an amplifier or room) driven with this pink noise would yield a close-to-flat response, which would make deviations from flatness very obvious.

Signal Distortion

Our discussion of the signals encountered in audio is incomplete without acknowledging what can happen to the signal in passing through transducers, amplifiers, and various forms of signal processing gear. Here is a list of some possible forms of distortion:

- **Bandwidth limitation** If the passband of an amplifier attenuates low or high frequencies, the signal output differs from the input by this bandwidth limitation.
- **Nonuniform response** Peaks and valleys within the passband also alter the signal waveshape.
- **Phase distortion** Any introduced phase shifts affect the time relationship between signal components.

- **Dynamic distortion** A compressor or expander changes the original dynamic range of a signal.

- **Crossover distortion** In class-B amplifiers, in which the output devices conduct for only half of the cycle, any discontinuities near zero output result in what is called crossover distortion.

- **Nonlinear distortion** If an amplifier is truly linear, there is a one-to-one relationship between input and output. Feedback helps to control nonlinear tendencies. The human ear is not linear. When a pure tone is impressed on the ear, harmonics can be heard. If two loud tones are presented simultaneously, sum and difference tones are generated in the ear itself, and these tones can be heard as can their harmonics. A cross-modulation test on an amplifier does essentially the same thing. If the amplifier (or the ear) were perfectly linear, no sum or difference tones or harmonics would be generated. The production of frequency elements that were not present in the input signal is the result of nonlinear distortion.

- **Transient distortion** Strike a bell and it rings. Apply a steep wavefront signal to an amplifier and it might ring too. For this reason, signals such as piano notes are difficult to reproduce. Tone burst test signals analyze the transient response characteristics of equipment. Transient intermodulation (TIM) distortion, slew induced distortion, and other measuring techniques evaluate transient forms of distortion.

Harmonic Distortion

The harmonic distortion method of evaluating the effects of circuit nonlinearities is universally accepted. In this method the device under test is driven with a sine wave of high purity. If the signal encounters any nonlinearity, the output waveshape is changed, that is, harmonic components appear that were not in the pure sine wave. A spectral analysis of the output signal measures these harmonic distortion products. A wave analyzer may use a constant passband width of 5 Hz, which is swept through the audio spectrum. Figure 5-18 shows results of such a measurement. The wave analyzer is first tuned to the fundamental, $f_0 = 1$ kHz, and the level is set for a convenient 1.00 V. The wave analyzer shows that the $2f_0$ second harmonic at 2 kHz measures 0.10 V. The third harmonic at 3 kHz gives a reading of 0.30 V, the fourth a reading of 0.05 V, and so on. The data are assembled in Table 5-3.

The total harmonic distortion (THD) may be found from the expression:

$$\text{THD} = \frac{\sqrt{(e_2)^2 + (e_3)^2 + (e_4)^2 \ldots (e_n)^2}}{e_o} \times 100 \qquad (5\text{-}1)$$

where $e_2, e_3, e_4, \ldots, e_n$ = voltages of second, third, fourth, etc. harmonics
e_o = voltage of fundamental

In Table 5-3, the harmonic voltages have been squared and added together. Using the equation:

$$\text{THD} = \frac{\sqrt{0.143125}}{1.00} \times 100$$

$$= 37.8\%$$

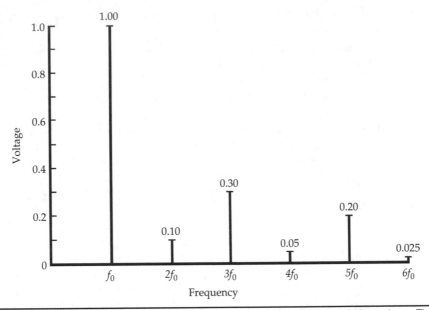

Figure 5-18 A distorted periodic wave is measured with a constant-bandwidth analyzer. The fundamental f_0 is set for some reference voltage, taken here as 1.00 V. Using a wave analyzer, the amplitude of the second harmonic at $2f_0$ is measured as 0.10 V. The wave analyzer similarly yields the amplitudes of each harmonic as shown. The root-mean-square of the harmonic voltages is then compared to the 1.00-V fundamental to find the total harmonic distortion expressed in percentage.

Harmonic	Volts	(Volts)^2
Second harmonic $2f_0$	0.10	0.01
Third harmonic $3f_0$	0.30	0.09
Fourth harmonic $4f_0$	0.05	0.0025
Fifth harmonic $5f_0$	0.20	0.04
Sixth harmonic $6f_0$	0.025	0.000625
Seventh and higher	(negligible)	——
		Sum 0.143125

Table 5-3 Harmonic Distortion Products (Fundamental $f_0 = 1$ kHz, 1.00 V amplitude)

A total harmonic distortion of 37.8% is a very high distortion that would make any amplifier sound poor on any type of signal, but the example has served our purpose.

A simple adaptation of the THD method can also be used. Consider Fig. 5-18 again. If the f_0 fundamental were adjusted to some known value and then a notch filter were

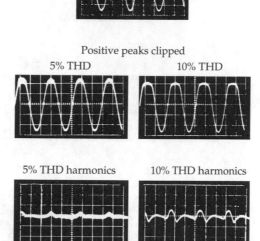

Undistorted sine wave

Positive peaks clipped

5% THD   10% THD

5% THD harmonics   10% THD harmonics

FIGURE 5-19 Oscillograms show an undistorted sine wave, which is applied to the input of an amplifer that clips the positive peaks of the signal. The appearance of the clipped sine wave for 5% and 10% total harmonic distortion is shown. If the fundamental is rejected by a notch filter, the summed harmonics appear as shown.

adjusted to f_0 essentially eliminating it, only the harmonics would be left. Measuring these harmonics together with an RMS (root-mean-square) meter accomplishes what was done in the square root portion of Eq. (5-1). Comparing this RMS measured value of the harmonic components with that of the fundamental and expressing it as a percentage gives the total harmonic distortion.

An undistorted sine wave is sent through an amplifier, which clips positive peaks, as shown in Fig. 5-19. On the left, the flattening of the positive peaks with 5% THD is evident, and shown below that is the combined total of all the harmonic products with the fundamental rejected. On the right is shown the effect of greater clipping to yield 10% THD. Figure 5-20 shows a sine wave passing through the amplifier and symmetrically clipped on both positive and negative peaks. The combined distortion products for symmetrical clipping have a somewhat different appearance, but they measure the same: 5% and 10% THD.

Consumer power amplifiers commonly have specifications listing total harmonic distortion nearer 0.05% rather than 5% or 10%. In a series of double-blind subjective tests, Clark found that 3% distortion was audible on different types of sounds. With carefully selected material (such as a flute solo), detecting distortions down to 2% or 1% might be possible. A distortion of 1% with sine waves is readily audible.

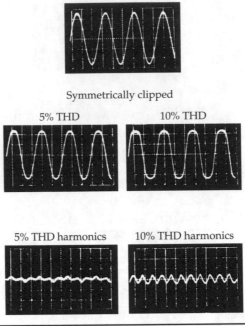

Undistorted sine wave

Symmetrically clipped

5% THD   10% THD

5% THD harmonics   10% THD harmonics

Figure 5-20   Oscillograms show an undistorted sine wave, which is applied to the input of an amplifer which clips both positive and negative peaks in a symmetrical fashion. The appearance of the clipped sine wave is shown for 5% and 10% total harmonic distortion. The appearance of the harmonics alone, with the fundamental filtered out, is also shown.

Resonance

The amplitude of vibration of any resonant system is maximum at the natural frequency or resonant frequency f_0 and is less at frequencies below and above that frequency. A modest excitation signal at that resonance frequency will result in a high amplitude. As shown in Fig. 5-21, the amplitude of vibration changes as the frequency of excitation is varied, going through a peak response at the frequency of resonance. Perhaps the simplest example of a resonant system is a weight on a spring.

Such resonance effects appear in a wide variety of systems: the interaction of mass and stiffness of a mechanical system such as a tuning fork, or the acoustical resonance of the air in a bottle, as the mass of the air in the neck of the bottle reacts with the springiness of the air entrapped in the body of the bottle.

Resonance effects are also present in electronic circuits as the inertia effect of an inductance reacts with the storage effect of a capacitance. An inductor (its electrical symbol is L) is commonly a coil of wire, and a capacitor (C) is made of sheets of conducting material separated by nonconducting sheets. Energy can be stored in the magnetic field of an inductor as well as in the electrical charges on the plates of a capacitor. The interchange of energy between two such storage systems can result in a resonance effect.

Figure 5-22 shows two circuits in which an inductor and a capacitor can exhibit resonance. Let us assume that an alternating current of constant amplitude, but varying frequency is flowing in a parallel resonant circuit (see Fig. 5-22A). As the frequency is

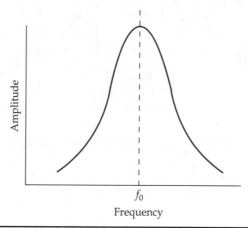

Figure 5-21 The amplitude of vibration of any resonant system is maximum at the natural frequency or resonant frequency f_0 and is less at frequencies below and above that frequency.

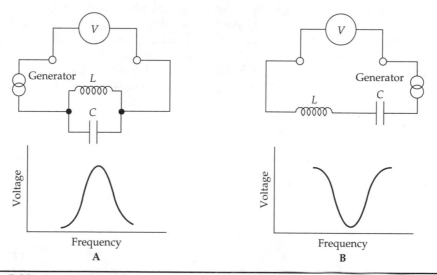

Figure 5-22 A comparison of (A) parallel resonance and (B) series resonance. For a constant alternating current flowing, the voltage across the parallel resonant circuit peaks at the resonance frequency while that of the series resonant circuit is a minimum.

varied, the voltage at the terminals reaches a maximum at the natural frequency of the *LC* system, falling off at lower and higher frequencies. In this way the typical resonance curve shape is developed. Another way of saying this is that the parallel resonant circuit exhibits maximum impedance (opposition to the flow of current) at resonance.

A series resonant circuit (see Fig. 5-22B) also uses an inductor *L* and a capacitor *C*. As the alternating current of constant magnitude and varying frequency flows in the circuit, the voltage at the terminals describes an inverted resonance curve in which the voltage is minimum at the natural frequency and rising at both lower and higher frequencies. It can also be said that the series resonant circuit presents minimum imped-ance at the frequency of resonance.

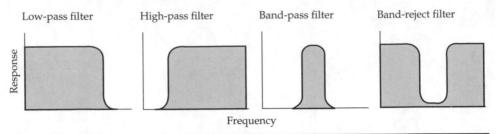

Figure 5-23   Frequency response characteristics for low-pass, high-pass, band-pass, and band-reject filters.

Audio Filters

Filters are used in equalizers, as well as loudspeaker crossovers. By adjusting the values of the resistors, inductors, and capacitors, any type of analog filter can be constructed to achieve almost any frequency and impedance matching characteristic desired. The common forms of filters are the low-pass filter, high-pass filter, band-pass filter, and band-reject filter. The frequency responses of these filters are shown in Fig. 5-23. Figure 5-24 shows how inductors and capacitors may be arranged in various passive circuits to form simple high- and low-pass filters. Filters of Fig. 5-24C will have much sharper cut-offs than the simpler ones in (A) and (B). There are many other highly specialized filters with specific features. With such filters, a wideband signal such as speech or music can be altered at will.

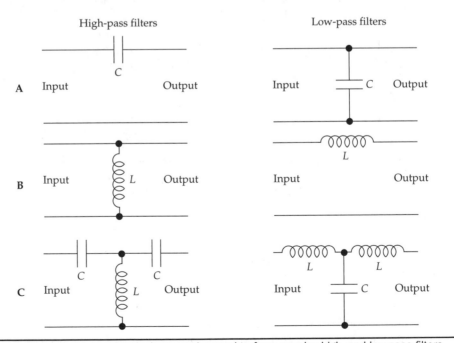

Figure 5-24   Inductors and capacitors can be used to form passive high- and low-pass filters. (A) Filters using capacitors. (B) Filters using inductors. (C) Filters using both capacitors and inductors; these have a sharper cut-off than those in A or B.

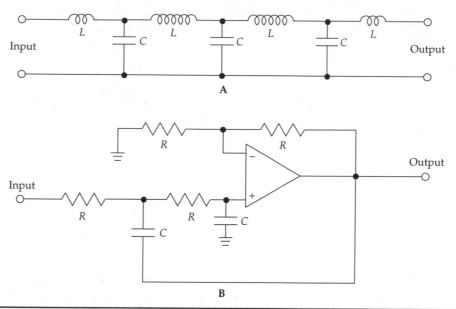

FIGURE 5-25 Two analog low-pass filters are shown. (A) A passive analog filter. (B) An active analog filter utilizing an integrated circuit.

Adjustable filters can be shifted to any frequency within their design band. One type is the constant bandwidth filter which offers the same bandwidth at any frequency. For example, a spectrum analyzer may have a 5-Hz bandwidth, whether it is tuned to 100 Hz or 10 kHz, or any other frequency within its operating band. Another adjustable filter offers a pass bandwidth that is a constant percentage of the frequency to which it is tuned. The ⅓-octave filter is such a device. If it is tuned to 125 Hz, the ⅓-octave bandwidth is 112 to 141 Hz. If it is tuned to 8 kHz, the ⅓-octave bandwidth is 7,079 to 8,913 Hz. The bandwidth is about 23% of the frequency to which it is tuned in either case.

Passive filters do not require any power source. Active filters depend on powered electronics such as discrete transistors or integrated circuits for their operation. A passive low-pass filter assembled from inductors and capacitors is shown in Fig. 5-25A. An active low-pass filter based in an operational amplifier integrated circuit is shown in Fig. 5-25B. Both passive and active filters are widely used; both offer advantages depending on the application.

Filters can be constructed in analog or digital form. All the filters discussed to this point have been of the analog type; they operate on a continuous analog signal. Digital filters perform numerical operations on digital audio signals sampled in discrete time. In many cases, digital filters are implemented as software programs running on microprocessors; analog-to-digital converters and digital-to-analog converters are used to transfer analog audio signals through the digital filter. An example of a digital filter is shown in Fig. 5-26. This example is a finite impulse response (FIR) filter, sometimes called a transversal filter. Digital samples are input to the filter and applied to the top section of z^{-1} blocks comprising a tapped delay line; the middle section shows multiplication by filter coefficients; these outputs are summed to produce the filter's output samples. In this way, for example, the signal's frequency response can be altered.

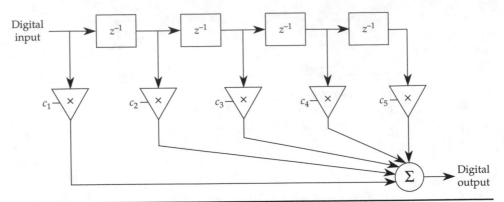

FIGURE 5-26 A finite impulse response (FIR) filter showing delay, multiplication, and summing to implement signal filtering.

Digital filters are part of a larger technology known as digital signal processing (DSP). DSP is widely used throughout the audio industry in many different applications ranging from processing of music signals for playback to signal analysis of room acoustics. For example, digital signal processing can be applied to loudspeaker-room-listener problems. A microphone placed at the listening position measures the frequency and phase response of a loudspeaker's output in a listening room, and from that data, inverse equalization is created to compensate for defects caused by the loudspeaker and the room acoustics.

CHAPTER 6

Reflection

I magine a sound source in a free field, or an approximation of one, such as an open meadow. Sound emanates from the source radially in all directions. Direct sound from the source moves past you, and never returns. Now consider the same source in a room. Direct sound moves past, but when it strikes room boundaries, it reflects from the boundary. A sound moves past you once in a direct path, and many additional times as many reflected paths, until it dies out. Sound comprising many reflections sounds dramatically different than sound in a free field. The reflections convey significant information about the room's size, shape, and boundary composition. Reflections help define the sound characteristic of a room.

Specular Reflections

The essential mechanism of reflection from a flat surface is simple. Figure 6-1 shows the reflection of point-source sound waves from a rigid, plane wall surface. The spherical wavefronts (solid lines) strike the wall and the reflected wavefronts (broken lines) are returned toward the source. This is called a specular reflection and behaves the same as light reflections from a mirror, described by Snell's law.

Sound follows the same rule as light: The angle of incidence is equal to the angle of reflection, as shown in Fig. 6-2. Geometry reveals that the angle of incidence θ_i equals the angle of reflection θ_r. Moreover, just as an image in a mirror, the reflected sound acts as though it originated from a virtual sound image. The virtual source is acoustically located behind the reflecting surface, just as viewing an image in a mirror. This image source is located the same distance behind the wall as the real source is in front of the wall. This is the simple case of a single reflecting surface.

When sound strikes more than one surface, multiple reflections will be created. For example, images of the images will exist. Consider the example of two parallel walls as shown in Fig. 6-3. Sound from the source will strike the left wall; this can be modeled as a virtual source located at I_L (a first-order image). Similarly, there is a virtual source located at I_R. Sound will continue to reflect back and forth between the parallel walls; for example, sound will strike the left wall, right wall, and left wall again, sounding as if there is a virtual source at location I_{LRL} (a third-order image). We observe that in this example, the walls are 15 distance units apart; thus the first-order images are 30 units apart, the second-order images are 60 units apart, the third-order images are 90 units apart, and so on. Using this modeling technique, we can ignore the walls themselves, and consider sound as coming from many virtual sources spaced away from the actual source, arriving at time delays based on their distance from the source.

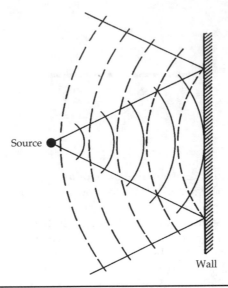

FIGURE 6-1 Reflection from a point source sound from a flat surface (incident sound, solid lines; reflected sound, broken lines). The reflected sound appears to be from a virtual image source.

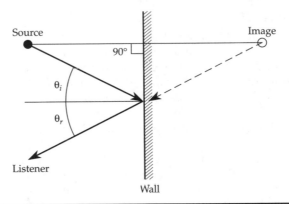

FIGURE 6-2 When sound strikes more than one surface, multiple reflections will be created. These can be viewed as virtual sound sources. Parallel walls can pose acoustical problems such as flutter echoes, reflections that are highly audible because of the regularity of the echo.

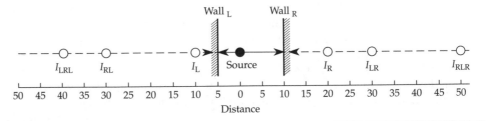

FIGURE 6-3 In a specular reflection, the angle of incidence (θ_i) is equal to the angle of reflection (θ_r).

In a rectangular room, there are six surfaces and the source has an image in all six surfaces, sending energy back to the receiver, resulting in a highly complex sound field. In computing the total sound intensity at a given receiving point, the contributions of all these images must be taken into consideration.

Flutter Echoes

Returning again to Fig. 6-3, we note that parallel walls such as these present an acoustic problem. If the distance between the walls is large enough so the time between reflections is outside the Haas fusion zone, a flutter echo is created as sound bounces back and forth from one wall to the other. Because of the regularity of these reflections, the ear is very sensitive to the effect; in fact, even if the time delays are otherwise in the fusion zone, the effect may still be audible as an echo. This echo can be very prominent in an otherwise diffuse sound field, and is highly undesirable. In theory, with perfectly reflective walls, there would be an infinite number of images. The acoustic effect is the same as being between two mirrors and seeing the series of images. In practice, successive images attenuate because of absorption or diffusion at the walls. Where possible, parallel walls should be avoided, and when unavoidable, they should be covered by absorbing or diffusing material. Splaying walls by a small amount of perhaps 5° or 10° can also avoid flutter echoes.

When sound strikes a boundary surface, some sound energy is transmitted or absorbed by the surface, and some is reflected. The reflected energy is always less than the incident energy. Surfaces that are made of heavy materials (measured by surface weight) are usually more reflective than lighter materials. Sound may undergo many reflections as it bounces around a room. The energy lost at each reflection results in the eventual demise of that sound.

Reflection depends partly on the size of the reflecting object. Sound is reflected from objects that are large compared to the wavelength of the impinging sound. Generally speaking, sound will be reflected from a rectangular panel if each of its two dimensions is five times the wavelength of sound. Thus, objects act as frequency-dependent reflectors. This book would be a good reflector for 10-kHz sound (wavelength about an inch). Moving the book in front of and away from your face would result in significant differences in high-frequency response because of acoustic shadowing. At the low end of the audible spectrum, 20-Hz sound (wavelength about 56 ft) would sweep past the book and the person holding it as though they did not exist, and without appreciable shadows.

Doubling of Pressure at Reflection

The sound pressure on a surface normal to an incident wave is equal to the energy density of the radiation in front of the surface. If the surface is a perfect absorber, the pressure equals the energy density of the incident radiation. If the surface is a perfect reflector, the pressure equals the energy density of both the incident and the reflected radiation. Thus the pressure at the face of a perfectly reflecting surface is twice that of a perfectly absorbing surface. In the study of standing waves, this pressure doubling takes on great significance.

Reflections from Convex Surfaces

Considering sound as rays is a simplified view. Each ray should really be considered as a beam of diverging sound with a spherical wavefront to which the inverse square law applies. Spherical wavefronts from a point source become plane waves at greater

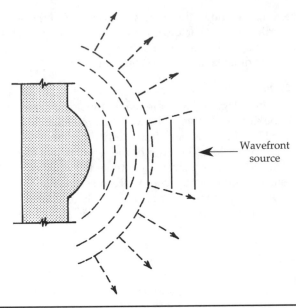

Wavefront
source

FIGURE 6-4   Plane sound waves impinging on a convex irregularity tend to be dispersed through a wide angle if the size of the irregularity is large compared to the wavelength of the sound.

distance from the source. For this reason, impinging sound on various surfaces can usually be thought of as plane wavefronts. Reflection of plane wavefronts of sound from a solid convex surface scatters the sound energy in many directions as shown in Fig. 6-4. This reflecting sound returns and diffuses the impinging sound. Polycylindrical sound-absorbing modules both absorb sound and contribute diffusion in the room. The latter is created by reflection from the cylindrically shaped convex surface of the modules.

Reflections from Concave Surfaces

Plane wavefronts striking a concave surface tend to focus to a point as illustrated in Fig. 6-5. The precision with which sound is focused is determined by the shape and relative size of the concave surface. Spherical concave surfaces are common because they are readily formed. They are often used to make a microphone highly directional by placing it at the focal point. Such microphones are frequently used to pick up field sounds at sporting events or record animal sounds in nature. The effectiveness of such concave reflectors depends on the size of the reflector with respect to the wavelength of sound. For example, a 3-ft-diameter spherical reflector will give good pickup directivity at 1 kHz (wavelength about 1 ft), but it is practically nondirectional at 200 Hz (wavelength about 5.5 ft). Concave surfaces, for example, domed ceilings and archways in churches or auditoriums, can be the source of serious problems as they produce concentrations of sound in direct opposition to the goal of uniform distribution of sound in rooms.

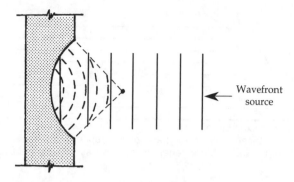

FIGURE 6-5 Plane sound waves impinging on a concave irregularity tend to be focused if the size of the irregularity is large compared to the wavelength of the sound.

Reflections from Parabolic Surfaces

A parabola generated by the equation $y = x^2$ has the characteristic of focusing sound precisely to a point. A very "deep" parabolic surface, such as that of Fig. 6-6, exhibits far better directional properties than a shallow one. Again, the directional properties depend on the size of the opening in terms of wavelengths.

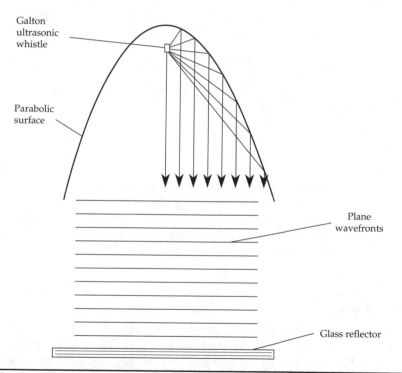

FIGURE 6-6 A parabolic surface can focus sound precisely at a focal point or, the converse, a sound source placed at the focal point can produce plane, parallel wavefronts. In this case, the source is an ultrasonic Galton whistle driven by compressed air.

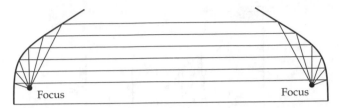

FIGURE 6-7   Graphic example of a whispering gallery showing symmetrical sound focusing points. A whisper directed tangentially to the paraboloid-shaped surface is readily heard by the receiver on the far side of the room. More generally, concave surfaces pose acoustical problems.

Plane waves striking such a reflector are brought to a focus at the focal point. Conversely, sound emitted at the focal point of the parabolic reflector generates plane wavefronts. For example, the parabola of Fig. 6-6 is used as a directional sound source with a small, ultrasonic Galton whistle pointed inward at the focal point. Standing waves are produced by reflections from a heavy glass plate. In a Galton whistle, the force exerted by the vibration of the air particles on either side of a node is sufficient to hold slivers of cork in levitation.

Whispering Galleries

St. Paul's Cathedral in London, St. Peter's Basilica in Vatican City, the Temple of Heaven in Beijing, Statuary Hall in the U.S. Capitol Building, Grand Central Station in New York City (in front of the Oyster Bar & Restaurant) and other structures contain a whispering gallery; the acoustical mechanism is diagrammed in Fig. 6-7. Reflections from the exterior surfaces of cylindrical shapes have been mentioned in the treatment of polycylindrical absorbers. In this case, the source and receiver are both inside a mammoth, hard-surfaced cylindrical room.

At the source, a whisper directed tangentially to the surface is clearly heard on the receiver side. The phenomenon is assisted by the fact that the walls are paraboloidshaped. This means that upward-directed components of the whispered sounds tend to be reflected downward and are conserved and transmitted rather than lost above. Although acoustically interesting, architectural configurations such as this are usually undesirable. Except for unique applications, concave surfaces such as sections of cylinders, spheres, paraboloids, and ellipsoids should not be used in acoustically important architecture. The sound focusing effects are exactly counter to the usual goal of providing uniform, diffuse sound throughout a space.

Standing Waves

The concept of standing waves directly depends on the reflection of sound. Assume two flat, solid parallel walls separated a given distance (as in Fig. 6-3). A sound source between them radiates sound of a specific frequency. As we observed, the wavefront striking the right wall is reflected back toward the source, striking the left wall where it is again reflected back toward the right wall, and so on. One wave travels to the right, the other toward the left. The two traveling waves interact to form a standing wave. Only the standing wave, the interaction of the two, is stationary. The frequency of the radiated sound establishes this resonant condition between the wavelength of the sound and the distance between the two surfaces. This phenomenon is entirely dependent on the reflection of sound at the two parallel surfaces. As discussed in other chapters, standing waves require careful design scrutiny, particularly in terms of a room's low-frequency response.

The Corner Reflector

We generally think of reflections as normal (perpendicular) reflections from surrounding walls, but reflections also occur at the corners of a room. Moreover, the reflections follow the source around the room. The corner reflector of Fig. 6-8 receives sound from the source at location 1, and sends a reflection directly back toward that source. If the angles of incidence and reflection are carefully noted, a sound at source location 2 will also send a direct, double-surface reflection back to that source. Similarly, a source at location 3 on the opposite side of the normal is subject to the same effect. A corner reflector thus has the property of reflecting sound back toward the source from any direction. Corner reflections suffer losses at two surfaces, tending to make them somewhat less intense than normal reflections at the same distance.

The corner reflector of Fig. 6-8 involves only two surfaces. The same follow-the-source principle applies to the four upper tri-corners of a room formed by ceiling and walls and another four formed by floor and wall surfaces. Following the same principles, sonar and radar systems have long employed targets made of three circular plates of reflecting material assembled so that each is perpendicular to the others.

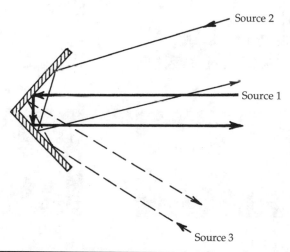

Source 2

Source 1

Source 3

Figure 6-8 A corner reflector has the property of reflecting sound back toward the source from any direction.

Mean Free Path

The average (mean) distance sound travels between successive reflections is called the mean free path. This distance is given by the expression:

$$\mathrm{MFP} = \frac{4V}{S} \qquad (6\text{-}1)$$

where MFP = mean free path
V = volume of the space, ft^3 or m^3
S = surface area of the space, ft^2 or m^2

For example, in a room measuring 25 × 20 × 10 ft, on average a sound travels a distance of 10.5 ft between reflections. Sound travels 1.13 ft/msec. At that speed it takes 9.3 msec to traverse the mean-free distance of 10.5 ft. Viewed another way, about 107 reflections take place in the space of a second.

Figure 6-9 shows echograms of reflections occurring during the first 0.18 sec in a recording studio having a volume of 16,000 ft^3 and a reverberation time of 0.51 sec at 500 Hz. The microphone was placed, successively, in four different locations in the room. The impulsive sound source was in a fixed position. The sound source was a pistol that punctured a paper with a blast of air, giving an intense pulse of sound of less than 1 msec in duration. The reflection patterns at the four positions show differences but, in each, scores of individual reflections are clearly resolved. These echograms define the transient sound field of the room during the first 0.18 sec as contrasted to the steady-state condition. These early reflections play a significant role in the perception of room acoustics.

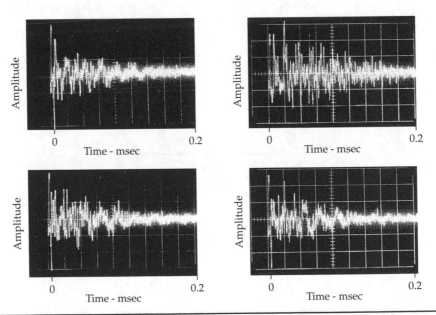

Figure 6-9 Individual reflections are resolved in these echograms taken at four different positions in a studio of 16,000 ft^3 volume and having a reverberation time of 0.51 sec. The horizontal time scale is 20 msec/div.

Perception of Sound Reflections

When reproducing recorded sound in a listening room, during the enjoyment of live music in a concert hall, or in any activity in any acoustical space, the sound falling on the ears of the listener is very much affected by reflections from the surfaces of the room. Our perception of these reflections is an important manifestation of sound reflection.

The Effect of Single Reflections

Research studies on the audibility of simulated reflections often use an arrangement of loudspeakers much like the traditional stereophonic playback configuration shown in Fig. 6-10. The observer is seated at the apex so that lines drawn to the two loudspeakers are approximately 60° apart (this angle varies with the investigator). The monaural signal is fed to one of the loudspeakers, which represents the direct signal. The signal to the other loudspeaker can be delayed any amount: This represents a lateral reflection. The two variables under study are the level of the reflection compared to that of the direct and the time delay of the reflection with respect to the direct signal.

Olive and Toole investigated listening conditions in small rooms, such as recording studio control rooms and home listening rooms. In one experiment, they investigated the effect of simulated lateral reflections in an anechoic environment with speech as the test signal. The work is summarized in Fig. 6-11. This graph plots reflection level against reflection delay, the two variables specified above. A reflection level of 0 dB means that the reflection is the same level as the direct signal. A reflection level of −10 dB means that the reflection level is 10 dB below the direct. In all cases, reflection delay is in milliseconds later than the direct signal.

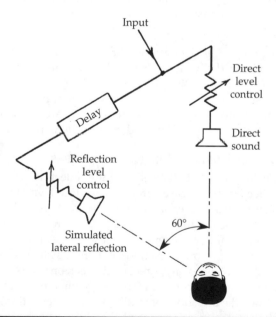

FIGURE 6-10 Typical equipment configuration used by many investigators to study the audible effect of a simulated lateral reflection on the direct signal. Reflection level (with respect to the direct) and reflection delay are the variables under control.

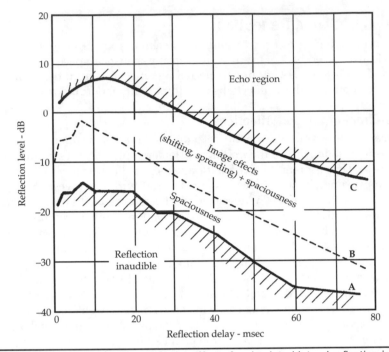

FIGURE 6-11 Results of investigations into the effect of a simulated lateral reflection in an anechoic environment and with speech as the test signal. Curve A is the absolute threshold of detection of the reflection. Curve B is the image-shift threshold. Curve C indicates the points at which the reflection is heard as a discrete echo. (*Composite of results: Curves A and B from Olive and Toole, Curve C from Meyer and Lochner*)

In Fig. 6-11, curve A is the absolute threshold of audibility of the echo. This means that at any particular delay, the reflection is not heard for reflection levels below this line. Note that for the first 20 msec, this threshold is essentially constant. At greater delays, progressively lower reflection levels are required for a just-audible reflection. For small room, delays in the 0- to 20-msec range are of greatest significance. In this range the reflection audibility threshold varies little with delay.

Perception of Spaciousness, Images, and Echoes

Assume a reflection delay of 10 msec with the reflection coming from the side. As the level of the reflection is increased from a very low level, the reflection is completely inaudible. As the level of the reflection is increased, it finally becomes audible as its level reaches about 15 dB below the direct signal. As the reflection level is increased beyond this point, the room takes on a sense of spaciousness; the anechoic room in which the tests were made sounds more like a normal room. The listener is not aware of the reflection as a discrete event, or of any directional effect, only this sense of spaciousness.

As the level of the reflection is increased further, other effects become audible. At about 10 dB above the threshold of audibility of the reflection, changes in the apparent size and location of the front auditory image become apparent. At greater delays, the image becomes smeared toward the reflection.

Reviewing what happens in the 10- to 20-msec delay range, as the reflection level is increased above the threshold of audibility, spatial effects dominate. As the reflection level is increased roughly 10 dB above the audibility threshold, image effects begin to enter, including image size and shifting of position of the image.

Reflections having a level another 10 dB above the image shift threshold introduce another perceptual threshold. The reflections are now discrete echoes superimposed on the central image. Such discrete echoes are damaging to sound quality. For this reason, reflection level/delay combinations that result in such echoes must be minimized in practical designs.

Lateral reflections provide important perceptual cues in a sound field. Lateral reflections can affect spaciousness and the size and position of the auditory image. Olive and Toole investigated a two-loudspeaker installation and found that the effects obtained from a single loudspeaker are correlated to the stereo case. This suggests that single-loudspeaker data can be applied to stereo playback.

Those interested in the reproduction of high-fidelity audio will see the practicality of the results of these reflection studies. The spaciousness of a listening room as well as the stereo image definition can be adjusted by careful manipulation of lateral reflections. However, lateral reflections can be used only after interfering early reflections are reduced.

Effect of Angle of Incidence, Signal Type, and Spectrum on Audibility of Reflection

Researchers have shown that the direction from which a reflection arrives has practically no effect on the perception of the reflection with one important exception. When the reflection arrives from the same direction as the direct signal, it can be up to 5 to 10 dB louder than the direct sound before it is detected. This is due to masking of the reflection by the direct signal. If the reflection is recorded along with the direct signal and reproduced over a loudspeaker, it will be masked by this 5- to 10-dB amount.

The type of signal has a major effect on the audibility of reflections. Consider the difference between continuous and noncontinuous sounds. Impulses, in the form of 2 clicks/sec, are of the noncontinuous type. Pink noise is an example of the continuous type. Speech and music lie between the two types. Figure 6-12 shows the differences in

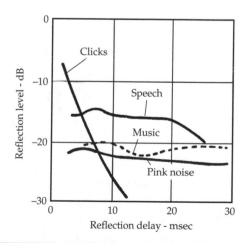

FIGURE 6-12   Absolute thresholds of perception of reflections of different types of signals, ranging from 2 clicks/sec (noncontinuous) to pink noise (continuous). The closeness of pink noise to classical music (Mozart) gives assurance that pink noise is a reasonable surrogate for music in measurements. (*Olive and Toole*)

audibility thresholds of continuous and noncontinuous sounds. Anechoic speech is closer to a noncontinuous sound than either music or pink noise. At delays of less than 10 msec, the level of impulses for threshold detection must be much higher than continuous sounds. The threshold curves for music (Mozart) and pink noise are very close together. This confirms the belief that pink noise is a reasonable surrogate for music for measurements.

Most researchers use sounds having the same spectrum for both the direct and reflected simulations. In real life, reflections depart from the original spectrum because sound-absorbing materials invariably absorb high frequencies more than low frequencies. In addition to this, off-axis loudspeaker response lowers the high-frequency content even more. Threshold audibility experiments show that radical low-pass filtering of the reflection signal produced only minor differences in thresholds. The conclusion is that alteration of reflection spectrum does not appreciably change audibility thresholds.

CHAPTER 7

Diffraction

From observation, we know that sound travels around obstacles and corners. For example, music reproduced in one room of a home can be heard down the hall and in other rooms. However, the character of the music heard in distant parts of the home is different. In particular, the bass notes are more prominent. This is partly because their longer wavelengths more readily diffract around obstacles and corners. Diffraction causes sound, which normally travels rectilinearly, to bend and travel in other directions.

Wavefront Propagation and Diffraction

Wavefronts of sound travel in straight lines. Sound rays, a concept applicable at mid and high audible frequencies, can be considered as beams of sound that travel in straight lines perpendicular to the wavefront. Sound wavefronts and sound rays travel in straight lines, except when something gets in the way. Obstacles can cause sound to change direction from its original rectilinear path. The mechanism by which this change of direction takes place is called diffraction. Incidentally, the word diffract is from the Latin word *diffringere*, which means to break into pieces.

Alexander Wood, a Cambridge acoustician, recalled how Newton pondered over the relative merits of the corpuscular and wave theories of light. Newton decided that the corpuscular theory was the correct one because light is propagated rectilinearly. Later it was demonstrated that light is not always propagated rectilinearly, that diffraction can cause light to change its direction of travel. In fact, all types of wave motion, including sound, are subject to diffraction, a phenomenon caused by phase interference.

Huygens formulated a principle that is the basis of the mathematical analysis of diffraction. The same principle also gives a simple explanation of how sound energy is diverted from a main beam into a shadow zone. Huygens' principle can be paraphrased: Every point on the wavefronts of sound that has passed through an aperture or passed a diffracting edge is considered a point source radiating energy back into the shadow zone. The sound energy at any point in the shadow zone can be mathematically obtained by summing the contributions of all of these point sources on the wavefronts.

Wavelength and Diffraction

Low-frequency sounds (long wavelengths) diffract (bend) more than high-frequency sounds (short wavelengths). Conversely, high frequencies diffract less than low frequencies. Diffraction is less noticeable for light than it is for sound because of the relatively short wavelengths of light. As a result, optical shadows are relatively more pronounced than acoustical shadows. You might not see the (diffracted) light from a room down the hallway, but you might easily hear the (diffracted) music from there.

Looked at in another way, the effectiveness of an obstacle in diffracting sound is determined by the acoustical size of the obstacle. Acoustical size is measured in terms of the wavelength of the sound. An obstacle is acoustically small if wavelengths are long, but the same object is acoustically large if wavelengths are short. This relationship is described in more detail below.

Diffraction of Sound by Obstacles

If an obstacle is acoustically small relative to wavelength, sound will easily diffract. Sound will bend around a small obstacle with little disturbance, creating little or no acoustical shadow. When an obstacle's dimensions are smaller or equal to the wavelength, virtually all sound will be diffracted. On the other hand, if an obstacle is acoustically large relative to wavelength, some sound will tend to be reflected by it. Thus we see that obstructions act as frequency-dependent reflectors.

As noted, the effectiveness of an obstacle in diffracting sound is determined by the acoustical size of the obstacle. Consider two objects in Fig. 7-1, and their behavior with the same wavelength of sound. In Fig. 7-1A, the obstacle is so small compared to the wavelength of the sound that it has no appreciable effect on the passage of sound. In Fig. 7-1B, however, if the obstacle is several wavelengths across, it casts a shadow behind the obstacle. Each wavefront passing the obstacle becomes a line of new point sources radiating sound into the shadow zone by diffraction. Although not shown, some sound waves are also reflected from the face of the barrier.

Another way of looking at the relationship of wavelength and diffraction is to change our basis of comparison, remembering that acoustical size depends on the wavelength of the sound. Consider Fig. 7-2A where the obstacle is the same physical size as that in Fig. 7-2B. However, the frequency of the sound in Fig. 7-2A is higher than that in Fig. 7-2B. We observe that the same obstacle casts a considerable shadow for high frequencies, but casts much less of a shadow for low frequencies.

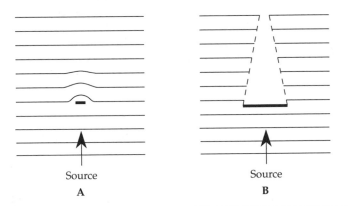

Figure 7-1 Diffraction is wavelength-dependent. (A) An obstacle very much smaller than the wavelength of sound allows the wavefronts to pass essentially undisturbed. (B) An obstacle larger than the wavelength casts a shadow that tends to be irradiated from sources on the wavefronts of sound that go past the obstacle.

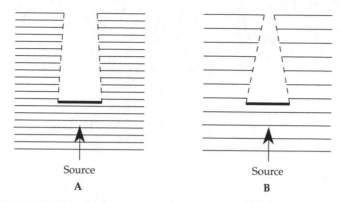

FIGURE 7-2 The same-sized obstacle will exhibit different degrees of diffraction, depending on the frequency (wavelength) of the sound. (A) When a high-frequency sound impinges on the obstacle, there is relatively little diffraction. (B) When a low-frequency sound impinges on the same obstacle, there is considerable diffraction. The edges act as new sources, radiating sound into the shadow zone.

As a final example, imagine an obstacle that is 0.1 ft across and an obstacle that is 1 ft across; if the frequency of the sounds impinging on the first obstacle is 1,000 Hz (wavelength 1.13 ft), and the frequency impinging on the second is 100 Hz (wavelength 11.3 ft), both would exhibit the same diffraction.

A highway noise barrier is a common sight along busy roadways, and an example of an obstacle that is designed to isolate listeners (e.g., in homes) from impinging traffic sounds, as shown in Fig. 7-3. The spacing of the sound wavefronts indicates relatively higher or lower frequency from the source (such as traffic). At higher frequencies, as shown in Fig. 7-3A, the barrier becomes effectively large and successfully shields a listener on the other side. Even if there is not complete isolation, at least higher-frequency traffic noise will be attenuated. At lower frequencies, as shown in Fig. 7-3B, the barrier becomes acoustically smaller. Low frequencies diffract over the barrier and are audible to the listener. Because of the change in the frequency response of the noise, the listener essentially hears bass-heavy traffic noise.

The sound reflected from the wall should be noted. It is as though the sound were radiated from a virtual image on the far side of the wall. The wavefronts passing the top edge of the wall can be considered as lines of point sources radiating sound. This is the source of the sound penetrating the shadow zone.

Figure 7-4 illustrates the effectiveness of highway barriers, showing the attenuation of sound in the shadow of a high, massive wall. The center of the highway is taken to be 30 ft from the wall on one side, and the home or other sensitive area is considered to be 30 ft on the other side of the wall (the shadow side). The 20-ft-high wall yields about 25 dB of attenuation from the highway noise at 1,000 Hz. However, the attenuation of the highway noise at 100 Hz is only about 15 dB. The wall is less effective at lower frequencies than at higher frequencies. The shadow zone behind the wall tends to be shielded from the high-frequency components of the highway noise. The low-frequency components penetrate the shadow zone by diffraction. To be effective, barriers must be sufficiently tall, and also long enough to prevent flanking around the end of the barrier. The effectiveness of any barrier is strictly frequency dependent.

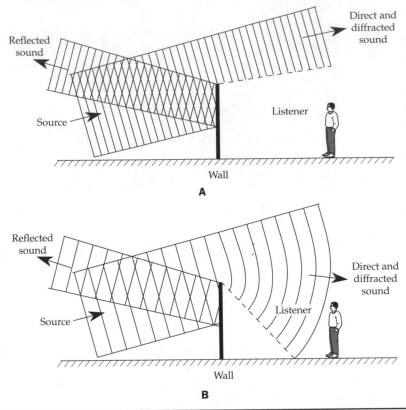

FIGURE 7-3 The sound striking a solid traffic barrier will be partly reflected, and partly diffracted. (A) High-frequency traffic sounds are successfully attenuated on the other side of the barrier because of limited diffraction. (B) Low-frequency traffic sounds are less attenuated because of more prominent diffraction. Sound passing the top edge of the barrier acts as though the wavefronts are lines of sources, radiating sound energy into the shadow zone.

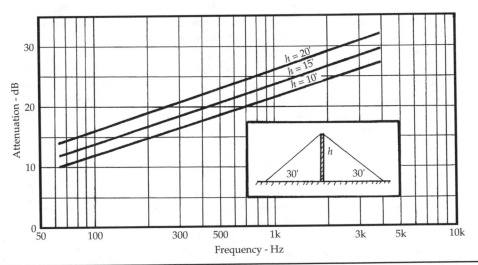

FIGURE 7-4 An estimation of the effectiveness of a traffic barrier in terms of sound (or noise) attenutation as a function of frequency and barrier height. (*Rettinger*)

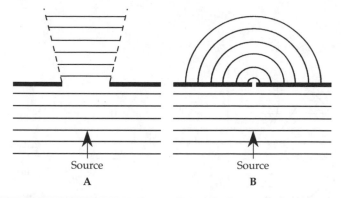

Source
A

Source
B

Figure 7-5 When plane waves of sound strike a barrier with an aperture, diffraction depends on the relative size of the opening. (A) An aperture large in terms of wavelength allows wavefronts to go through with little disturbance. These wavefronts act as lines of new sources, radiating sound energy into the shadow zone. (B) If the aperature is small compared to wavelength, the small wavefronts which penetrate the opening act as point sources, radiating a hemispherical field of sound into the shadow zone.

Diffraction of Sound by Apertures

Diffraction through apertures depends on the size of the opening and the wavelength of the sound. The amount of diffracted sound relative to total sound passing through an aperture increases as the opening size decreases. As with diffraction around obstacles, diffraction through apertures is wavelength dependent. Diffraction increases as frequency decreases. Thus, an opening is more acoustically transparent at low frequencies.

Figure 7-5A illustrates the diffraction of sound by an aperture that is many wavelengths wide. The wavefronts of sound strike a solid obstacle; some sound passes through the wide aperture and although not shown, some sound is reflected. By diffusion, some of the energy in the main beam is diverted into the shadow zone. Each wavefront passing through the aperture becomes a row of point sources radiating diffracted sound into the shadow zone. The same principle holds for Fig. 7-5B, except that the aperture is small compared to the wavelength of sound. Most of the sound energy is reflected from the wall surface and only a small amount of energy passes through it. The points on the limited wavefront going through the aperture are so close together that their radiations take the form of a hemisphere. Following Huygens' principle, the sound emerges from the aperture as a hemispherical wave that diverges rapidly. Most of the energy passing through a small aperture is through diffraction. Because of diffraction, even a small opening can transmit a relatively large amount of sound energy, particularly at low frequencies.

Diffraction of Sound by a Slit

Figure 7-6 diagrams a classical experiment first performed by Pohl, and described more recently by Wood. The equipment layout of Fig. 7-6A is very approximate. The source/slit arrangement rotated about the center of the slit and the measuring radiometer was at a

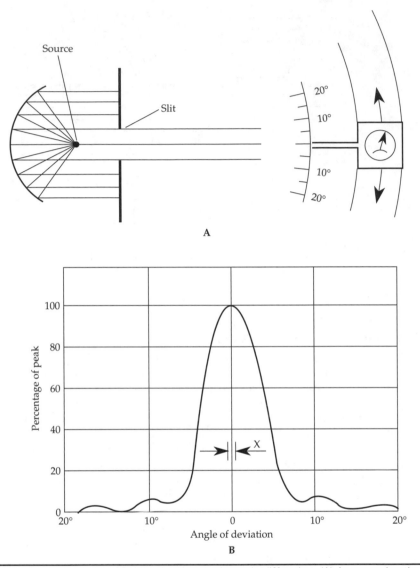

FIGURE 7-6 A consideration of Pohl's classic experiment in diffraction. (A) An approximation of the equipment arrangement showing source and slit. (B) Diffraction causes the beam to broaden with a characteristic pattern. The narrower the slit, the greater this broadening of the beam. (*Wood*)

distance of 8 m. The slit width was 11.5 cm wide, and the wavelength of the measuring sound was 1.45 cm (23.7 kHz). Figure 7-6B shows the intensity of the sound versus the angle of deviation. The dimension X indicates the geometrical boundaries of the ray. Any response wider than X is caused by diffraction of the beam by the slit. A narrower slit would yield correspondingly more diffraction and a greater width of the beam. The increase in width of the beam, that is, the extent of diffraction, is the striking feature of this experiment.

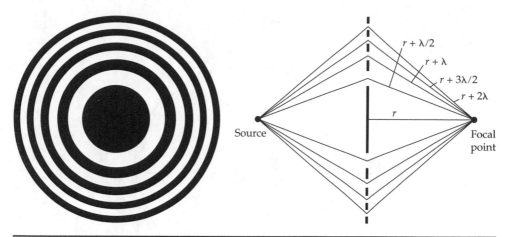

Figure 7-7 The zone plate acts as an acoustic lens. The slits are arranged so that the several path lengths differ by multiples of a half wavelength of the sound so that all diffracted rays arrive at the focal point in phase, combining constructively. (*Olson*)

Diffraction by the Zone Plate

The zone plate shown in Fig. 7-7 can be considered an acoustic lens. It consists of a circular plate with a set of concentric, annular slits of specific radii. If the focal point is at a distance of r from the plate, the next longer path must be $r + \lambda/2$, where λ is the wavelength of the sound falling on the plate from the source. Successive path lengths are $r + \lambda$, $r + 3\lambda/2$, and $r + 2\lambda$. These path lengths differ by $\lambda/2$, which means that the sound through all the slits will arrive at the focal point in phase which, in turn, means that they add constructively, intensifying the sound at the focal point.

Diffraction around the Human Head

Figure 7-8 illustrates the diffraction caused by a sphere roughly the size of the human head. This diffraction by the head as well as reflections and diffractions from the shoulders and upper torso influences human perception of sound. In general, for sound of frequency 1 to 6 kHz arriving from the front, head diffraction tends to increase the sound pressure in front and decrease it behind the head. As expected, for lower frequencies, the directional pattern tends to become circular.

Diffraction by Loudspeaker Cabinet Edges

Loudspeaker cabinets are notorious for diffraction effects. If a loudspeaker is placed near a wall and aimed away from the wall, the wall is still illuminated with sound diffracted from the corners of the cabinet. Reflections of this sound can affect the quality of the sound at the listener's position. Vanderkooy and Kessel computed the magnitude of loudspeaker cabinet edge diffraction. The computations were made on a box loudspeaker cabinet with front baffle having the dimensions 15.7×25.2 in and depth of 12.6 in. A point source of sound was located symmetrically at the top of the baffle, as shown in Fig. 7-9. The sound from this point source was computed at a distance from the box.

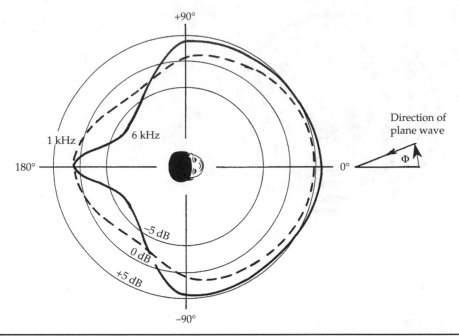

Figure 7-8 Diffraction occurs around a solid sphere about the size of a human head. For sound in the 1- to 6-kHz range, sound pressure is generally increased in the front hemisphere and reduced in the rear. (*Muller, Black and Davis, as reported by Olson*)

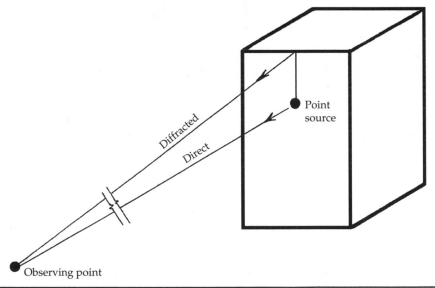

Figure 7-9 Experimental set-up for the measurement of loudspeaker cabinet edge diffraction, as shown in Fig. 7-10. Sound arriving at the observation point is the combination of the direct sound plus the edge diffraction. (*Vanderkooy and Kessel*)

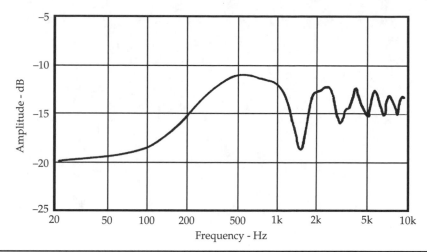

Figure 7-10   The calculated effects of loudspeaker edge diffraction on the direct signal in set-up of Fig. 7-9. There is a significant change in overall frequency response of the system. (*Vanderkooy and Kessel*)

The sound arriving at the observation point is the combination of the direct sound plus the edge diffraction. The resulting response is shown in Fig. 7-10. Fluctuations due to edge diffraction for this particular experiment approached ±5 dB. This is a significant change in overall frequency response of a reproduction system.

This cabinet diffusion effect can be controlled by setting the loudspeaker face flush in a much larger baffling surface. Diffusion can be reduced by rounding cabinet edges and using foam or other absorbing material around the front of the cabinet.

Diffraction by Various Objects

Early sound level meters were simply boxes with a microphone. Diffraction from the edges and corners of the box seriously affected the accuracy of the readings at high frequencies. Modern sound level meters have cases with carefully rounded contours, and the microphone is mounted on a smooth, slender, rounded stalk to remove it from the meter case.

Similarly, diffraction from the casing of a studio microphone can cause deviations from the desired flat response. This must be taken into account when designing a microphone.

When measuring sound absorption in large reverberation chambers, the common practice is to place the material to be measured in an 8×9 ft frame on the floor. Diffraction from the edges of this frame may result in absorption coefficients greater than unity. In other words, diffraction of sound makes the sample appear larger than it really is. Similarly, in practice, the absorption of materials is increased because of diffraction at their edges. For this reason, and others, it is preferable to space absorbing panels apart from each other, rather than join them into one section. This takes advantage of increased absorption because of edge diffraction.

Small cracks around observation windows or back-to-back microphone or electrical service boxes in partitions can destroy the hoped-for isolation between studios or between studio and control room. The sound emerging on the other side of the hole or slit is spread in all directions by diffraction. For this reason, when designing for sound isolation, any opening in a partition must be sealed.

CHAPTER 8

Refraction

A t the turn of the twentieth century, Lord Rayleigh was puzzled because some powerful sound sources, such as cannon fire, could be heard only short distances sometimes and very great distances at other times. He calculated that if all the power from a siren driven by 600 horsepower was converted into sound energy and spread uniformly over a hemisphere, the sound should be audible to a distance of 166,000 miles, more than six times the circumference of the earth. However, such sound propagation is never experienced and a maximum range of a few miles is typical.

When dealing with sound propagation, particularly outdoors, refraction plays a large role. Refraction is a change in the direction of sound propagation that occurs when there is a change in the transmission medium. In particular, the change in the medium changes the speed of sound propagation, and therefore the sound bends.

There are many reasons why Rayleigh's estimate was wrong, and why sound is not heard over great distances. For one thing, refraction in the atmosphere will profoundly affect the propagation of sound over distance. In addition, the efficiency of sound radiators is usually quite low; not much of that 600 horsepower would actually be radiated as sound. Energy is also lost as wavefronts drag across the rough surface of the earth. Another loss is dissipation in the atmosphere, particularly affecting high frequencies. Early experiments by Rayleigh and others accelerated research on the effects of temperature and wind gradients on the transmission of sound. This chapter will help to advance the understanding of refraction effects.

The Nature of Refraction

The difference between absorption and reflection of sound is obvious, but sometimes there is confusion between refraction and diffraction (and possibly diffusion, the subject of the next chapter).

Refraction is the change in the direction of travel of sound because of differences in the velocity of propagation. Diffraction is the change in the direction of travel of sound as it encounters sharp edges and physical obstructions. Of course, in practical situations, it is entirely possible for both effects to simultaneously affect the same sound.

Figure 8-1 recalls a common observation of the apparent bending of a stick as one end is immersed in water. This is an illustration of the refraction of light, caused by the different refractive indices of air and water, and hence different speeds of propagation. The refraction of sound, which is another wave phenomenon, is similar. In the case of water and air, the change in the refractive indices is abrupt, as is the bending of light. Sound refraction can occur abruptly, or gradually, depending on how the transmission mediums affect its speed.

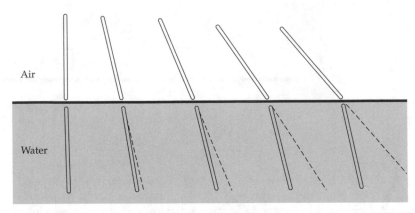

FIGURE 8-1   Partially immersing a stick in water illustrates refraction of light, created because the speeds of propagation are different in air and water. Sound is another wave phenomenon that is also refracted because of changes in media sound speed.

Refraction of Sound in Solids

The sound ray concept is helpful in considering direction of propagation. Rays of sound are always perpendicular to sound wavefronts. Figure 8-2 shows two sound rays passing from a dense medium to a less dense medium. The sound speed in the denser medium is greater than that in the less dense one, as shown in Table 8-1. As one ray

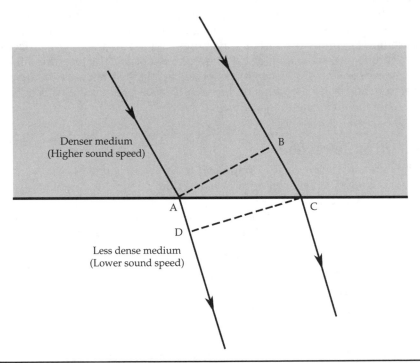

FIGURE 8-2   Rays of sound are refracted when traveling from a denser medium having a higher sound speed into a less dense medium having a lower sound speed. The wavefront A-B is not parallel to wavefront C-D because the direction of the wave is changed due to refraction.

Medium	Speed of Sound ft/sec	m/sec
Air	1,130	344
Seawater	4,900	1,500
Wood, fir	12,500	3,800
Steel bar	16,600	5,050
Gypsum board	22,300	6,800

TABLE 8-1 Speed of Sound

reaches the boundary between the two media at A, the other ray still has some distance to go. In the time it takes one ray to travel from B to C, the other ray has traveled a shorter distance from A to D in the new medium. Wavefront A-B represents one instant of time as does wavefront D-C an instant later. But these two wavefronts are no longer parallel. The rays of sound have been refracted at the interface of the two media having unequal sound speeds. In particular, in this example, the speed of sound is faster in the denser medium and slower in the less dense medium.

An analogy may be useful. Assume that the shaded area of Fig. 8-2 is paved and the unshaded area is ploughed. Assume also that the wavefront A-B is a line of soldiers. The line of soldiers A-B, marching in military order, makes good progress on the pavement. As soldier A reaches the ploughed ground he or she slows down and begins plodding over the rough surface. Soldier A travels to D on the ploughed surface in the same time that soldier B travels the distance B to C on the pavement. This tilts the column of soldiers off in a new direction, which is the definition of refraction. In any homogeneous medium, sound travels rectilinearly (in the same direction). If a medium of another density is encountered, the sound is refracted.

Refraction of Sound in the Atmosphere

The earth's atmosphere is anything but a stable, uniform medium for the propagation of sound. Sometimes the air near the earth is warmer than the air at greater heights; sometimes it is colder. At the same time this vertical layering exists, other changes may be taking place horizontally. It is a wondrously intricate and dynamic system, challenging acousticians (and meteorologists) to make sense of it.

In the absence of thermal gradients, a sound ray may be propagated rectilinearly, as shown in Fig. 8-3A. As noted, rays of sound are perpendicular to sound wavefronts.

In Fig. 8-3B, a thermal gradient exists between the cool air near the surface of the earth and the warmer air above. This affects the wavefronts of sound. Sound travels faster in warm air than in cool air causing the tops of the wavefronts to go faster than the lower parts. The tilting of the wavefronts directs the sound rays downward. Under such conditions, sound from source S is continually bent down toward the surface of the earth and may follow the earth's curvature and be heard at relatively great distances.

In Fig. 8-3C, the thermal gradient is reversed as the air near the surface of the earth is warmer than the air higher up. In this case the lower parts of the wavefronts travel faster than the tops, resulting in an upward refraction of the sound rays. The same

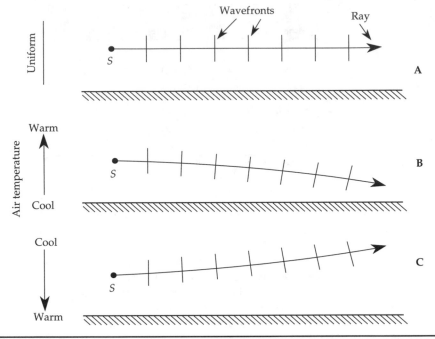

FIGURE 8-3   Refraction of sound paths resulting from temperature gradients in the atmosphere. (A) Air temperature constant with height. (B) Cool air near the surface of the earth and warmer air above. (C) Warm air near the earth and cooler air above.

sound energy from source S would now be dissipated in the upper reaches of the atmosphere, reducing the chances of it being heard at any great distance at the surface of the earth.

Figure 8-4A presents a distant view of the downward refraction scenario of Fig. 8-3B. Sound traveling directly upward from the source S penetrates the temperature gradient at right angles and would not be refracted. It would speed up and slow down slightly as it penetrates the warmer and cooler layers, but would still travel in a vertical direction. All sound rays, except the vertical, would be refracted downward. The amount of this refraction varies materially: the rays closer to the vertical are refracted much less than those more or less parallel to the surface of the earth.

Figure 8-4B is a distant view of the upward refraction scenario of Fig. 8-3C. Shadow zones are to be expected in this case. Again, the vertical ray is the only one escaping refractive effects.

Wind can have a significant effect on sound propagation and plays an important role in analyzing noise pollution over large distances. For example, it is a common experience to hear sound better downwind than upwind. As a result, for example, on a windy day, sound can pass over a traffic barrier that would normally block traffic noise. However, this is not because wind is blowing sound toward the listener. Rather, wind speed is generally slower close to the ground than at greater heights. This wind gradient may affect the propagation of sound. Plane waves of sound from a distant source traveling with the wind will bend down toward the earth. Plane waves traveling against the wind will bend upward. Figure 8-5 illustrates the effect of wind on the downward

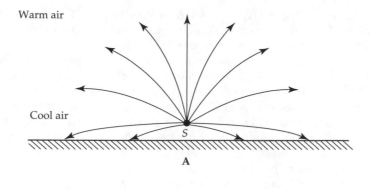

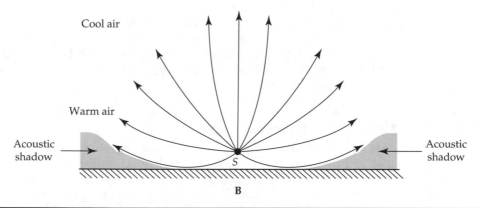

FIGURE 8-4 Comprehensive illustration of refraction of sound from source. (A) Cool air near the ground and warmer air above. (B) Warm air near the ground and cooler air above; note that acoustic shadow areas result from the upward refraction.

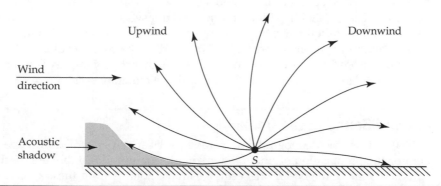

FIGURE 8-5 Wind gradients can affect the propagation of sound. A shadow sound is created upwind while good listening conditions exist downwind.

refraction case of Fig. 8-4A. An upwind shadow is created while downwind sound is helpfully redirected downward. This is not true refraction but the effect is the same, and it can be as significant as temperature-related refraction.

To a lesser extent, if wind moves the air at a certain speed, it is to be expected that the speed of sound will be affected. For example, if sound travels 1,130 ft/sec and a 10 mi/hr (about 15 ft/sec) wind prevails, upwind the sound speed with respect to the earth would be increased about 1%, and downwind it would be decreased by the same amount. This is a small change, but it may also affect sound propagation.

It is possible, under unusual circumstances, that sound traveling upwind may actually be favored. For instance, upwind sound is kept above the surface of the ground, minimizing losses at the ground surface.

In some cases, there may be both temperature and wind changes that simultaneously affect sound propagation. They might combine to produce a larger effect, or tend to negate one another. Thus, results are unpredictable. For example, a lighthouse foghorn might be clearly audible near the lighthouse and also at a distance far away, but mysteriously, not be audible at points between. Interestingly, in 1862, during a Civil War battle in Iuka, Mississippi, it is believed that two divisions of Union soldiers did not enter battle because a wind-related acoustic shadow prevented them from hearing the sounds of the conflict raging six miles downwind.

Refraction of Sound in Enclosed Spaces

Refraction has an important effect on sound outdoors and sound traveling great distances. Refraction plays a much less significant role indoors. Consider a multiuse gymnasium that also serves as an auditorium. With a normal heating and air-conditioning system, efforts are made to avoid large horizontal or vertical temperature gradients. If there is temperature uniformity, and no troublesome drafts, sound refraction effects should be reduced to inconsequential levels.

Consider the same gymnasium also used as an auditorium but with less sophisticated air conditioning. Assume that a large heater is ceiling-mounted. The unit would produce hot air near the ceiling and rely on slow convection currents to move some of the heat down to the audience level.

This reservoir of hot air near the ceiling and cooler air below could have a minor effect on the transmission of sound from the sound system and on the acoustics of the space. The feedback point of the sound system might shift. The standing waves of the room might change slightly as longitudinal and transverse sound paths are increased in length because of their curvature due to refraction. Flutter echo paths may also shift. With a sound system mounted high at one end of the room, sound paths could be curved downward. Such downward curvature might actually improve audience coverage, depending somewhat on the directivity of the radiating system.

Refraction of Sound in the Ocean

In 1960, a team of oceanographers headed by Heaney performed an experiment to monitor the propagation of underwater sound. Charges of 600 lb were discharged at various depths in the ocean off Perth, Australia. Sounds from these discharges were detected near Bermuda, a distance of over 12,000 miles. Sound in seawater travels 4.3 times faster than in air, but it still took 3.71 hours for the sound to make the trip.

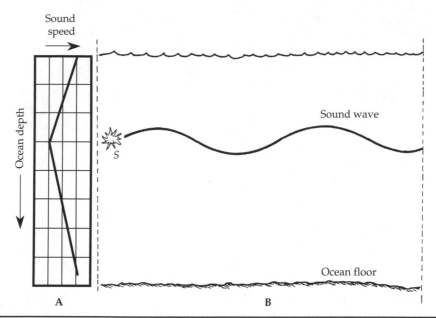

FIGURE 8-6   Schematic showing how oceanic refraction can affect sound propagation underwater. (A) Sound speed decreases with depth in the upper reaches of the ocean (temperature effect) and increases at greater depths (pressure effect), creating a sound channel at the inversion depth (about 700 fathoms). (B) A ray of sound is kept in this sound channel by refraction. Sound travels great distances in this channel because of the lower losses.

Oceanic refraction played a significant role in this experiment. The depth of the ocean may be over 5,000 fathoms (30,000 ft). At about 700 fathoms (4,200 ft) an interesting effect occurs. The sound speed profile shown in Fig. 8-6A very approximately illustrates the principle. In the upper reaches of the ocean, the speed of sound decreases with depth because temperature decreases. At greater depths the pressure effect prevails, causing sound speed to increase with depth because of the increase in density. The V-shaped profile changeover from one effect to the other occurs near the 700-fathom (4,200-ft) depth.

A sound channel is created by this V-shaped sound-speed profile. A sound emitted in this channel tends to spread out in all directions. Any ray traveling upward will be refracted downward, while any ray traveling downward will be refracted upward. Thus, sound energy in this channel is propagated great distances with modest losses.

Refraction in the vertical plane is prominent because of the vertical temperature/ pressure gradient. There is relatively little horizontal sound speed gradient and therefore very little horizontal refraction. Sound tends to spread out in a thin sheet at this 700-fathom depth. Spherical divergence in three dimensions is changed to two-dimensional propagation at this unique depth.

These long-distance sound channel experiments have suggested that such measurements can be used to monitor global climate change by detecting changes in the average temperature of the oceans. The speed of sound is a function of the temperature of the ocean. Accurate measurements of time of transit over a given course yield information on the temperature of that ocean.

CHAPTER 9
Diffusion

To make their calculations easier, theorists often assume a completely diffuse sound field, one that is isotropic and homogeneous. That is, a sound field where at any point, sound can arrive from any direction, and a field which is the same throughout the room. In practice, that rarely occurs, especially in small rooms. Instead, the characteristics of sound are markedly different throughout most rooms. In some cases, this is welcome, because it may, for example, help a listener to localize the source of the sound. In most room designs, diffusion is used to more effectively distribute sound and to provide a more equal response throughout a room that immerses the listener in the sound. It is often difficult to provide sufficient diffusion, particularly at low frequencies and in small room, because of the room's modal response. The goal of most room designs is to obtain sound energy across the audible frequency range that is uniformly distributed throughout the room. This is unattainable, but diffusion greatly assists in the effort.

The Perfectly Diffuse Sound Field

Even though unattainable, it is instructive to consider the characteristics of a diffuse sound field. Randall and Ward have suggested the following:

- The frequency and spatial irregularities obtained from steady-state measurements must be negligible.
- Beats in the decay characteristic must be negligible.
- Decays must be perfectly exponential (they will appear as straight lines on a logarithmic scale).
- Reverberation time will be the same at all positions in the room.
- The character of the decay will be essentially the same for all frequencies.
- The character of the decay will be independent of the directional characteristics of the measuring microphone.

These six factors are observation oriented. More theoretically, a diffuse sound field would be defined in terms such as energy density, energy flow, and superposition of an infinite number of plane progressive waves. However, these six characteristics point us to practical ways of judging the diffuseness of the sound field of a given room.

Evaluating Diffusion in a Room

To obtain the frequency response of an amplifier, a variable-frequency signal is input and the output is observed to measure the response. The same approach can be applied

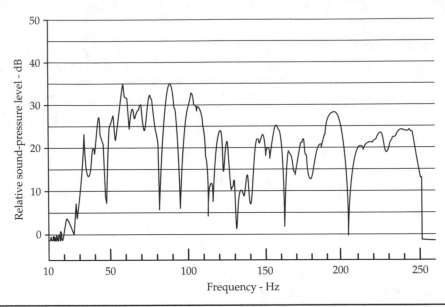

FIGURE 9-1   A slowly swept sine-wave sound-transmission response of a 12,000-ft³ studio. Fluctuations of this magnitude, which characterize many studios, are evidence of nondiffusive conditions.

to a room's playback system by reproducing a variable-frequency signal through a loudspeaker and measuring the acoustical signal picked up by a microphone located in the room. However, the acoustical response of playback systems in rooms are never as flat as the response of electrical devices. These deviations are partly due to nondiffusive conditions in the room. As noted, diffusion is welcome because it helps envelop the listener in the sound field. On the other hand, too much diffusion can make it difficult to localize the sound source.

Steady-State Measurements

Figure 9-1 shows a recorder tracing of the steady-state response of a studio with a volume of 12,000 ft³. In this example, the loudspeaker was placed in one lower tricorner of the room, and the microphone was at the upper diagonal tricorner about 1 ft from each of the three surfaces. These positions were chosen because all room modes terminate in the corners and all modes should be represented in the trace. The fluctuations in this response cover a range of about 35 dB over the linear 30- to 250-Hz sweep. The nulls are very narrow, and the narrow peaks show evidence of being single modes because the mode bandwidth of this room is close to 4 Hz. The wider peaks are the combined effect of several adjacent modes. The rise from 30 to 50 Hz is due primarily to loudspeaker response and the 9-dB peak between 50 and 150 Hz is due to radiating into $\frac{1}{4}$ space; therefore, these are testing artifacts that are not part of the room response.

The response of Fig. 9-1 is typical of even the best studios. Such variations in response are evidence of a sound field that is not perfectly diffused. A steady-state response such as this taken in an anechoic room would still show variations, but of lower amplitude. A very live room, such as a reverberation chamber, would show even greater variations.

Fixed measurements are one way to obtain the steady-state response of a room. Another method of appraising room diffusion is to rotate a highly directional microphone in various planes while holding the loudspeaker frequency response constant and recording the microphone's output to the constant excitation of the room. This method is used with some success in large spaces, but the method is ill adapted to smaller rooms, where diffusion problems are greatest. In principle, however, in a totally homogeneous sound field, a highly directional microphone pointed in any direction should pick up a constant signal. Both fixed- and rotating-measurement methods reveal the same deviations from a truly homogeneous sound field. We see that the criteria of negligible frequency and spatial irregularities are not met in any practical room.

Decay Beats

By referring ahead to Fig. 11-9, we can compare the smoothness of the reverberation decay for the eight octaves from 63 Hz to 8 kHz. In general, the smoothness of the decay increases as frequency is increased. The reason for this, as explained in Chap. 11, is that the number of modes within an octave span increases greatly with frequency, and the greater the mode density, the smoother their average effect. Conversely, in this example, beats in the decay are greatest at 63 and 125 Hz. If all decays have the same character at all frequencies and that character is smooth decay, complete diffusion prevails. In practice, decays (such as those of Fig. 11-9) with significant changes in character are more common, especially for the 63-Hz and 125-Hz decays.

The beat information on the low-frequency reverberation decay makes possible a judgment on the degree of diffusion. The decays of Fig. 11-9 indicate that the diffusion of sound in this particular studio is about as good as can be achieved by traditional means. Reverberation-time measuring devices that yield information only on the average slope and not the shape of the decay omit information that most consultants consider important in evaluating the diffuseness of a space.

Exponential Decay

A truly exponential decay can be viewed as a straight line on a level versus time plot, and the slope of the line can be described either as a decay rate in decibels per second or as a reverberation time in seconds. The decay of the 250-Hz octave band of noise pictured in Fig. 9-2 has two exponential slopes. The initial slope gives a reverberation time of 0.35 second and the final slope a reverberation time of 1.22 seconds. The latter slow decay when the level is low is probably due to a specific mode or group of modes encountering low absorption either by striking the absorbent at grazing angles or striking where there is little absorption. This is typical of one type of nonexponential decay, or stated more precisely, of a dual exponential decay.

Another type of nonexponential decay is illustrated in Fig. 9-3. The deviations from the straight-line slope are considerable. This is a decay of an octave band of noise centered on 250 Hz in a 400-seat chapel, poorly isolated from an adjoining room. Decays taken in the presence of acoustically coupled spaces are characteristically concave upward (such as in Fig. 9-3) and often the deviations from the straight line are even greater. When the decay traces are nonexponential, that is, when they depart from a straight line in a level versus time plot, we conclude that true diffuse conditions do not prevail.

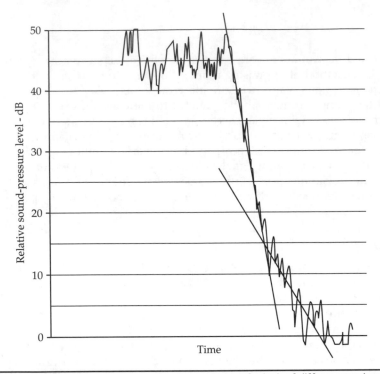

FIGURE 9-2 A typical double-slope decay, showing evidence of a lack of diffuse sound conditions. The slower decaying final slope is probably due to modes that encounter lower absorption.

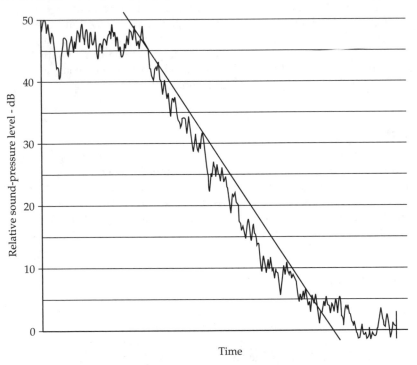

FIGURE 9-3 The nonexponential form of this decay is attributed to acoustically coupled spaces. The absence of a diffuse sound field is indicated.

Spatial Uniformity of Reverberation Time

When reverberation time for a given frequency is reported, it is most accurately stated as the average of multiple observations at each of several positions in the room. This accounts for the fact that reverberatory conditions differ from place to place in the room. Figure 9-4 shows the results of measurements in a small video studio of 22,000 ft³ volume. The multiple uses of the space require variable reverberation time, which is accomplished by hinged wall panels that can be closed, revealing absorbent sides, or opened, revealing reflecting sides. Multiple reverberation decays were recorded at the same three microphone positions for both "panels-reflective" and "panels-absorptive" conditions. The open and filled circles show the average values for the reflective and absorptive conditions, and the solid, dashed and dotted lines represent the average reverberation time at each of the three microphone positions. It is evident that there is considerable variation, which means that the sound field of the room is not completely homogeneous during this transient decay period. Inhomogeneities of the sound field are one reason why reverberation times vary from point to point in the room, but there are other factors as well. Uncertainties in fitting a straight line to the decay also contribute to the spread of the data, but this effect should be relatively constant from one position to another. It seems reasonable to conclude that spatial variations in reverberation time are related, at least partially, to the degree of diffusion in the space.

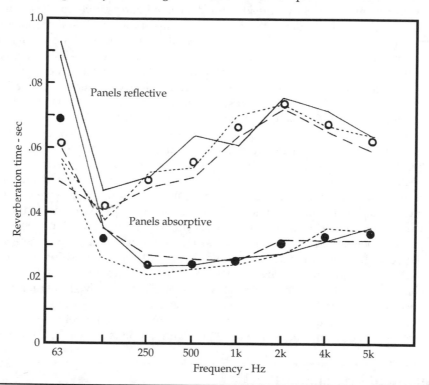

FIGURE 9-4   Reverberation time characteristics of a 22,000-ft³ studio with acoustics adjustable by hinged panels, absorbent on one side and reflective on the other. At each frequency, the variation of the average reverberation time at each of the three positions indicates nondiffuse conditions. This is particularly evident at low frequencies.

Octave-Band Center Frequency (Hz)	Panels Reflective			Panels Absorptive		
	RT_{60}	Standard Deviation	Standard Deviation % of Mean	RT_{60}	Standard Deviation	Standard Deviation % of Mean
63	0.61	0.19	31	0.69	0.18	26
125	0.42	0.05	12	0.32	0.06	19
250	0.50	0.05	10	0.24	0.02	8
500	0.56	0.06	11	0.24	0.01	4
1,000	0.67	0.03	5	0.26	0.01	4
2,000	0.75	0.04	5	0.31	0.02	7
4,000	0.68	0.03	4	0.33	0.02	6
8,000	0.63	0.02	3	0.34	0.02	6

TABLE 9-1   Reverberation Time of a Small Video Studio

Standard deviations of the reverberation times give us a measure of the spread of the data as measured at different positions in a room. When we calculate an average value, we lose all evidence of the spread of the data going into the average. The standard deviation can be used to quantify the data spread. Plus or minus one standard deviation from the mean value embraces 68% of the data points if the distribution is normal (Gaussian), and reverberation data should qualify reasonably well. Table 9-1 shows an analysis of the reverberation times for the small studio plotted in Fig. 9-4. For the "panels reflective" condition at 500 Hz, the mean reverberation time is 0.56 second with a standard deviation of 0.06 second. For a normal distribution, 68% of the data points would fall between 0.50 and 0.62 second. That 0.06 standard deviation is 11% of the 0.56 mean. The percentages listed in Table 9-1 give us a rough appraisal of the precision of the mean.

The columns of percentage in Table 9-1 are plotted in Fig. 9-5. Variability of reverberation time values at the higher frequencies settles down to reasonably constant values of around of 3% to 6%. Because we know that each octave at high frequencies contains a large number of modes that results in smooth decays, we conclude that at higher audible frequencies essentially diffuse conditions exist, and that the 3% to 6% variability is normal experimental measuring variation. At low frequencies, however, the high percentages (high variabilities) are the result of greater mode spacing producing considerable variation in reverberation time from one position to another. These high percentages include the uncertainty in fitting a straight line to the uneven decay characteristic of low frequencies. However, as Fig. 9-4 showed, there are great differences in reverberation time between the three measuring positions. Therefore, for this studio for two different conditions of absorbance (panels open/closed), diffusion is poor at 63 Hz, somewhat better at 125 Hz, and reasonably good at 250 Hz and above.

Geometrical Irregularities

Studies have been made regarding what type of wall protuberances provide the best diffusing effect. Somerville and Ward reported that geometrical diffusing elements reduced fluctuations in a swept-sine steady-state transmission test. The depth of such

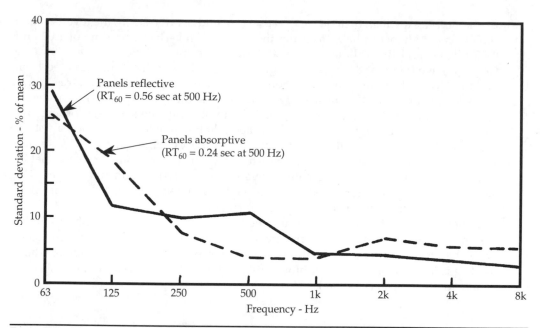

FIGURE 9-5 Closer examination of the reverberation time variations of the data of Table 9-1. The standard deviation, expressed as a percentage of the mean value, shows lack of diffusion, especially below 250 Hz.

geometrical diffusers must be at least one-seventh of a wavelength before their effect is felt. They studied cylindrical, triangular, and rectangular elements and found that the straight sides of the rectangular-shaped diffuser provided the greatest effect for both steady-state and transient phenomena. Other experience indicates superior subjective acoustical properties in studios and concert halls in which rectangular ornamentation in the form of coffering is used extensively.

Absorbent in Patches

Applying all the absorbent in a room on one or two surfaces does not result in a diffuse condition, nor is the absorbent used most effectively. Consider the results of an experiment showing the effect of distributing the absorbent. The experimental room is approximately a 10-ft cube and the room was tiled (not an ideal recording or listening room, but acceptable for this experiment). For test 1, reverberation time for the bare room was measured and found to be 1.65 seconds at 2 kHz. For test 2, a common commercial absorber was applied to 65% of one wall (65 ft²), and the reverberation time at the same frequency was found to be about 1.02 seconds. For test 3, the same area of absorber was divided into four sections, one piece mounted on each of four of the room's six surfaces. This brought the reverberation time down to about 0.55 second.

The area of the absorber was identical between tests 2 and 3; the only difference was that in test 3 it was in four pieces, one on each of three walls and one piece on the floor.

By the simple expedient of dividing the absorbent and distributing it, the reverberation time was cut almost in half. Inserting the values of reverberation time of 1.02 and 0.55 second and the volume and area of the room into the Sabine equation (see Chap. 11), we find that the average absorption coefficient of the room increased from 0.08 to 0.15 and the number of absorption units from 48 to 89 sabins. This extra absorption is due to an edge effect related to diffraction of sound that makes a given sample appear to be much larger acoustically. Stated another way, the sound-absorbing efficiency of 65 ft² of absorbing material is only about half that of four 16-ft² pieces distributed about the room, and the edges of the four pieces total about twice that of the single 65-ft² piece. So, one advantage of distributing the absorbent in a room is that its sound-absorbing efficiency is greatly increased, at least at certain frequencies. The above statements are true for 2 kHz, but at 700 Hz and 8 kHz, the difference between one large piece and four distributed pieces is small.

Another significant result of distributing the absorbent is that it contributes to diffusion of sound. Patches of absorbent with reflective walls showing between the patches have the effect of altering wavefronts, which improves diffusion. Sound-absorbing modules placed along a wall distribute the absorbing material and simultaneously contribute to the diffusion of sound.

Concave Surfaces

A concave surface such as that in Fig. 9-6A tends to focus sound energy and consequently should be avoided because focusing is the opposite of the diffusion we are seeking. The radius of curvature determines the focal distance; the flatter the concave surface, the greater the distance at which sound is concentrated. Such surfaces often cause problems in microphone placement. Concave surfaces might produce some awe-inspiring effects in a whispering gallery, but they are to be avoided in listening rooms and small studios.

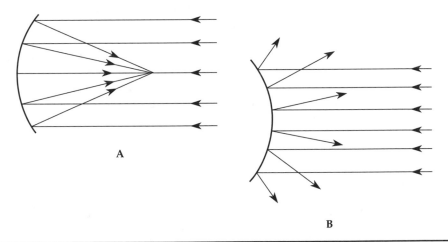

A

B

FIGURE 9-6 Generally, concave surfaces are undesirable, while convex surfaces are very desirable. (A) Concave surfaces tend to focus sound. Concave surfaces should be avoided if the goal is to achieve well-diffused sound. (B) Convex surfaces tend to diffuse sound.

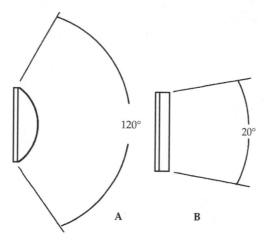

Figure 9-7 Polycylindrical diffusers, when properly designed, are very effective at providing wideband diffusion. (A) A polycylindrical diffuser reradiates sound energy not absorbed through an angle of about 120°. (B) A similar flat element reradiates sound in a much smaller angle of about 20°.

Convex Surfaces: The Polycylindrical Diffuser

The polycylindrical diffuser (poly) is an effective diffusing element, and one relatively easy to construct; it presents a convex section of a cylinder. Three things can happen to sound falling on such a cylindrical surface made of plywood or hardboard: The sound can be reflected and thereby dispersed as in Fig. 9-6B; the sound can be absorbed; or the sound can be reradiated. Such cylindrical elements serve as absorbers in the low-frequency range where absorption and diffusion are often needed in small rooms. The reradiated portion, because of the diaphragm action, is radiated almost equally throughout an angle of roughly 120°, as shown in Fig. 9-7A. A similar flat element reradiates sound in a much narrower angle, about 20°, as shown in Fig. 9-7B. Therefore, reflection, absorption, and reradiation characteristics favor the use of the cylindrical surface. Some very practical polys and their absorption characteristics are presented in Chap. 12. The dimensions of such diffusers are not critical, although to be effective their size must be comparable to the wavelength of the sound being considered. The wavelength of sound at 1,000 Hz is a bit over 1 ft, at 100 Hz about 11 ft. A poly element 3 or 4 ft across would be effective at 1,000 Hz, much less so at 100 Hz. In general, a poly base or chord length of 2 to 6 ft with depths of 6 to 18 in meet most needs. Axes of symmetry of the polys on different room surfaces should be mutually perpendicular.

It is important that diffusing elements be characterized by randomness. A wall lined with polys, all of 2-ft chord and of the same depth, might be beautiful to behold, but is not very effective for diffusion. The regularity of the structure would cause it to act as a diffraction grating, affecting one particular frequency in a much different way than other frequencies, which is counter to the ideal of wide-frequency diffusion.

Plane Surfaces

Geometrical sound diffusing elements made up of two flat surfaces to give a triangular cross section, or of three or four flat surfaces to give a polygonal cross section, may also be used. In general, their diffusing qualities are inferior to the cylindrical section.

CHAPTER 10

Comb-Filter Effects

The effect of a delayed reflection on the frequency response of a signal is often called a comb filter. Comb filtering is a steady-state phenomenon. It has limited application to music and speech, which are highly transient phenomena. With transient sounds, the audibility of a delayed replica is more the result of successive sound events. A case might be made for combing effects during brief periods of speech and music that approach steady state. However, the study of the audible effects of delayed reflections is better handled with the generalized threshold approach of sound reflection. Still, it is important to understand the nature of comb filtering, and know when it will, and will not, pose an acoustical problem.

Comb Filters

A filter changes the frequency response or transfer function of a signal. For example, an electrical filter might attenuate low frequencies of a signal to reduce unwanted vibration or noise. A filter could also be a system of pipes and cavities used to change an acoustical signal, such as is used in some microphones to adjust the pick-up pattern.

In the days of multitrack tape recording, multiple-head tape recorders were used to provide delayed replicas of sounds that were then mixed with the original sound to produce phasing and flanging effects. These same effects can also be created electronically or algorithmically. Whatever the means, these audible effects are the result of comb filters.

Superposition of Sound

Imagine a laboratory with a large tank of shallow water. Two stones are dropped in the tank simultaneously. Each stone causes circular ripples to flow out from the drop points. Each set of ripples expands through the other ripple pattern. We note that at any point in the water, the net effect is the combination of both ripple patterns at that point. As we will see later, both constructive and destructive interference will result. This is an example of superposition.

The principle of superposition states that every infinitesimal volume of a medium is capable of transmitting many discrete disturbances in many different directions, all simultaneously and with no detrimental effect on other disturbances. If you were able to observe and analyze the motion of a single air particle at a given instant under the influence of several disturbances, you would find that its motion is the vector sum of the various particle motions required by each of the disturbances passing by. At that instant, the air particle moves with amplitude and direction of vibration to satisfy the requirements of each disturbance just as a water particle responds to each of several disturbances in the ripple tank.

At a given point in space, assume an air particle responds to a passing disturbance with amplitude A and 0° direction. At the same instant another disturbance requires the same amplitude A, but with a 180° direction. This air particle satisfies both disturbances at that instant by not moving at all.

Tonal Signals and Comb Filters

A microphone is a passive instrument. Its diaphragm responds to fluctuations in air pressure that occur at its surface. If the rate of such fluctuations (frequency) falls within its operating range, it generates an output voltage proportional to the magnitude of the pressure. For example, if a 100-Hz acoustical tone actuates the diaphragm of a microphone in free space, a 100-Hz voltage appears at the microphone terminals. If a second 100-Hz tone, identical in pressure but 180° out of phase with the first signal, strikes the microphone diaphragm, one acoustically cancels the other, and the microphone voltage falls to zero. If an adjustment is made so that the two 100-Hz acoustical signals of identical amplitude are in phase, the signals reinforce each other, and the microphone delivers twice the output voltage, an increase of 6 dB. The microphone responds to the pressures acting on its diaphragm. That is, the microphone simply responds to the vector sum of air pressure fluctuations impinging upon it. This characteristic of the microphone can help us understand acoustical comb-filter effects.

A 500-Hz sine tone is shown as a frequency component in Fig. 10-1A. All of the energy in this pure tone is located at this frequency. Figure 10-1B shows an identical signal except it is delayed by 0.5 msec with respect to the signal of A. The signal has the

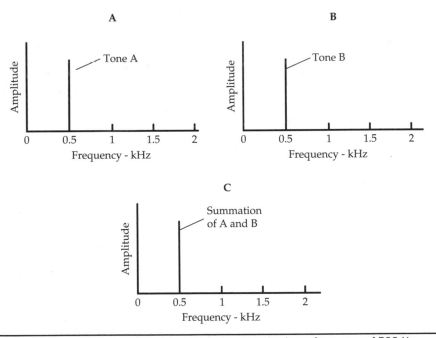

Figure 10-1   Tonal signals and time delay. (A) A sine wave having a frequency of 500 Hz. (B) Another sine wave of 500 Hz that is delayed 0.5 msec from A. (C) The summation of A and B. The 500-Hz signal and its delayed counterpart reach their peaks at slightly different times, but adding them together simply yields another sine wave; there is no comb filtering. A linear frequency scale is used.

same frequency and amplitude, but the timing is different. Consider both A and B as acoustical signals combining at the diaphragm of a microphone. Signal A could be a direct signal and B a reflection of A from a nearby side wall. What is the nature of the combined signal output by the microphone?

Because signals A and B are 500-Hz sine tones, both vary from a positive peak to a negative peak 500 times per second. Because of the 0.5-msec delay, these two tonal signals will not reach their positive or negative peaks at the same instant. Often along the time axis, both are positive or both are negative, and at times one is positive while the other is negative. When the sine wave of sound pressure representing signal A and the sine wave of sound pressure representing signal B combine (with due respect to positive and negative signs), they produce another sine wave of the same frequency, but of different amplitude.

Figure 10-1 shows the two 500-Hz tones as lines in the frequency domain. Figure 10-2 shows the same 500-Hz direct tone and the delayed tone in the time domain. The delay is accomplished by applying the 500-Hz tone to a delay device and combining the original and the delayed tones.

In Fig. 10-2A, the direct 500-Hz tone is shown originating at zero time. One cycle of a 500-Hz tone (1/500 = 0.002 second) takes 2 msec. One cycle is also equivalent to 360°. The 500-Hz signal, e, is plotted according to the time and degree scales at the bottom of the figure.

A delay of 0.1 msec is equivalent to 18°; a delay of 0.5 msec is equivalent to 90°; a delay of 1 msec is equivalent to 180°. The effect of these three delays on the tonal signals is shown in Fig. 10-2B. (Later the same delays will be compared with music and speech signals). The combination of e and e_1 reaches a peak of approximately twice that of e (+6 dB). A shift of 18° is a small shift, and e and e_1 are practically in phase. The curve $e + e_2$, at 90° phase difference has a lower amplitude, but is still a sine form. Adding e to e_3 (delay 1 msec, shift of 180°) yields zero amplitude as adding two waves of identical amplitude and frequency but with a phase shift of 180° results in cancellation of one by the other.

Adding direct and delayed sine waves of the same frequency results in other sine waves of the same frequency. Adding direct and delayed sine waves of different frequencies gives periodic waves of irregular wave shape. Adding direct and delayed periodic waves does not create comb filtering. Comb filtering requires signals having distributed energy such as speech, music, and pink noise.

Combing of Music and Speech Signals

The spectrum of Fig. 10-3A can be considered an instantaneous segment of music, speech, or any other signal having a distributed spectrum. Figure 10-3B is essentially the same spectrum but delayed 0.1 msec from Fig. 10-3A. Considered separately, the delay difference is inconsequential but their summation produces a new result. Figure 10-3C is the acoustical combination of the A and B sound pressure spectra at the diaphragm of a microphone. The resulting response of Fig. 10-3C is different from the result of combining tonal signals. This response shows comb filtering with characteristic peaks (constructive interference) and nulls (destructive interference) in the frequency response. Plotted on a linear frequency scale, the pattern looks like a comb; hence the term, comb filter.

Combing of Direct and Reflected Sound

The 0.1-msec delay in Fig. 10-3 could have been from a digital-delay device, or a reflection from a wall or other object. The spectral shape of a signal will be changed somewhat upon reflection, depending on the angle of incidence, the acoustical characteristics

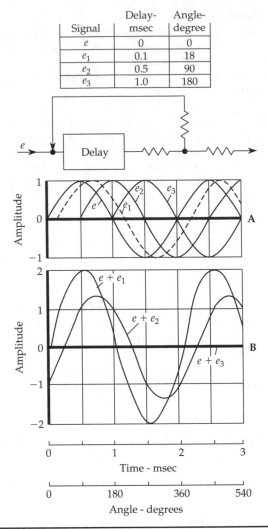

Signal	Delay-msec	Angle-degree
e	0	0
e_1	0.1	18
e_2	0.5	90
e_3	1.0	180

FIGURE 10-2 An exercise to demonstrate the result from combining sine waves. (A) 500-Hz sine waves are displayed with delays of 0.1, 0.5, and 1.0 msec to conform to the distributed spectrum cases in Fig. 10-4. (B) Combining sine waves such as these does not yield comb filtering, but simply other sine waves. A distributed spectrum is required for the formation of comb filtering. A linear frequency scale is used.

of the surface, and so on. When a direct sound is combined with its reflection, a comb filter is produced, with characteristic nulls (also called notches) in the frequency response. Nulls result when two signals are out of phase; they are one-half wavelength apart in time. The frequency of the nulls (and peaks) is determined by the delay between the direct and reflected sound. The frequency of the first null occurs where the period is twice the delay time. This is given by $f = 1/(2t)$, where t is the delay in seconds. Each successive null occurs at odd multiples thus: $f = n/(2t)$, where $n = 1, 3, 5, 7$, and so on. The first

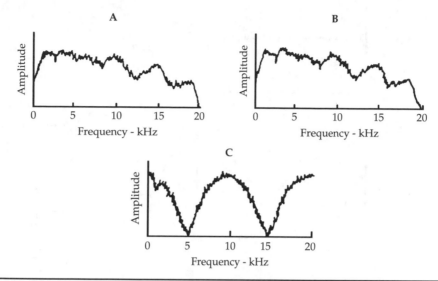

Figure 10-3 Comb filtering of signals having distributed spectra. (A) Instantaneous spectrum of music signal. (B) A replica of A, which is delayed 0.1 msec from A. (C) A summation of A and C showing typical comb filtering. A linear frequency scale is used.

peak occurs at $f = 1/t$ and successive peaks occur at $f = n/t$, where $n = 1, 2, 3, 4, 5$, and so on. The spacing between nulls or between peaks is $1/t$.

A reflection delayed 0.1 msec will have traveled (1,130 ft/sec) (0.001 sec) = 1.13 ft further than the direct signal. This difference in path length, only about 1.34 in, could result from a grazing angle with both source and listener, or microphone, close to the reflecting surface. Greater delays are expected in more normal situations such as those of Fig. 10-4. The spectrum of Fig. 10-4A is from a random noise generator driving a loudspeaker and received by an omnidirectional microphone in free space. Noise of this type is used widely in acoustic measurements because it is a continuous signal, its energy is distributed throughout the audible frequency range, and it is closer to speech and music signals than sine or other periodic waves.

In Fig. 10-4B, the loudspeaker faces a reflective surface; the microphone diaphragm is placed about 0.7 in from the reflective surface. Interference occurs between the direct sound the microphone picks up from the loudspeaker, and the sound reflected from the surface. The output of the microphone shows the comb-filter pattern characteristic of a 0.1-msec delay.

Placing the microphone diaphragm about 3.4 in from the reflective barrier, as shown in Fig. 10-4C, yields a 0.5-msec delay, which results in the comb-filter pattern shown. Increasing the delay to 0.5 msec has increased the number of peaks and the number of nulls fivefold. In Fig. 10-4D, the microphone is 6.75 in from the reflective barrier, giving a delay of 1.0 msec. Doubling the delay has doubled the number of peaks and nulls.

Increasing the delay between the direct and reflected components increases the number of constructive and destructive interference events proportionally. Starting with the flat spectrum of Fig. 10-4A, the spectrum of B is distorted by the presence of a reflection delayed by 0.1 msec. An audible response change would be expected. One might suspect that the distorted spectrum of D might be less noticeable because the multiple, closely spaced peaks and narrow nulls tend to average out the overall response aberrations.

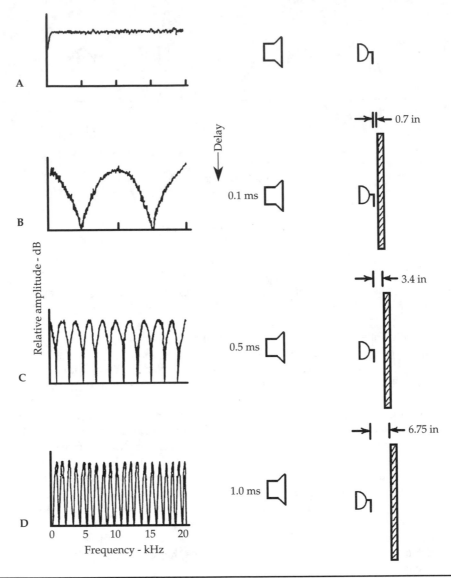

Figure 10-4   A demonstration of comb filtering in which direct sound from a loudspeaker is acoustically combined with a reflection from a surface at the diaphragm of a microphone. (A) No surface, no reflection. (B) Placing the microphone 0.7 in from the surface creates a delay of 0.1 msec, and the combination of the direct and the reflected rays shows cancellations at 5 and 15 kHz and every 10 kHz. (C) A delay of 0.5 msec creates cancellations much closer together. (D) A delay of 1 msec results in cancellations even more closely together. If t is taken as the delay in seconds, the first null is $1/(2t)$ and spacing between nulls or between peaks is $1/t$. A linear frequency scale is used.

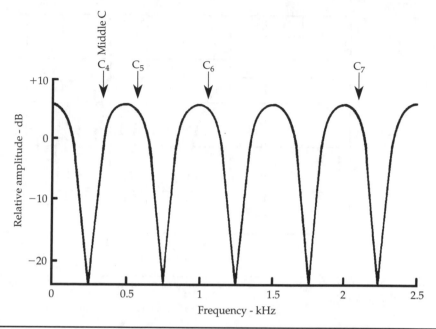

Figure 10-5 Passing a music signal through a 2-msec comb filter affects the components of that signal. Components spaced one octave can be boosted 6 dB at a peak or essentially eliminated at a null, or can be given values between these extremes. A linear frequency scale is used.

Reflections following closely after the arrival of the direct component are expected in small rooms because the dimensions of the room are limited. Conversely, reflections in large spaces would have greater delays, which generate more closely spaced comb-filter peaks and nulls. Thus, comb-filter effects resulting from reflections are more commonly associated with small room acoustics. The size of various concert halls and auditoriums renders them relatively immune to audible comb-filter distortions; the peaks and nulls are so numerous and packed so closely together that they merge into an essentially uniform response. Figure 10-5 illustrates the effect of passing a music signal through a 2-msec comb filter. The relationship between the nulls and peaks of response is related to several musical notes as indicated. Middle C, (C_4), has a frequency of 261.63 Hz, and is close to the first null of 250 Hz. The next higher C, (C_5), has a frequency twice that of C_4 and is treated favorably with a +6-dB peak. Other Cs up the keyboard will be either discriminated against with a null, or favored with a peak in response, or something in between. Whether viewed as fundamental frequencies or a series of harmonics, the timbre of the sound suffers.

The comb filters illustrated in Figs. 10-3, 10-4, and 10-5 are plotted to a linear frequency scale. In this form, the comb appearance and visualization of the delayed effects are most graphic. A logarithmic-frequency scale, however, is more common in the electronics and audio industry, and is more representative of what we hear. A comb filter resulting from a delay of 1 msec plotted to a logarithmic frequency scale is shown in Fig. 10-6.

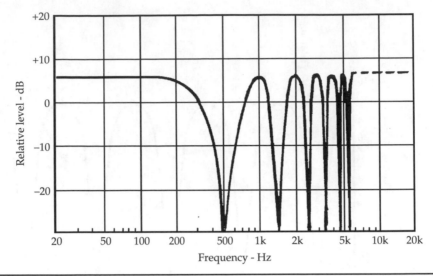

FIGURE 10-6   Plotting to the more familiar logarithmic scale helps estimate the effects of comb filtering on a signal.

Comb Filters and Critical Bands

One way to evaluate the relative audibility of comb-filter effects is to consider the critical bands of the human ear. The critical bandwidths at representative frequencies are shown in Table 10-1. The bandwidths of the critical bands vary with frequency. For example, the critical bandwidth of the human ear at 1 kHz is about 128 Hz. A peak-to-peak comb-filter frequency of 125 Hz corresponds to a reflection delay of about 8 msec $1 / 0.008 = 125$ Hz , which corresponds to a difference in path length between the direct and reflected components of about 9 ft (1,130 ft/sec × 0.008 sec = 9.0 ft). This case for an 8-msec delay is plotted in Fig. 10-7B. Figure 10-7A shows a shorter delay of 0.5 msec. Figure 10-7C shows a longer delay of 40 msec.

Center Frequency (Hz)	Width of Critical Band* (Hz)
100	38
200	47
500	77
1,000	128
2,000	240
5,000	650

*Calculated equivalent rectangular band as proposed by Moore and Glasberg.

TABLE 10-1   Auditory Critical Bands

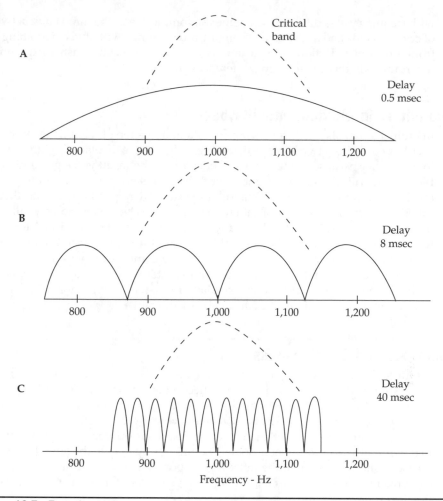

FIGURE 10-7   To estimate the perceptual importance of comb filters, they are compared to the auditory critical band effective at a frequency of 1,000 Hz. (A) With a delay of 0.5 msec, the width of the auditory critical band is comparable to one comb peak. (B) With a delay of 8 msec, two comb peaks fall within a critical band. (C) At a delay of 40 msec, the width of the critical band is large, relatively, so that no analysis of the comb filter is possible. This seems to confirm the observation that in large spaces (long delays) comb filters are inaudible, while they often are very troublesome in small spaces (short delays). Also, critical bands are much narrower at low frequencies, suggesting that comb-filter effects are more audible at low frequencies. A linear frequency scale is used.

The relative coarseness of the critical bands suggests that the ear is relatively insensitive to the peaks and nulls resulting from a 40-msec delay (Fig. 10-7C). Therefore, the human ear may not interpret response aberrations such as timbral changes resulting from 40-msec combing, or combing from longer delays. On the other hand, the combing resulting from the 0.5-msec delay (Fig. 10-7A) could be delineated by the ear's critical band at 1,000 Hz, resulting in a perceived timbre of the signal. Figure 10-7B illustrates an intermediate example in which the ear may be marginally able to analyze the 8-msec combed signal. The width of the critical bands of the auditory system increases rapidly with frequency. It is difficult to imagine the complexity of the interaction between a set of critical bands and a constantly changing music signal, with diverse combing patterns from a host of reflections. Only carefully controlled psychoacoustics experiments can determine whether the resulting differences are audible.

Comb Filters in Multichannel Playback

In multichannel playback, for example, in standard stereo playback, the input signals to each ear come from two loudspeakers. These signals are displaced in time with respect to each other because of the loudspeaker spacing; the result is the generation of comb filters. Blauert indicated that comb-filter distortion is not generally audible. As the perception of timbre is formed, the auditory system disregards these distortions; this is called binaural suppression of differences in timbre. However, no generally accepted theory exists to explain how the auditory system accomplishes this. Distortion can be heard by plugging one ear; however, this destroys the stereo effect. Comparing the timbre of signals from two loudspeakers (producing comb-filter distortion) and one loudspeaker (that does not) will demonstrate that stereo comb-filter distortion is barely audible. The timbre of the two is essentially the same. Furthermore, the timbre of the stereo signal changes little as the head is turned.

Reflections and Spaciousness

A reflected sound reaching the ear of a listener is always somewhat different from the direct sound. The characteristics of the reflecting wall vary with frequency. By traveling through the air, both the direct and reflected components of a sound wave are altered slightly, due to the air's absorption of sound, which varies with frequency. The amplitude and timing of the direct and reflected components differ. The human ear responds to the frontal, direct component somewhat differently than to the lateral reflection from the side. The perception of the reflected component is always different from the direct component. The amplitudes and timing will be related, but with an interaural correlation less than maximum.

Weakly correlated input signals to the ears contribute to the impression of spaciousness. If no reflections occur, such as when listening outdoors, there is no feeling of spaciousness. If a room supplies "correct" input signals to the ears, the perception of the listener is that of being completely enveloped and immersed in the sound. The lack of strong correlation is a prerequisite for the impression of spaciousness.

Comb Filters in Microphone Placement

When two microphones separated in space pick up a single sound at slightly different times, their combined output will be similar to the single microphone with delayed reflections. Therefore, spaced microphone stereo-pickup arrangements are susceptible to comb-filter problems. Under certain conditions the combing is audible, imparting phasiness to the overall sound reproduction, interpreted by some as room ambience. It is not ambience, however, but distortion of the time and intensity cues presented to the microphones. It is evident that some people find this distortion pleasing, so spaced microphone pickups are favored by many producers and listeners.

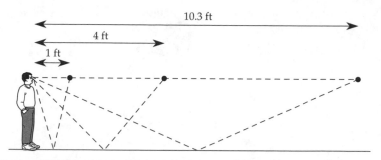

FIGURE 10-8   Common microphone placements can produce comb-filter effects (see Table 10-2). At a distance of 1 ft, the reflection level of −20 dB yields minimum comb-filter problems. At a distance of 4 ft, the reflection level of −8 dB may yield some comb-filter problems. (C) At a distance of 10.3 ft, the reflection level almost equals the direct level, and thus comb-filter problems are certain.

Path Length (ft)		Difference		First Null 1/(2*t*)	Peak/Null Spacing 1/*t*	Reflection Level
Direct	Reflected	Distance (ft)	Time (msec)	(Hz)	(Hz)	(dB)
1.0	10.1	9.1	8.05	62	124	−20
4.0	10.0	6.0	5.31	94	189	−8
10.3	11.5	1.2	1.06	471	942	−1

TABLE 10-2   Comb-Filter Effects from Microphone Placement (Refer to Fig. 10-8)

Comb-Filter Effects in Practice: Six Examples

Example 1   Figure 10-8 shows three microphone placements that produce comb filters of varying degree. A hard, reflective floor is assumed in each case, and other room reflections are not considered. In a close source-to-microphone setup, the direct component travels 1 ft and the floor-reflected component travels 10.1 ft (see Table 10-2). The difference between these (9.1 ft) means that the floor reflection is delayed 8.05 msec (9.1/1130 = 0.00805 second). The first null is therefore at 62 Hz with subsequent null and peak spacing of 124 Hz. The level of the reflection is −20 dB referred to the direct component (20 log 1.0/10.1 = 20 dB). The direct component is thus 10 times stronger than the floor reflection. The effect of the comb filter would be negligible in this case.

Similar calculations for two more source-to-microphone distances are included in Table 10-2. A source-to-microphone distance of 4 ft presents an intermediate case, where the reflection level is 8 dB below the direct-signal level; the comb-filter effect would be marginal. With a source-to-microphone distance of 10.3 ft, the reflection-level difference of only 1 dB; the reflection is almost as strong as the direct signal; the comb-filter effect would be considerable. In contrast to these microphone placements, consider what would happen if the microphone was placed on the floor some distance from the source. A small floor reflection may occur, but this technique essentially eliminates the difference between the direct and reflected path length.

Example 2   Figure 10-9 shows two microphones on a podium. Stereo reproduction systems are relatively rare in auditoriums. The chances are very good that the two microphones are fed into a monaural system and thus become an excellent producer of comb-filter effects. The common excuse for two microphones is to give the speaker greater freedom of movement, or to provide a spare microphone in case of failure of one. Assuming the microphones are properly polarized and the talker is dead center, there would be a helpful 6-dB boost in level. Assume also that the microphones are 24 in apart and the talker's lips are 18 in from a line drawn through the two microphones and on a level with the

FIGURE 10-9   An example of comb-filter production, two microphones feeding into the same monaural amplifier with a sound source that moves about.

microphones. If the talker moves laterally 3 in, a 0.2-msec delay is introduced, attenuating important speech frequencies. If the talker does not move, the speech quality would probably not be good, but it would be stable. Normal talker movements shift nulls and peaks up and down the frequency scale with quite noticeable shifts in quality.

Example 3   Figure 10-10 shows comb-filter possibilities in a singing group with each singer holding a microphone. Each microphone is fed to a separate channel but ultimately mixed together. Each singer's voice is picked up by all microphones but only adjacent singers may create noticeable comb filters. For example, the voice of singer A is picked up by both microphones and mixed, and may produce a comb-filter response resulting from the path difference. However, if singer A's mouth is at least three times farther from singer B's microphone than from A's own microphone, the comb-filter effects are minimized. This "3:1 rule" works because maintaining this distance means that delayed replicas are at least 9 dB below the main signal. Comb-filter peaks and nulls are 1 dB or less in amplitude and thus essentially imperceptible.

Example 4   Figure 10-11 shows dual monaural loudspeakers, for example, one on stage left and the other on stage right. Two sources radiating identical signals create comb filters over the audience area. On the line of symmetry (often down the center aisle), both signals arrive at the same time and no comb filters are produced. Equi-delay contours range out from stage center over the audience area, the 1-msec-delay contour nearest the center line of symmetry, and greater delays as the sides of the auditorium are approached.

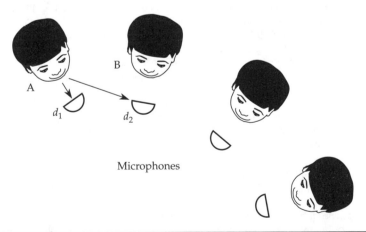

FIGURE 10-10   For group singing, if d_2 is at least 3 times as great as d_1, the comb-filter effect is minimized.

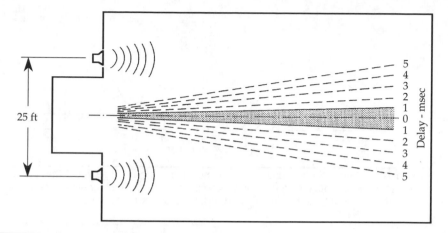

FIGURE 10-11   In the common split system in which two loudspeakers radiate identical signals, zones of constructive and destructive interference result which degrade sound quality in the audience area.

Example 5   Figure 10-12 shows the frequency-response bands in a three-way loudspeaker. Frequency f_1 is radiated by both low-frequency and midrange drivers, and both outputs may be essentially equal in magnitude; moreover, the two radiators are physically displaced. These are the ingredients for comb-filtering. The same process is at work at f_2 between the midrange and high-frequency drivers. Only a narrow band of frequencies is affected, the width of which is determined by the relative amplitudes of the two radiations. The steeper the crossover curves, the narrower the frequency range affected.

Example 6   Figure 10-13 shows how a permanently mounted microphone may be flush mounted on a surface. One advantage is an approximate 6-dB gain in sensitivity due to the pressure rise at the table surface. Another advantage is minimizing comb-filter distortions. A direct signal from the source strikes the microphone diaphragm; since the diaphragm is flush with the surface, there can be no reflections from the surface.

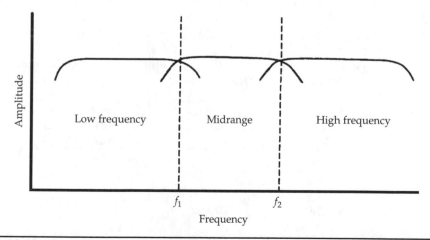

FIGURE 10-12   Comb-filter distortion can occur in the crossover regions of a multielement loudspeaker because the same signal is radiated from two physically separated drivers.

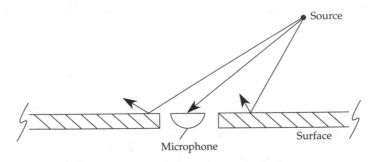

FIGURE 10-13 With a flush-mounted microphone, sounds from the source *S* that strike the surface do not reach the microphone, thus avoiding comb-filter effects. Another advantage of this mounting is an increase of level due to the pressure buildup near the reflecting surface.

Estimating Comb-Filter Response

A few simple relationships can be used to estimate the effect of comb filters on the response of a system. If the delay is *t* seconds, the spacing between peaks and the spacing between nulls is $1/t$ Hz. For example, a delay of 0.001 second (1 msec) spaces the peaks at 1,000 Hz ($1/0.001 = 1,000$ Hz) and the nulls will also be spaced the same amount, as shown in Table 10-3.

As noted, the frequency at which the first null (i.e., the null of lowest frequency) will occur is $1/(2t)$ Hz. For the same delay of 1 msec, the first null will occur at $1/(2 \times 0.001) = 500$ Hz. For this 1-msec delay, the first null is at 500 Hz, peaks are spaced 1,000 Hz, and nulls are spaced 1,000 Hz apart. Of course, there is a peak between each adjacent pair of nulls at which the two signals are in phase. Adding two sine waves with the same frequency, the same amplitude, in phase, doubles the amplitude; yielding a peak 6 dB higher than either component by itself (20 log 2 = 6.02 dB). The nulls will be at a theoretical minimum of minus infinity as they cancel at phase opposition. In this way, the entire response curve can be sketched as the phase of the two waves alternates between the in-phase and the phase-opposition condition through the spectrum.

An important point to observe is that the $1/(2t)$ expression above gives a null at 500 Hz, which robs energy from any distributed signal subject to that delay. A music or speech signal passing through a system having a 1-msec delay will have important components removed or reduced. This is comb-filter distortion.

If the mathematics of the $1/t$ and the $1/(2t)$ functions seem too laborious, Figs. 10-14 and 10-15 are included as graphical solutions.

Delay (msec)	Frequency of Lowest Null (Hz)	Spacing between Nulls and Spacing between Peaks (Hz)
0.1	5,000	10,000
0.5	1,000	2,000
1.0	500	1,000
5.0	100	200
10.0	50	100
50.0	10	20

TABLE 10-3 Comb-Filter Peaks and Nulls

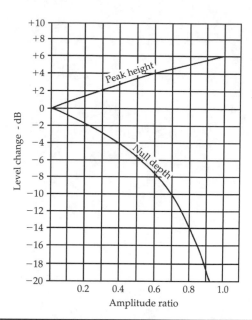

FIGURE 10-14   The effect of amplitude ratios on comb-filter peak height and null depth.

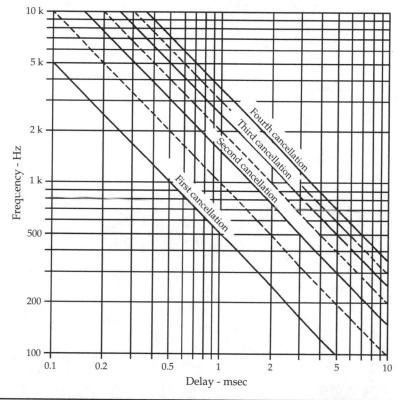

FIGURE 10-15   The magnitude of the delay determines the frequencies at which destructive interference (cancellations) and constructive interference (peaks) occur. The broken lines indicate the peaks between adjacent cancellations.

CHAPTER 11

Reverberation

If you press the accelerator pedal of an automobile, the vehicle accelerates to a certain speed. If the road is smooth and level, this speed will remain constant. With a constant force on the accelerator, the engine produces enough horsepower to overcome frictional and aerodynamic losses, and a balanced (steady-state) condition results. If you take your foot off the accelerator, the car will gradually slow, and come to a stop.

Sound in a room behaves similarly. When a loudspeaker is turned on, it emits noise in a room that quickly grows to a certain level. This level is the steady-state or equilibrium point at which the sound energy radiated from the loudspeaker is enough to overcome losses in the air and at the room boundaries. A greater sound energy radiated from the loudspeaker will result in a higher equilibrium level, while less energy to the loudspeaker will result in a lower equilibrium level.

When the loudspeaker is turned off, it takes a finite length of time for the sound level in the room to decay to inaudibility. This aftereffect of the sound in a room, after the excitation signal has been removed, is reverberation and it has an important bearing on the acoustic quality of the room.

A symphony orchestra recorded in a large anechoic chamber, with almost no room reverberation, would yield a recording of very poor quality for normal listening. This recording would be even thinner, weaker, and less resonant than most outdoor recordings of music, which are noted for their flatness. Clearly, symphonic and other music requires reverberation to achieve an acceptable sound quality. Similarly, many music and speech sounds require a room's reverberant assistance to sound natural, because we are accustomed to hearing them in reverberant environments.

Formerly, reverberation was considered the single most important characteristic of an enclosed space for speech or music. Today, reverberation is considered one of several important and measurable parameters that define the sound quality of an acoustic space.

Growth of Sound in a Room

When a sound is initiated in a room, the room will contain the energy, as it builds to a final, steady-state value. The time required to reach this final value is determined by the rate of growth of sound in the room.

Let us consider a source S and a listener L in a room, as shown in Fig. 11-1A. As source S is suddenly energized, sound travels outward from S in all directions. Sound travels a direct path to the listener L and we shall consider zero time (Fig. 11-1B) as that time at which the direct sound reaches the ears of the listener. The sound pressure at L instantly jumps to a value less than that which left S due to spherical divergence and small losses in the air. The sound pressure at L stays at this value until reflection R_1 arrives and

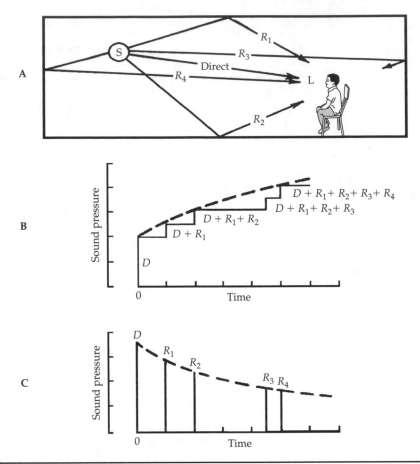

Figure 11-1   The growth and decay of sound in a room. (A) The direct sound arrives first at time $t = 0$, reflected components arriving later. (B) The sound pressure at L grows stepwise. (C) The sound decays exponentially after the source ceases.

then the sound pressure suddenly jumps to the $D + R_1$ value. Shortly thereafter R_2 arrives, causing the sound pressure to increase a bit more. The arrival of each successive reflected component causes the level of sound to increase stepwise. These additions are, in reality, vector additions involving both magnitude and phase, but we are keeping things simple for the purpose of illustration.

Sound pressure at the listener grows step by step as one reflected component after another adds to the direct component. The reason the sound pressure at L does not instantly go to its final value is that sound travels by paths of varying length. Given the speed of sound, reflected components are delayed by an amount proportional to the difference in distance between the reflected path and the direct path. The growth of sound in a room is thus relatively slow due to finite transit time, but in practice, sound growth is so fast as to be perceived as being instantaneous by a listener. On the other hand, in comparison, sound decay is very slow and is usually readily heard (as reverberation) by a listener. Therefore, the characteristics of sound decay are more important in practical room acoustic design.

The ultimate level of sound in the room is determined by the energy from the source S. The energy it radiates is dissipated as heat in wall reflections and other boundary losses, along with a small loss in the air itself. With a constant input to S, the sound-pressure level grows, as shown in Fig. 11-1B, to a steady-state equilibrium. Increasing the input to the source S yields a new equilibrium of room-sound-pressure level versus room losses.

Decay of Sound in a Room

After turning off source S, the room is momentarily still filled with sound, but stability is destroyed because the losses are no balanced by energy from S. Support is cut off to the rays of sound moving through the room.

What is the fate, for example, of the ceiling reflected component R_1 (Fig. 11-1)? As S is cut off, R_1 is on its way to the ceiling. It loses energy at the ceiling reflection and heads toward L. After passing L it strikes the rear wall, then the floor, the ceiling, the front wall, the floor again, and so on, losing energy at each reflection. Soon it is so weak it can be considered dead. The same thing happens to R_2, R_3, R_4, and a multitude of others not shown. Figure 11-1C shows the exponential decrease of the first reflection components, which would also apply to the wall reflections not shown and to the many multiple reflection components. The sound in the room thus dies away, but it takes a finite time to do so because of the speed of sound, losses at reflections, the damping effect of the air, and divergence.

Idealized Growth and Decay of Sound

From the view of geometrical (ray) acoustics, the decay of sound in a room, as well as its growth, is a stepwise phenomenon. However, in the practical world, the great number of small steps involved results in smooth growth and decay of sound. The idealized forms of growth and decay of sound in a room are shown in Fig. 11-2A. Here the sound pressure is shown on a linear scale and is plotted against time. Figure 11-2B shows the same growth and decay, except that sound-pressure level is plotted in decibels, that is, on a logarithmic scale.

During the growth of sound in a room, power is being applied to the sound source. During decay, the power to the source is cut off, hence the difference in the shapes of the growth and decay curves. The decay of Fig. 11-2B is a straight line in this idealized form, and this becomes the basis for measuring the reverberation time of an enclosure.

Calculating Reverberation Time

Reverberation time (RT) is a measure of the rate of decay of sound. It is defined as the time in seconds required for sound intensity in a room to drop 60 dB from its original level. This represents a change in sound intensity or sound power of 1 million (10 log 1,000,000 = 60 dB), or a change of sound pressure or sound-pressure level of 1 thousand (20 log 1,000 = 60 dB). This reverberation time measurement is referred to as RT_{60}. The 60-dB figure was chosen arbitrarily, but it roughly corresponds to the time required for a loud sound to decay to inaudibility. The ear is most sensitive to the initial part of most reverberation decays. Working in the 1890s, Wallace Clement Sabine, a Harvard University physics professor, devised the first reverberation-time equation. He used a portable

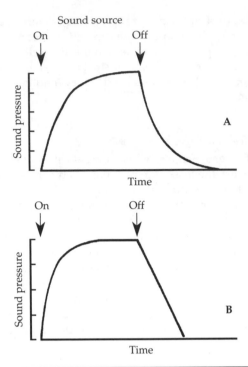

Sound source

On   Off

Time

A

On   Off

Time

B

FIGURE 11-2 The growth and decay of sound in a room. (A) Sound-pressure on vertical scale is measured in linear units. (B) Sound-pressure level on vertical scale is measured in logarithmic units (decibels).

wind chest and organ pipes as a sound source, a stopwatch, and keen ears to measure the time from the interruption of the source to inaudibility. Today we have better technical measuring facilities, but we can only refine our understanding of the basic concept. The equation used to calculate reverberation, often called the Sabine equation, is given below.

This approach to measuring reverberation time is illustrated in Fig. 11-3A. Using a recording device that gives a trace of the decay, it is simple to measure the time required for the 60-dB decay. While simple to measure in theory, in practice, problems can be encountered. For example, obtaining a straight decay spanning 60 dB or more as in Fig. 11-3A is a difficult practical problem. Background noise suggests that a higher source level is needed. This may occur if the background noise level is 30 dB (as in Fig. 11-3A), because source levels of 100 dB are quite attainable. However, if the noise level is near 60 dB, as shown in Fig. 11-3B, a source level greater than 120 dB is required. If a 100-W (watt) amplifier driving a loudspeaker gives a sound-pressure level of 100 dB at the required distance, doubling the power of the source increases the sound-pressure level only 3 dB; hence 200 W gives 103 dB, 400 W gives 106 dB, 800 W gives 109 dB, and so on. The limitations of size and cost can set a ceiling on the maximum levels in a practical case.

The situation of Fig. 11-3B is commonly encountered, where circumstances yield a trace showing less than the desired 60 dB. The solution is to extrapolate the usable initial portion of the decay. For example, the time required for sound pressure to fall 30 dB is extended to yield an estimate for time required to fall 60 dB.

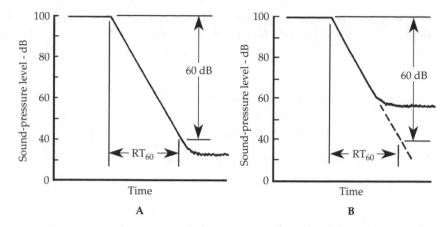

Figure 11-3 The length of the decay depends on the strength of the source, and the noise level. (A) Rarely do practical circumstances allow a full 60-dB decay. (B) The slope of the limited decay is extrapolated to determine the reverberation time.

Actually, it is important to strive for the greatest decay range possible because we are interested in the entire extent of the decay. For example, it has been demonstrated that in evaluating the quality of speech or music, the first 20 or 30 dB of decay is the most important to the human ear. On the other hand, the significance of double-slope phenomena is revealed only near the end of the decay. In practice, the highest amplitude of sound source reasonably attainable is used, and filters can be incorporated to improve the signal-to-noise ratio.

Sabine Equation

Sabine's reverberation equation, published in 1900, was developed in a strictly empirical fashion. He had two lecture halls at his disposal and by adding or removing seat cushions of a uniform kind, he was able to clarify the role of absorption in room reverberation. He observed that reverberation time depends on room volume, and absorption. The greater the absorption, the shorter the reverberation time. Likewise, the greater the room volume, the longer the reverberation time because sound will strike the absorbing room boundaries less often.

The quantity $4V/S$ describes the average distance a sound travels between two successive reflections; it is sometimes called the mean free path. In the equation, V is the room volume, and S is the room surface area. For example, consider a room measuring $23.3 \times 16 \times 10$ ft. The volume is 3,728 ft^3 and the surface area is 1,533 ft^2. The mean free path for this room is $4V/S$ or $(4)(3,728)/1,533 = 9.7$ ft. Since the speed of sound is 1,130 ft/sec, on average a sound ray will travel 8.5 msec before striking another room surface. It might take between four to six bounces before the sound energy in this sound ray is mainly lost, a decay process that will take 42.5 msec. But the number of bounces actually required will depend on how absorptive the room is. For example, the decay process will require more bounces in a live room, and thus take longer. In addition, because the mean free path is longer in large rooms, the decay process will take longer in large rooms.

Using this kind of statistical theory and geometrical ray acoustics, Sabine devised his equation for room reverberation. In particular, he devised the following relationship:

$$RT_{60} = \frac{0.049V}{A} \qquad (11\text{-}1)$$

where RT_{60} = reverberation time, sec
V = volume of room, ft³
A = total absorption of room, sabins

The Sabine equation can also be expressed in metric units:

$$RT_{60} = \frac{0.161V}{A} \qquad (11\text{-}2)$$

where RT_{60} = reverberation time, sec
V = volume of room, m³
A = total absorption of room, metric sabins

Using the Sabine equation is a straightforward process, but a few notes are worth considering. The quantity A is the total absorption in the room. For example, this accounts for the absorption of room surfaces. (In many cases, audience absorption must be considered as well, and if desired, a value for air absorption may be included, as described below). It would be easy to obtain the total absorption if all the room surfaces were uniformly absorbing, but this condition rarely exists. Walls, floor, and ceiling may be covered with quite different materials, and the doors and windows must be considered separately as well.

The total absorption A can be found by considering the absorption contributed by each type of surface. To obtain the total absorption A of the room, it is necessary to combine the respective absorptions of the various materials lining the room by multiplying the square-foot area S_i of each type of material by its respective absorption coefficient α_i, and summing the result to obtain total absorption. In particular, $A = \Sigma S_i \alpha_i$, where i represents each of the surface areas and its respective absorption coefficient. The quantity $\Sigma S_i \alpha_i / \Sigma S_i$ represents the average absorption coefficient $\alpha_{average}$.

For example, let us say that an area S_1 (expressed in square feet or square meters) is covered with a material having an absorption coefficient α_1, as obtained from the table in the appendix. This area then contributes $(S_1)(\alpha_1)$ absorption units, in sabins, to the room. Likewise, another area S_2 is covered with another kind of material with absorption coefficient α_2, and it contributes $(S_2)(\alpha_2)$ sabins of absorption to the room. The total absorption in the room is $A = S_1\alpha_1 + S_2\alpha_2 + S_3\alpha_3 + \cdots$, etc. With a figure for A in hand, it is a simple matter to use Eq. (11-1) or (11-2) and calculate the reverberation time. This is further demonstrated in the examples at the end of the chapter.

The absorption coefficients of practically all materials vary with frequency. For this reason, it is necessary to calculate total absorption at different frequencies. A typical reference frequency for reverberation time is 500 Hz, and 125 Hz and 2 kHz are used as well. To be precise, any reverberation time should be accompanied by an indication of frequency. For example, a reverberation time at 125 Hz might be quoted as $RT_{60/125}$. When there is no frequency designation, the reference frequency is assumed to be 500 Hz. Note that 1 sabin = 0.093 metric sabin.

A word about the limitations of the Sabine equation is in order. For live rooms, the assumed statistical conditions prevail and Sabine's equation gives accurate results. However, in very absorbent rooms, the equation produces erroneous results. For example, consider again the room measuring $23.3 \times 16 \times 10$ ft. This room has a volume of 3,728 ft³ and a total surface area of 1,532 ft². If we further assume that all of its surfaces are perfectly absorbent ($\alpha = 1.0$), we see that total absorption is 1,532 sabins. Substituting these values:

$$RT_{60} = \frac{(0.049)(3,728)}{1,532} = 0.119\,\text{second}$$

Clearly, since the correct RT_{60} value in a perfectly absorbing room should be zero, the equation is in error. Perfectly absorbing walls would allow no reflections. This paradox results from assumptions upon which the Sabine equation is derived; in particular, it assumes that sound in a room is diffuse, as in a live room. As a result, the Sabine equation is most accurate in a live room where the average absorption coefficient is less than 0.25.

Eyring-Norris Equation

Eyring-Norris and others derived equations which overcame this difficulty and can be used in more absorptive rooms. For average absorption coefficients of 0.25 or less, the other equations are basically equivalent to the Sabine equation.

In particular, Eyring and Norris proposed an alternate equation for absorptive rooms:

$$RT_{60} = \frac{0.049V}{-S\ln(1-\alpha_{average})} \tag{11-3}$$

where V = room volume, ft³
          S = total surface area of room, ft²
         $\ln$ = natural logarithm (to base "e")
   $\alpha_{average}$ = average absorption coefficient ($\Sigma S_i \alpha_i / \Sigma S_i$)

Young points out that the absorption coefficients published by materials manufacturers (such as the list in the appendix) are Sabine coefficients and can be applied directly in the Sabine equation. Young recommends that Eq. (11-1) or (11-2) be used for engineering computations rather than the Eyring-Norris equation or its several derivatives. Two unassailable reasons for this are simplicity and consistency. Many technical writings offer the Eyring-Norris or other equations for studio use. There is authoritative backing for using Eyring-Norris for more absorbent spaces, but commonly available coefficients apply only to Sabine. For this reason, we use the Sabine equation in this volume. Other researchers have suggested alternate reverberation equations; these include Hopkins-Striker, Millington, and Fitzroy.

Air Absorption

In large rooms, where sound travels over long path lengths, the passage through air can effectively add absorption to the room, lowering reverberation times. Air absorption is only significant at higher frequencies—above 2 kHz. Air absorption is not a significant factor in small rooms, and can be neglected. When accounted for, a term $4mV$ is added to the denominator of reverberation-time equations, where m is the air attenuation coefficient in sabins per foot (or sabins per meter) and V is room volume in cubic feet (or cubic meters). For example, the Sabine and Eyring-Norris equations, respectively, are modified as:

$$RT_{60} = \frac{0.049V}{A + 4mV} \tag{11-4}$$

$$RT_{60} = \frac{0.049V}{-S\ln(1-\alpha_{average}) + 4mV} \tag{11-5}$$

Some values of m in sabins per foot are: 0.003 at 2 kHz, 0.008 at 4 kHz, and 0.025 at 8 kHz; and in sabins per meter, 0.009, 0.025, and 0.080, respectively. The value of m depends on relative humidity of air, and the values given apply to RH between 40% and 60%. Air absorption increases at low humidity.

Measuring Reverberation Time

There are many approaches to measuring the reverberation time of a room, and many instruments are available to serve those who have an interest in reverberation effects. For example, sound contractors need to know the approximate reverberation time of the spaces in which they are to install a sound-reinforcement system, and measuring it may be more conclusive than calculating it. Measurements can be more accurate than calculations because of inherent uncertainty in absorption coefficients. Acoustical consultants called upon to correct a problem space or verify a carefully designed and newly constructed space, generally lean toward the method of recording many sound decays. These sound decays give a wealth of detail meaningful to the practiced eye.

Impulse Sources

Both impulse sources and steady-state sources can be used to examine a room's response. Any sound source used to excite an enclosure must have enough energy throughout the spectrum to ensure decays sufficiently raised above the noise to give the required accuracy. Examples of impulse sound sources are powerful electrical spark discharges, pistols firing blanks, and balloon bursts. For large spaces, even small cannons have been used as impulse sources to provide adequate energy, especially at lower frequencies. Whatever the actual method of producing an impulse, the impulse decay, the room's response measured over time, can be used to examine the temporal characteristics of sound in a room. For example, the more diffuse the room's sound field, the smoother the decay response. Conversely, for example, a room with echoes will yield a nonuniform decay response as energy is concentrated at the times when echoes appear.

The impulse decays for a small studio room are shown in Fig. 11-4. The sound source was an air pistol that ruptures paper discs. The peak sound-pressure level at 1-meter

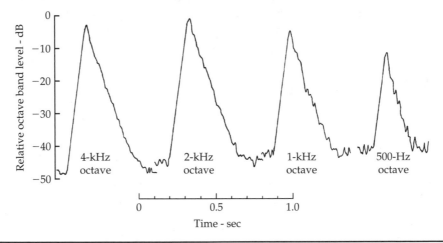

FIGURE 11-4 Reverberatory decays produced by impulse excitation of a small studio. The upgoing left side of each trace is instrumentation limited; the downgoing right side is the reverberatory decay.

distance is 144 dB, and the duration of the major pulse is less than 1 msec. It is ideal for recording echograms; the decays of Fig. 11-4 were made from impulses recorded for that purpose.

In Fig. 11-4, the straight, upward traveling segment on the left side is the same slope for all decays because it is a result of instrumentation limitation. The useful measure of reverberation is the downward traveling, more irregular slope on the right. This slope yields a reverberation time after the manner of Fig. 11-3. Note that the octave-band noise level is higher for the lower frequency bands. The impulse is barely above the noise floor for the 250 Hz and lower octaves. This is often a major limitation of the method.

Steady-State Sources

As noted, steady-state sources can be used to examine a room's response; for example, as noted, Sabine used a wind chest and organ pipes for his early measurements. However, care must be taken to choose a steady-state source that provides accurate response data. Sine-wave sources providing energy at a single frequency give highly irregular decays that are difficult to analyze. Warbling a tone, which spreads its energy over a narrow band, is an improvement over the fixed tone, but random noise sources are preferred. Bands of random (white or pink) noise give a steady and dependable indication of the average acoustical effects taking place within that particular part of the spectrum. Octave and $\frac{1}{3}$-octave bands of random noise are commonly used. Steady-state sources are useful for measuring the growth of sound in a room, as well as the decay response.

Measuring Equipment

The equipment layout shown in Fig. 11-5 can be used to obtain reverberation decays. A wideband pink-noise signal is amplified and used to drive a loudspeaker. A switch for interrupting the noise excitation is provided. By aiming the loudspeaker into a corner of the room (especially in smaller rooms), all resonant modes are excited, because all modes terminate in the corners.

An omnidirectional microphone is positioned on a tripod, usually at ear height for a listening room, or microphone height for a room used for recording. The smaller the microphone capsule, generally the less its directional effects. Some larger capsule microphones (e.g., 1-in-diameter diaphragms) can be fitted with random incidence correctors, but using a smaller microphone (e.g., $\frac{1}{2}$-in-diameter diaphragm) is preferred for more uniform sensitivity to sound arriving from all angles. In Fig. 11-5A, the microphone is a high-quality condenser microphone, part of the Brüel & Kjaer (B&K) sound-level meter, but separated from it by an extension cable. This device provides a preamplifier, octave filters, a calibrated method, and a line-level output signal for recording.

Measurement Procedure

The measurement procedure begins when a the room is filled with loud wide-band pink noise, usually loud enough to require the use of ear protectors for everybody in the room. When the noise is topped, the sound in the room decays into silence. The microphone, at its selected position, picks up this decay, which is recorded for later analysis.

The signal-to-noise ratio determines the length of the reverberatory decay available for study. As noted, it is often not possible to realize the entire 60-dB decay used in the definition of RT_{60}. It is quite possible, however, to obtain 45- to 50-dB decays with the

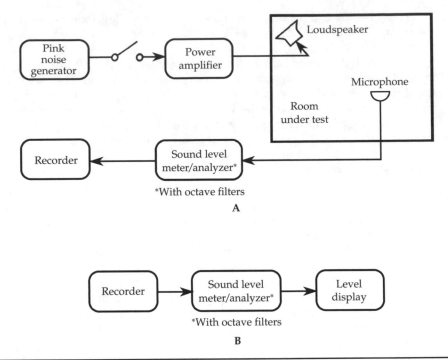

FIGURE 11-5   Equipment configuration for measuring the reverberation time of an enclosure. (A) Recording decays on location. (B) Later decay playback for analysis.

equipment shown in Fig. 11-5 by the simple expedient of double filtering. For example, to obtain an RT_{60} measurement at 500 Hz, the octave filter centered on 500 Hz in the sound-level meter is used both in recording and in later playback.

The analysis procedure outlined in Fig. 11-5B uses recorder playback and a B&K sound-level meter, with the addition of a graphic-level display. The line output of the recorder is connected to the input of the sound-level meter circuit through a 40-dB attenuating pad. To do this, the microphone of the sound-level meter is removed and a special fitting is screwed in its place. The output of the sound-level meter is connected directly to the graphic-level display input, completing the equipment interconnection. The appropriate octave filters are switched in as the decay recording is played back.

Reverberation and Normal Modes

The natural resonances of a room are revealed in its normal modes, as described in Chap. 13. It is necessary to somewhat anticipate this topic to understand the relationship of these natural resonances and the reverberation of the room. For now, simply stated, most rooms have preferred resonance frequencies where sound energy at those frequencies is accentuated.

The Sabine equation or alternatives are widely used to characterize a room's reverberation time. However, single-number calculations or measurement cannot completely describe the behavior of room reverberation, particularly when considering the effects of normal modes of rooms. Room modes pose particular difficulties when characterizing the reverberation time of small rooms.

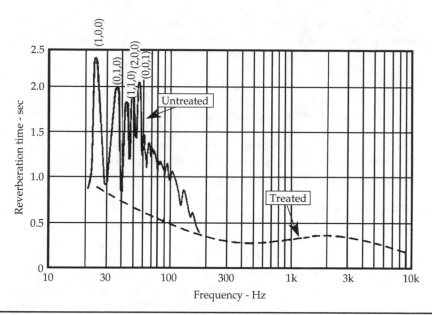

FIGURE 11-6 Reverberation time measured with pure sine signals at low frequencies reveals slow sound decay (long reverberation time) at the modal frequencies. These peaks apply only to specific modes and are not representative of the room as a whole. High modal density, resulting in uniformity of distribution of sound energy and randomizing of directions of propagation, is necessary for reverberation equations to apply. (*Beranek and Schultz*)

Consider an untreated, small studio room. Starting with an oscillator set to about 20 Hz below the first axial mode, the acoustics of the room do not load the loudspeaker and a relatively weak sound is produced with the amplifier gain turned up full (even assuming the use of a good subwoofer). As the oscillator frequency is adjusted upward, however, the sound becomes very loud as the (1, 0, 0) mode (24.18 Hz in this example) is energized, as shown in Fig. 11-6. Adjusting the oscillator upward we go through a weak valley but at the frequency of the (0, 1, 0) mode (35.27 Hz) there is high-level sound once more. Similar peaks are found at the (1, 1, 0) tangential mode (42.76 Hz), the (2, 0, 0) axial mode (48.37 Hz), and the (0, 0, 1) axial mode (56.43 Hz).

Now that the loudness of peaks and valleys of the room's modes have been noted, let's examine the decay of sound. After exciting the (1, 0, 0) mode at 24.18 Hz to steady-state conditions, the sound source is switched off; the decay time of the resulting reverberation is measured as 2.3 seconds. Similar decays are observed at 35.27, 42.76, 48.37, and 56.43 Hz with faster decays (shorter reverberation times) at frequencies in between those modes. The long decay times at the modal frequencies are decay rates characteristic of individual modes, not of the room as a whole.

Long reverberation time implies low absorbance, and short reverberation time implies high absorbance. It is interesting that the sound absorbing qualities of the walls, floor, and ceiling can vary considerably within a frequency range of a few hertz. For the (1, 0, 0) mode, only the absorbance of the two ends of the room comes into play; the four other surfaces are not involved. For the (0, 0, 1) mode, only the floor and ceiling are involved. In this low-frequency range, we have measured the decay rate of individual modes, but not the average condition of the room.

We see now why it is difficult to apply the concept of reverberation time to small rooms having dimensions comparable to the wavelength of sound. Schultz states that reverberation time is a statistical concept "in which much of the mathematically awkward details are averaged out." In small rooms these details are not averaged out.

The reverberation-time equations of Sabine, Eyring-Norris, and others are based on the assumption of an enclosed space with a highly uniform distribution of sound energy and random direction of propagation of sound. At low frequencies in the room presented in Fig. 11-6, energy is distributed very unevenly and direction of propagation is far from random. After the room is treated, reverberation time measurements would follow the broken line, but statistical randomness still does not prevail below 200 Hz even though modal frequencies are brought under some measure of control.

Analysis of Decay Traces

An octave band of pink noise viewed on an oscilloscope shows a trace that looks very much like a sine wave, except that it is constantly shifting in amplitude and phase, because of the property of random noise. This characteristic of random noise has its effect on the shape of the reverberatory decay trace. Consider what this constantly shifting random noise signal does to the normal modes of a room. When the axial, tangential, and oblique resonant modes are considered, they are quite close together in frequency. For example, the number of modes included within an octave band centered on 63 Hz are: 4 axial, 6 tangential, and 2 oblique modes between the −3-dB points. These are shown in Fig. 11-7 in which the taller lines represent the more dominant axial modes, the intermediate height the tangential modes, and the shorter lines the oblique modes.

When the excitation noise from the loudspeaker energizes the room, it excites one mode, and an instant later another mode. While the response shifts to the second mode, the first mode begins to decay. Before it decays very far, however, the random-noise instantaneous frequency is once more back at the first mode, giving it another boost. All the modes of the room are in constant agitation, alternating between high and somewhat lower levels, as they begin to decay in between stimuli. It is strictly a random situation, but it can be said with confidence that each time the excitation noise is stopped, the modal excitation pattern will be somewhat different. For example, the 12 modes in the 63-Hz octave will all be highly energized, but each to a

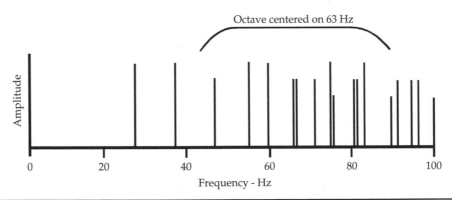

FIGURE 11-7 The normal modes included in an octave centered on 63 Hz (−3-dB points). The tallest lines indicate axial modes, the intermediate lines indicate tangential modes, and the shortest lines indicate oblique modes.

somewhat different level the instant the noise is stopped. This helps to address, but does not completely solve, the effect that room normal modes have on measured reverberation time in small rooms.

Mode Decay Variations

To illustrate this discussion, let's consider measurements in a real room. The room is a rectangular studio for voice recording having the dimensions 20 ft 6 in × 15 ft × 9 ft 6 in, with a volume of 2,921 ft³. The measuring equipment is that described in Fig. 11-5, and the technique is that described above. Four successive 63-Hz octave decays from the graphic-level recorder are shown in Fig. 11-8A. These decays are not identical, and differences can be attributed to the random nature of the noise signal. In particular, the fluctuations in the decays result from beats between closely spaced modes. Because the excitation level of the modes is constantly shifting, the form and degree of the beat pattern shifts from one decay to another depending on where the random excitation happens to be the instant the noise is stopped.

Even though the four decays are similar, fitting a straight line to evaluate the reverberation time of each can be affected by the beat pattern. For this reason, it is good practice to record perhaps five decays for each octave for each microphone position in a room. With eight octaves (63 Hz to 8 kHz), five decays per octave, and three microphone positions, this means 120 separate decay readings for each room, which is laborious.

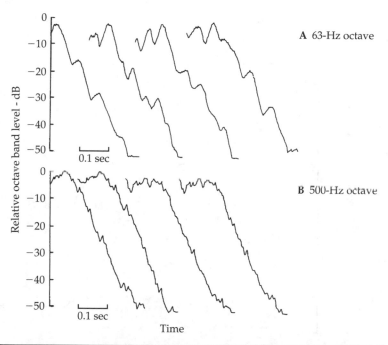

FIGURE 11-8   Decays of random noise recorded in a small studio having a volume of 2,921 ft³. (A) Four successive 63-Hz octave decays recorded under identical conditions. (B) Four successive 500-Hz octave decays also recorded under identical conditions. The differences noted result from the differences in random-noise excitation the instant the source is turned off to start the decay.

However, this approach yields a statistically significant view of the variation with frequency. A handheld reverberation-time measuring device could accomplish this with less work, but it may not give detail of the shape of each decay. There is much information in each decay, and acoustical flaws can often be identified from aberrant decay shapes.

Four decays at 500 Hz are also shown in Fig. 11-8B for the same room and the same microphone position. The 500-Hz octave (354 to 707 Hz) embraces about 2,500 room modes. With such a high modal density, the 500-Hz octave decay is much smoother than the 63-Hz octave with only a dozen modes. Even so, the irregularities for the 500-Hz decay of Fig. 11-8B result from the same cause. Remembering that some modes die away faster than others, the decays in Fig. 11-8 for both octaves are composites of all modal decays included.

Frequency Effect

Typical decays for octave bands of noise from 63 Hz to 8 kHz are shown in Fig. 11-9. The greatest fluctuations are in the two lowest bands, the least in the two highest. This is what we would expect, given that the higher the octave band, the greater the number of

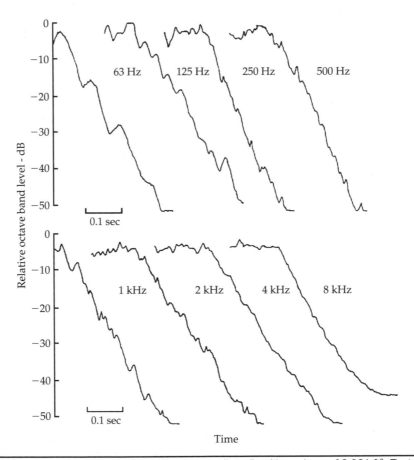

FIGURE 11-9   Decay of octave bands of noise in a small studio with a volume of 2,921 ft³. Fluctuations due to modal interference are greatest for low-frequency octaves containing fewer modes.

normal modes included, and the greater the statistical smoothing. However, we should not necessarily expect the same decay rate because reverberation time is different for different frequencies. In this particular voice studio (Fig. 11-9) a uniform reverberation time with frequency was the design goal, which was approximated in practice.

Reverberation Characteristic

Rooms absorb sound at different rates at different frequencies, producing different reverberation times, and this can greatly influence the room's sonic characteristic. For example, a room with longer reverberation times at higher frequencies, but shorter reverberation times at low frequencies may be described as sounding "thin" or "harsh." Because reverberation time will vary at different frequencies, it is common practice to measure reverberation times at different octaves. Rather than plot amplitude versus time as in a reverberation time graph, this analysis plots reverberation time versus frequency; this is sometimes called the reverberation characteristic.

This reverberation-time analysis is quite different from a real-time analysis that shows level versus frequency. For example, consider a room that exhibits excessive reverberation time at low frequencies, causing poor bass clarity. If the problem is mistakenly addressed through equalization, by rolling off low frequencies, a real-time analysis might show a flat low-frequency response. But reverberation time measurements would show that reverberation time is still too long at low frequencies and bass clarity is still poor; the problem has not been addressed. That is because equalization will not change the reverberation characteristic. To solve the problem of frequency balance, acoustical treatment must be applied to change the reverberation characteristic.

To measure the reverberation characteristic, a pink-noise source is band-pass filtered to produce octave-band noise at various center frequencies. For example, these center frequencies may be used: 63; 125; 250; 500; 1,000; 2,000; 4,000; and 8,000 Hz. The source is activated to produce a steady-state level, then turned off. The RT_{60} is the time required for the level to drop 60 dB. As noted, because of dynamic-range constraints, the first portion of the decay can be extrapolated to produce the full measurement. The entire measurement is conveniently performed by recording the steady-state noise and the decay. Because of fluctuations in level at low frequencies, measurements must be repeated. Because only the time decay of a narrow band is relevant, the frequency response of the source loudspeaker and microphone are relatively unimportant. Also, because of room modes, measurements at low frequencies are affected by position of the source and the microphone.

As noted, the results are plotted as reverberation time versus frequency. Figure 11-10 shows an example of a room's reverberation characteristic before and after room treatment. As shown, the room exhibited a significant rise in reverberation time in the upper bass and lower midrange. After treatment, reverberation time was much flatter, with a desirable moderate increase in reverberation time at low frequencies. Generally, rooms are preferred with a smooth characteristic, with somewhat higher RT_{60} at low frequencies. Undue bumps in the reverberation-time characteristic can be reduced by adding absorption treatment in that frequency range. A practical example of room treatment to obtain desirable reverberation times is given at the end of this chapter.

Reverberation Time Variation with Position

In most rooms, there is enough variation of reverberation time from one physical position to another to justify taking measurements at several positions. The average then

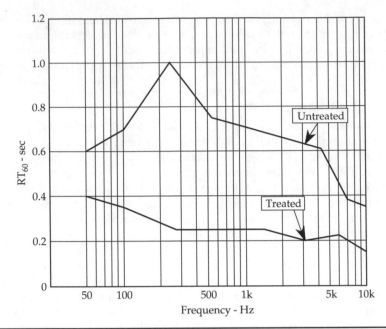

FIGURE 11-10   An example of a room's reverberation characteristic before and after room treatment. A significant rise in reverberation time in the upper bass and lower midrange is changed to a flatter characteristic with a moderate increase in reverberation time at low frequencies.

gives a better statistical picture of the behavior of the sound field in the room. If the room is symmetrical, it may be efficient to spot all measuring points on one side of the room to increase the effective coverage with a given effort.

Decay Rate and the Reverberant Field

The definition of reverberation time is based on uniform distribution of energy and random directions of propagation. These idealized conditions do not exist in small, absorptive rooms because of room mode effects; so what we measure, technically, should not be called reverberation time. It is more properly termed decay rate. A reverberation time of 0.3 second is equivalent to a decay rate of 60 dB/0.3 sec = 200 dB/sec. Speech and music in small, absorptive rooms certainly decay, even though the modal density is too low to meet the requirements of the definition of reverberation time.

In small and relatively dead rooms such as many studios, control rooms, and home listening rooms, the direct sound from the source usually dominates. A true reverberant field may be below the ambient noise level. However, the reverberation time equations have been derived for conditions that exist only in the reverberant field. In this sense, then, the concept of reverberation time is inapplicable to small, relatively dead rooms. We measure something very much like the reverberation time in large, more live spaces. But in small, dead rooms, more accurately, what we measure is the decay rate of the normal modes of the room.

Each axial mode decays at its own rate determined by the absorbance of a pair of walls and their spacing. Each tangential and oblique mode has its own decay rate

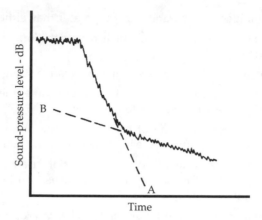

FIGURE 11-11 Reverberatory decay with a double slope due to acoustically coupled spaces. The shorter reverberation time represented by slope A is that of the main room. A second, highly reflective space is coupled through an open doorway; its longer reverberation time is represented by slope B. Those seated near the doorway are subjected first to the main-room response and then to the decay of the coupled space.

determined by distance traveled, the number of surfaces involved, the variation of the absorption coefficient of the surfaces with angle of incidence, and so on. Whatever average decay rate is measured for an octave of random noise will be representative of the average decay rate at which that octave of speech or music signals would die away. Although the applicability of computing reverberation time from the equations based on reverberant field conditions might be questioned because of the lack of reverberant field, the measured decay rates most certainly apply to this space and to these signals.

Acoustically Coupled Spaces

The shape of a reverberation decay can reveal acoustical problems in the space. One common effect that alters the shape of the decay is due to acoustically coupled spaces. This is quite common in large public gathering spaces, but is also found in offices, homes, and other small spaces. The principle involved is illustrated in Fig. 11-11. In this example, the main space, perhaps an auditorium, is acoustically quite dead and has a reverberation time corresponding to slope A. An adjoining hall with hard surfaces opens into the main room and has a reverberation time corresponding to slope B. A person seated in the main hall near the connecting opening could very well experience a double-slope reverberation decay. After the sound level in the main room falls to a fairly low level, the main room reverberation would be dominated by sound fed into it from the slowly decaying sound in the adjoining hall. Assuming the reverberation time described by slope A is acoustically correct for the main room, persons subjected to slope B would hear degraded sound.

Electroacoustically Coupled Spaces

What is the overall reverberant effect when music recorded in a studio with one reverberation time is reproduced in a listening room with a different reverberation time? The sound in the listening room is indeed affected by the reverberation of both studio and listening room. This is demonstrated as follows:

- The combined reverberation time is greater than either alone.
- The combined reverberation time is nearer the longer reverberation time of the two rooms.
- The combined decay departs somewhat from a straight line.
- If one room has a very short reverberation time, the combined reverberation time will be very close to the longer one.
- If the reverberation time of each of the two rooms alone is the same, the combined reverberation time is 20.8% longer than one of them.
- The character and quality of the sound field transmitted by a stereo system conforms more closely to the mathematical assumptions of the above than does a monaural system.
- The first five items can be applied to the case of a studio linked to an echo chamber as well as a studio linked to a listening room.

Eliminating Decay Fluctuations

The measurement of reverberation time by the classical method described above requires the recording of many decays for each condition. Schroeder published an alternate method by which the equivalent of the average of a great number of decays can be obtained in a single decay. One practical, but clumsy, method of accomplishing the required mathematical steps is to:

1. Record the decay of an impulse (noise burst or pistol shot) by the normal method.
2. Play back that decay reversed.
3. Square the voltage of the reversed decay as it grows.
4. Integrate the squared signal with a resistance-capacitance circuit.
5. Record this integrated signal as it grows during the reversed decay. Reverse it and this trace will be mathematically identical to averaging an infinite number of traditional decays.

Influence of Reverberation on Speech

Consider what happens to the spoken word "back" in a reverberant space. The word starts abruptly with a "ba . . ." sound and ends with the consonant ". . . ck" which is much lower in level. As measured on a graphic-level recorder, the "ck" sound is nominally about 25 dB below the peak level of the "ba" sound and reaches a peak nominally about 0.32 second after the "ba" peak.

Both the "ba" and "ck" sounds are transients that grow and decay after the manner of Fig. 11-2. Sketching these to scale yields Fig. 11-12. The "ba" sound grows to a peak at an arbitrary level of 0 dB at time $t = 0$, after which it decays according to the RT_{60} reverberation time of the room, which is assumed to be 0.5 second. The "ck" consonant sound, peaking 0.32 second later, is 25 dB below the "ba" sound peak. It too decays at the same rate as the "ba" sound according to the assumed 0.5-second RT_{60}. Under the influence of the 0.5-second reverberation time, the "ck" consonant sound is not masked by the reverberation time of "ba." If the reverberation time is increased to 1.5 seconds, as shown by

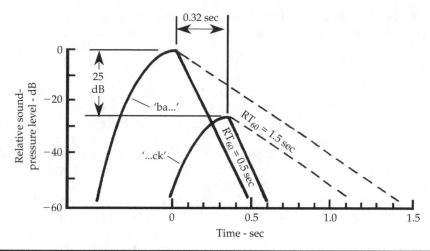

FIGURE 11-12 An illustration of the effects of reverberation on the intelligibility of speech. Understanding the word "back" depends on apprehending the later, lower level consonant ". . . ck," which can be masked by reverberation if the reverberation time is too long.

the broken lines, the consonant "ck" is completely masked by reverberation. Likewise, a syllable that ends one word might mask the opening syllable of the next word.

Excessive reverberation can impair the intelligibility of speech by masking the lower level consonants. In the word "back," the word is unintelligible without clear audibility of the "ck" part. Understanding the "ck" ending is the only way to distinguish "back" from bat, bad, bass, ban, or bath. In this simplified way, we observe the effect of reverberation on the understandability of speech and the reason why speech is more intelligible in rooms having lower reverberation times. Similarly, in a highly reverberant room, a speaker may have to speak slowly to be understood. It is a straightforward procedure to predict with reasonable accuracy the intelligibility of speech in a space from geometrical factors and a knowledge of reverberation time.

On the other hand, a completely dead room, or an outdoor environment, is not optimal for speech intelligibility; at even a moderate distance away, the sound level from a speaker may be too low to clearly hear speech. The reverberation in live rooms add acoustic power to original sound sources, reinforcing them and making them more audible, as well as more pleasant sounding. A dead room adds less power to the source, and the overall sound is softer. A good room design must therefore balance acoustic gain with intelligibility. In part, this means that different reverberation times determine whether a room is best suited for speech or music.

Influence of Reverberation on Music

The effect of reverberation on speech can be easily gauged according to the intelligibility of speech. The effect of hall resonance or reverberation on music is intuitively grasped but is more difficult to quantify. For example, a reverberation time that seems suitable for one type of music may be unsuitable for another type of music. In any case, the optimal reverberation time for music is both variable and subjective. This subject has received much attention from scientists as well as musicians. Beranek attempted to summarize knowledge and to pinpoint essential characteristics of concert and opera halls around the world, but our understanding of the problem is still incomplete. Suffice it to say that the reverberation decay of a music hall is an important factor among many, another being the echo pattern of the early sound of a room. It is beyond the scope of this book to treat this subject in great detail, but two commonly overlooked points are discussed briefly.

Normal modes have great basic importance in any room's acoustical response, and they are also active in concert halls and listening rooms. An interesting phenomenon is pitch change during reverberant decay. For example, in reverberant churches, organ tones have been observed to change pitch as much as a semitone during decay. In searching for an explanation for this phenomenon, two factors have been suggested: shift of energy between normal modes and the perceptual dependence of pitch on sound intensity. Balachandran demonstrated the physical (as opposed to psychophysical) reality of the effect using the Fast Fourier Transform (FFT) technique on a reverberant field created by 2-kHz pulses. In his case study, he showed the existence of a primary 1,992-Hz spectral peak, and curiously, another peak at 3,945 Hz. Because a 6-Hz change would be just perceptible at 2 kHz, and a 12-Hz change perceptible at 4 kHz, we see that the 39-Hz shift from the octave of 1,992 Hz would give an impression of pitch change. The reverberation time of the hall in which this effect was recorded was about 2 seconds.

Optimum Reverberation Time

Considering the full range of possible reverberation times, seemingly there must be an optimum time between the too-dry condition of the outdoors and anechoic chambers, and the obvious problems associated with excessively long reverberation times as in a masonry cathedral. There is great disagreement as to what the optimal value is in any given room because it is a subjective problem, with different cultural and aesthetic expectations, and differences in opinions must be expected. The optimal value depends not only on the one making the judgment, but also on the type of sound sources being considered.

In general, long reverberation time yields lack of definition in music and loss of intelligibility in speech. Spaces for speech particularly require shorter reverberation times than for music because of the importance of the clarity provided by direct sound. In dead spaces in which reverberation time is very short, loudness and tonal balance in music may suffer. It is not possible to specify precisely optimum reverberation times for different services, but Figs. 11-13 through 11-15 show approximate recommendations given by experts who do not always agree with each other.

Reverberation chambers, which are used for measuring absorption coefficients, are carefully designed for the longest practical RT_{60} to achieve the maximum accuracy. The optimum here is the maximum attainable.

The best reverberation time for a space in which music is played depends on the size of the space and the type of music. No single optimum universally fits all types of music; the best that can be done is to establish a range based on subjective judgment. Slow, solemn, melodic music, such as some organ music, is best served by long reverberation time. Quick rhythmic music requires a shorter reverberation time from chamber music.

The reverberation times for churches in Fig. 11-13 range from highly reverberant liturgical churches and cathedrals to the shorter ranges of the lower shaded area characteristic of more sermon-oriented churches. Churches generally represent a compromise between music and speech.

Figure 11-14 represents the range of recommended reverberation times for different concert halls. Symphony orchestras are near the top and lighter music somewhat lower. The lower shaded area applies for opera and chamber music.

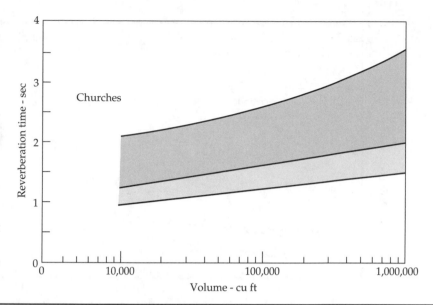

FIGURE 11-13 Range of optimum reverberation times for churches. The upper area applies to the more reverberant liturgical churches and cathedrals, the lower to churches having services more oriented to speech. A compromise between music and speech is required in most churches.

Recording studios present still other problems that do not conform to simple rules. Multitrack recording in which musical instruments are recorded on separate tracks for later mixdown generally requires quite dead spaces to realize adequate acoustical separation between tracks. Music producers often require different reverberation times for different instruments; hence hard areas and absorptive areas may be found in the same

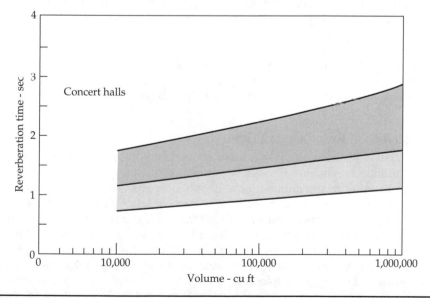

FIGURE 11-14 Range of optimum reverberation times for concert halls. Symphony orchestras are near the top of the shaded areas; lighter music is lower. The lower shaded area applies to opera and chamber music.

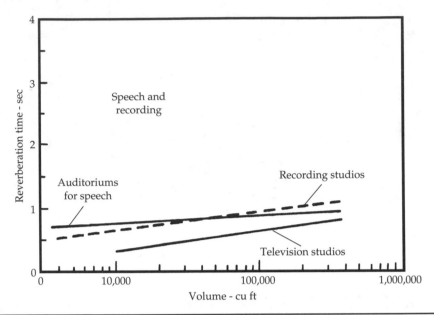

FIGURE 11-15   Spaces designed for speech and music recording require shorter reverberation times.

studio. The range of reverberation time realized in this manner is limited, but proximity to reflective surfaces does affect local conditions.

Generally, spaces used primarily for speech and recording require close to the same reverberation times, as shown in Fig. 11-15. Television studios have even shorter reverberation times to deaden the sounds associated with some video equipment, dragging cables, and other production noises. It should also be remembered that acoustics in television are dominated by the setting and local furnishings. In many of the spaces represented in Fig. 11-15, speech reinforcement is employed. Movie theatres are usually quite absorptive; adequate reverberation is already recorded in the soundtrack, and in particular added room reverberation would reduce the intelligibility of dialogue. For hearing-impaired listeners, reverberation times should be decreased for better intelligibility. Many spaces that are used for multiple functions can benefit from adjustable acoustic methods.

Bass Rise of Reverberation Time

The goal in voice studios is to achieve a reverberation time that is the same throughout the audible spectrum. This can be difficult to realize, especially at low frequencies. Adjustment of reverberation time at high frequencies is easily accomplished by adding or removing relatively inexpensive absorbers. At low frequencies, the situation is quite different as absorbers are bulky, difficult to install, and sometimes unpredictable.

Researchers at the British Broadcasting Corporation (BBC) observed that subjective judgments indicated a tolerance for a certain amount of bass rise of reverberation time. Investigating this in controlled tests, Spring and Randall found that bass rise to the extent indicated in Fig. 11-16 was tolerated by the test subjects for voice signals. Taking the 1-kHz value as reference, rises of 80% at 63 Hz and 20% at 125 Hz were found to be acceptable. These tests were made in a studio $22 \times 16 \times 11$ ft (volume about 3,900 ft^3) for which the mid-band reverberation time was 0.4 second (which agrees fairly well with Fig. 11-15).

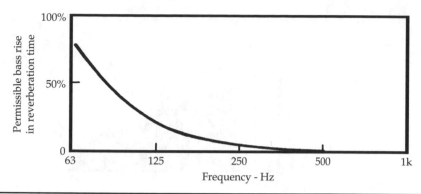

FIGURE 11-16 Permissible bass rise of reverberation time for voice studios derived by subjective evaluation in controlled tests by BBC researchers. (*Spring and Randall*)

Bass rise in reverberation time for music performance has traditionally been accepted to add sonority and warmth to the music in concert halls. A bass rise may help overcome the ear's relative insensitivity at low frequencies, or may simply be a cultural preference. Presumably, somewhat greater bass rise than that for speech would be desirable in listening rooms designed for classical music. One metric used to define a bass boost is the bass ratio (BR), where $BR = (RT_{60/125} + RT_{60/250})/(RT_{60/500} + RT_{60/1,000})$. In other words, the sum of RT_{60} at 125 and 250 Hz is divided by the sum of RT_{60} at 500 and 1,000 Hz; a value greater than unity shows that reverberation time is longer at low frequencies. Some designers recommend a bass ratio of 1.1 to 1.45 for halls with a reverberation time less than 1.8 seconds, and a bass ratio of 1.1 to 1.25 for halls with a longer reverberation time. A bass ratio of less than 1.0 is not recommended. This is explored more fully in Chap. 22.

Initial Time-Delay Gap

One important characteristic of natural reverberation in concert halls was revealed by Beranek's study of halls around the world. At a given seat, the direct sound arrives first because it follows the shortest path. Shortly after the direct sound, the reverberant sound arrives. The time between the two is called the initial time-delay gap (ITDG), as shown in Fig. 11-17. If this gap is less than about 40 msec, the ear integrates the direct and the reverberant sound successfully. In addition to all of the reflections responsible for achieving reverberation density, the initial time-delay gap is another important delay that must be considered. In particular, this gap is important in concert-hall design (and in artificial reverberation algorithms) because it is the cue that gives the ear information on the size of the hall.

Listening Room Reverberation Time

The reverberation characteristic of a typical home listening room is of interest to the audiophile, the broadcaster, and the recording engineer. The monitoring room of the broadcast or recording studio should have a reverberation time similar to that of the listening room in which the final product will be heard. Generally, such rooms should be deader than the listening room, which will add its own reverberation to that of recording or broadcast studio.

Figure 11-18 shows the average reverberation time of 50 British living rooms measured by Jackson and Leventhall using octave bands of noise. The average reverberation

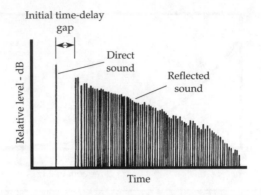

FIGURE 11-17   The introduction of an initial time-delay gap plays an important role in room reverberation. That time gap between arrival of the direct sound, and the first reflected sounds, helps identify the size of the room.

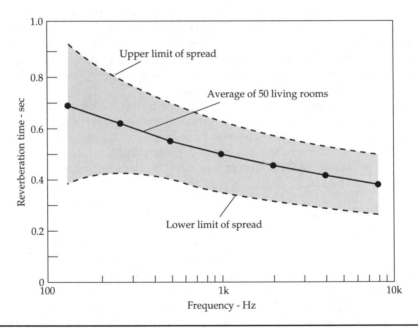

FIGURE 11-18   Average reverberation time for 50 British living rooms. (*Jackson and Leventhall*)

time decreases from 0.69 second at 125 Hz to 0.4 second at 8 kHz. This is considerably higher than earlier measurements of 16 living rooms made by BBC engineers in which average reverberation times ranged from 0.35 to 0.45 second. Apparently, the living rooms measured by the BBC engineers were better furnished than those measured by Jackson and Leventhall.

The 50 rooms of the Jackson-Leventhall study were of varying sizes, shapes, and degree of furnishing. The sizes varied from 880 to 2,680 ft³, averaging 1,550 ft³. An optimum reverberation time for speech for rooms of this size is about 0.3 second (Fig. 11-15). Only those living rooms near the lower limit approach this, and in them we would expect to find much heavy carpet and overstuffed furniture. These reverberation measurements tell us little or nothing about the possible presence of timbral defects. The BBC engineers checked for timbral changes and reported problems in a number of the living rooms studied.

Artificial Reverberation

Artificial reverberation is considered a necessity in audio-signal processing. Recordings of music played in dry (nonreverberant) studios lack the richness of the room effect contributed by the concert hall. The addition of artificial reverberation to such recordings has become standard practice, and there is great demand for equipment that will provide natural-sounding artificial reverberation at a reasonable cost.

There are many ways of generating artificial reverberation, but the challenge is finding that method which mimics actual music halls and does not introduce frequency-response aberrations into the signal. Historically, a dedicated reverberation room was employed. The program is played into this room, picked up by a microphone, and the reverberated signal mixed back into the original in the amount to achieve the desired effect. Small reverberation rooms are afflicted with serious timbral problems because of widely spaced modes. Large rooms are expensive. Even though the three-dimensional reverberation room approach has certain desirable qualities, the problems outweigh the advantages, and they are now a thing of the past.

Most artificial reverberation is created digitally using digital hardware or software programs. The principle of digital reverberation is illustrated in the signal flow schematic of Fig. 11-19. The input signal is delayed, and a portion of the delayed signal is fed back and mixed with the incoming signal, the mixture being delayed again, and so on.

Schroeder found that approximately 1,000 echoes per second are required to avoid a flutter effect and to sound natural to the ear. With a 40-msec delay, only $1/0.04 = 25$ echoes are produced each second, a far cry from the 1,000 per second desired. One solution is to arrange many simple reverberators in parallel. Four such simple reverberators, arranged in parallel, might produce $4 \times 25 = 100$ echoes per second. It would require 40 such reverberators in parallel to achieve the required echo density.

One approach to producing the necessary echo density, and simultaneously, a flat frequency response is illustrated in Fig. 11-20. Numerous delays feed back on themselves, combining to feed other delays, which in turn recirculate back to the first delay. The + signs in Fig. 11-20 indicate mixing (addition), and the × signs indicate gain (multiplication). Multiplying by a fraction less than unity gives a gain of less than unity—in other words, attenuation. The digital reverberator of Fig. 11-20 only suggests how greater echo density along with good frequency response might be achieved. The digital-reverberation algorithms used today are far more complicated than this. The resulting reverberation available today has a high echo density, flat frequency response, and a natural sound.

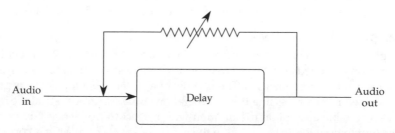

Figure 11-19 A simple digital reverberation algorithm using a delay line and feedback.

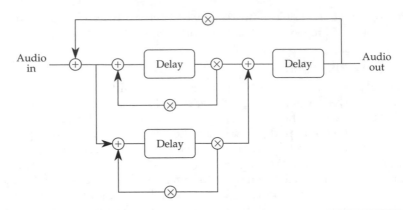

Figure 11-20   The required echo density is achieved in a reverberation algorithm that incorporates numerous delays and recirculation of the signal. Actual digital reverberation algorithms are more complicated than this one.

Reverberation Time Calculations

As noted, reverberation time is defined as the number of seconds required for sound intensity in a room to drop 60 dB from its original level. The Sabine equation, presented earlier in the chapter as Eq. (11-1), can be used to calculate reverberation time.

Example 1: Untreated Room

This example illustrates the implementation of Sabine's equation. The dimensions of an untreated room are $23.3 \times 16 \times 10$ ft. The room has a concrete floor and the walls and the ceiling are of frame construction with $\frac{1}{2}$-in gypsum board (drywall) covering. As a simplification, the door and a window will be neglected as having minor effect. Figure 11-21 shows the untreated condition. The concrete floor area of 373 ft^2 and the gypsum board area of 1,159 ft^2 are entered in the table. The appropriate absorption coefficients α are entered from the table in the appendix for each material and for the six frequencies. Multiplying the concrete floor area of $S = 373$ ft^2 by the coefficient $\alpha = 0.01$ gives a floor absorption of 3.7 sabins. This is entered under $S\alpha$ for 125 and 250 Hz. The absorption units (sabins) are then figured for both materials and for each frequency. The total number of sabins at each frequency is obtained by adding that of the concrete floor to that of the gypsum board. The reverberation time for each frequency is obtained by dividing $0.049\ V = 182.7$ by the total absorption for each frequency.

To visualize the variation of reverberation time with frequency, the values are plotted in Fig. 11-22A. The peak reverberation time of 3.39 seconds at 1 kHz is excessive and would yield poor sound conditions. Two persons separated 10 ft would have difficulty understanding each other as the reverberation of one word masks the next word.

Example 2: Modified Room

The goal now is to correct the reverberation of curve A of Fig. 11-22. It is evident that much absorption is needed at midband frequencies, a modest amount at higher frequencies, and very little at lower frequencies. The need is for a material having an absorption characteristic shaped more or less like the reverberation curve A. An acoustical tile $\frac{3}{4}$-in thick provides a suitable absorption distribution with respect to frequency. Giving no thought at this point to how it is to be placed in the room, what area of this tile is correct?

		Size		23.3 × 16 × 10 ft								
		Treatment		None								
		Floor		Concrete								
		Walls/ceiling		Gypsum board, 1/2", on frame construction								
		Volume		(23.3)(16)(10) = 3,728 cu ft								

Material	S sq ft	125 Hz		250 Hz		500 Hz		1 kHz		2 kHz		4 kHz	
		α	$S\alpha$	α	$S\alpha$	α	$S\alpha$	α	$S\alpha$	α	$S\alpha$	α	$S\alpha$
Concrete	373	0.01	3.7	0.01	3.7	0.015	5.6	0.02	7.5	0.02	7.5	0.02	7.5
Gypsum board	1,159	0.29	336.1	0.10	115.9	0.05	58.0	0.04	46.4	0.07	81.1	0.09	104.3
Total absorption (sabins)		339.8		119.6		63.6		53.9		88.6		111.8	
Reverberation time (seconds)		0.54		1.53		2.87		3.39		2.06		1.63	

S = area of material
α = absorption coefficient for that material and for that frequency
$A = S\alpha$ absorption units, sabins

$$RT_{60} = \frac{(0.049)(3728)}{A} = \frac{182.7}{A}$$

Example: For 125 Hz, $RT_{60} = \frac{182.7}{339.8} = 0.54$ second

FIGURE 11-21 Room conditions and calculations for reverberation example 1.

Figure 11-23 organizes the calculations. Everything is identical to Fig. 11-21, except that the 3/4-in acoustical tile has been added with coefficients from the appendix. In Fig. 11-21, there are a total of 53.9 sabins at the peak reverberation time at 1 kHz and 339.8 sabins at 125 Hz at which the reverberation time is a reasonable 0.54 second. How much 3/4-in acoustical tiles would be required to add 286 sabins at 1 kHz? The absorption coefficient

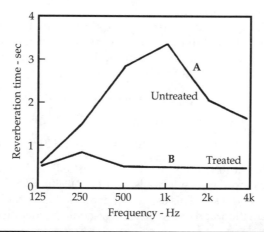

FIGURE 11-22 The calculated reverberation characteristics of a room measuring 23.3 × 16 × 10 ft. (A) The untreated condition of example 1. (B) The treated condition of example 2.

		Size	23.3 × 16 × 10 ft								
		Treatment	Acoustical tile								
		Floor	Concrete								
		Walls/ceiling	Gypsum board, ½", on frame construction								
		Volume	(23.3)(16)(10) = 3,728 cu ft								

| Material | S | 125 Hz | | 250 Hz | | 500 Hz | | 1 kHz | | 2 kHz | | 4 kHz | |
	sq ft	α	Sα	α	Sα	α	Sα	α	Sα	α	Sα	α	Sα
Concrete	373	0.01	3.7	0.01	3.7	0.015	5.6	0.02	7.5	0.02	7.5	0.02	7.5
Gypsum board	1,159	0.29	336.1	0.10	115.9	0.05	58.0	0.04	46.4	0.07	81.1	0.09	104.3
Acoustical tile	340	0.09	30.6	0.28	95.2	0.78	265.2	0.84	285.6	0.73	248.2	0.64	217.6
Total absorption (sabins)		370.4		214.8		328.8		339.5		336.8		329.4	
Reverberation time (seconds)		0.49		0.85		0.56		0.54		0.54		0.55	

S = area of material
α = absorption coefficient for that material and for that frequency
$A = S\alpha$ absorption units, sabins
$$RT_{60} = \frac{(0.049)(3728)}{A} = \frac{182.7}{A}$$

FIGURE 11-23   Room conditions and calculations for reverberation example 2.

of this material is 0.84 at 1 kHz. To get 286 sabins at 1 kHz with this material would require 286/0.84 = 340 ft² of the material. This is entered in Fig. 11-23 and the calculations extended. Plotting these reverberation time points gives the curve of Fig. 11-22B, showing reverberation time that is uniform across the band. The overall precision of coefficients and measurements is limited, so the deviation of curve B from flatness is insignificant.

An average reverberation time of 0.54 second in a room of 3,728 ft³ volume would be quite acceptable for a music listening room. If it were decided that the 0.85-second point at 250 Hz should be brought down closer to 0.54 second, it would require about 100 ft² of an absorber tuned to 250 Hz.

How should the 340 ft² of ³⁄₄-in acoustical tile be distributed? For maximum diffusion it should be applied in irregular patches and distributed to the three axes of the room. As a first approach, the 340 ft² could be distributed in proportion to areas on the three axes of the room as follows:

Area of N and S ends = (2)(10)(16) = 320 ft² (21%)

Area of E and W walls = (2)(10)(23.3) = 466 ft² (30%)

Area of floor/ceiling = (2)(16)(23.3) = 746 ft² (49%)

This would put (0.49)(340) = 167 ft² on the ceiling, (0.3)(340) = 102 ft² distributed between the east and west side walls, and (0.21)(340) = 71 ft² distributed between the north and south end walls. This recommendation should provide the desired results, but some experimentation, and critical listening, should be included in any such project.

CHAPTER 12

Absorption

The law of conservation of energy states that energy can neither be created nor destroyed. However, energy can be changed from one form to another. If there is excessive sound energy in a room, the energy itself cannot be eliminated but it can be transformed into an innocuous form. This is the function of sound-absorbing materials. Generally, sound absorbers can be considered as belonging to one of these types: porous absorbers, panel absorbers, and volume or resonance absorbers. Generally, porous absorbers are most effective at higher frequencies, while panel and volume absorbers are most effective at lower frequencies.

All of these types of absorbers operate in fundamentally the same way. Sound is the vibratory energy of air particles, and by using absorbers, the vibratory energy can be dissipated in the form of heat. Thus, sound energy is reduced. The amount of heat generated from sound absorption is minuscule. It would take the sound energy of millions of people talking to brew a cup of tea, so we must abandon any ecological hopes of warming our homes with sound—even the sound of heated arguments.

Dissipation of Sound Energy

A sound wave S traveling in air strikes a concrete block wall covered with an acoustical material, as shown in Fig. 12-1. What happens to the energy it contains? As a sound wave travels through air, there is first a small heat loss E from air absorption that is appreciable only at higher audio frequencies. When the sound wave hits a wall, there is a reflected component A returned to the air from the surface of the acoustical material.

More interestingly, some of the sound penetrates the acoustical material represented by the shaded layer in Fig. 12-1. The direction of travel of the sound is refracted downward because the acoustical material is denser than air. There is heat lost F by the frictional resistance the acoustical material offers to the vibration of air particles. As the sound ray strikes the surface of the concrete blocks, two things happen: a component B is reflected, and the ray is also bent strongly downward as it enters the much denser concrete blocks. There is further heat loss G within the concrete blocks. As the ray travels on, getting weaker all the time, it strikes the concrete-air boundary and undergoes another reflection C and emerges with refraction D with heat lost (I, J, and K) in all three media.

The sound ray S experiences many complex events during its travel through this barrier, and every reflection and passage through air or acoustical material dissipates some of its original energy. The refractions bend the ray but do not necessarily dissipate heat. Fortunately, this minutia is not involved in practical absorption problems. We usually consider only the aggregate of these individual actions.

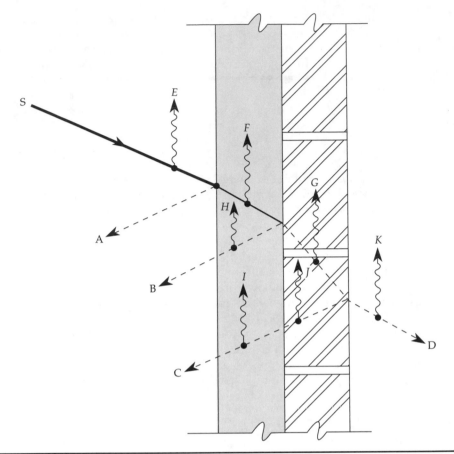

FIGURE 12-1   A sound ray impinging on an acoustical material on a masonry wall undergoes reflection from three different surfaces and absorption in the air and two different materials, with different degrees of refraction at each interface. In this chapter, the absorbed component is of chief interest.

Absorption Coefficients

Absorption coefficients are used to rate a material's effectiveness in absorbing sound. Absorption coefficients vary with the angle at which sound impinges upon the material. In an established diffuse sound field in a room, sound is traveling in every imaginable direction. In many calculations, we need sound absorption coefficients that are averaged over all possible angles of incidence. The random incidence absorption coefficient is a coefficient that is averaged over all incidence angles. This is usually referred to as the absorption coefficient of a material, designated as α. The absorption coefficient is a measure of the efficiency of a surface or material in absorbing sound. If 55% of the incident sound energy is absorbed at some frequency, the absorption coefficient α is said to be 0.55 at that frequency. A perfect sound absorber will absorb 100% of incident sound; thus α is 1.0. A perfectly reflecting surface would have α of 0.0.

Different references may use different symbols for the absorption coefficient; for example, *a* is sometimes used instead of α. Partly, this is because there are several different absorption coefficients. As noted, absorption varies according to the incident angle of sound striking the surface (absorption also varies according to frequency). One type of absorption coefficient measures absorption at a specific angle of incidence. Another type of absorption coefficient measures absorption from a diffuse sound field, that is, sound from a random distribution of angles. In this book, α refers to the absorption coefficient from a diffuse sound field (averaging all angles of incidence) at a given frequency. When an absorption coefficient at a particular angle is cited, it will be referred to as α_θ, where θ is the angle of incidence.

The sound absorption A provided by a particular area of material is obtained by multiplying its absorption coefficient by the surface area of the material exposed to sound. Therefore:

$$A = S\alpha \tag{12-1}$$

where A = absorption units, sabins or metric sabins
$\quad\quad S$ = surface area, sq ft or sq m
$\quad\quad \alpha$ = absorption coefficient

Sound absorption A is measured in sabins, in honor of Wallace Sabine. An open window is considered a perfect absorber because sound passing through it never returns to the room. An open window would have an absorption coefficient of 1.0 and by definition; a 1-ft^2 open window provides 1 sabin of sound absorption. Ten square feet of open window would thus give 10 sabins of absorption. As another example, suppose that carpet has an absorption coefficient of 0.55; 20 ft^2 of carpet would therefore provide 11 sabins of absorption.

Either sabins or metric sabins can be used. A metric sabin is the absorption from a 1-m^2 open window. Since 1 m^2 equals 10.76 ft^2, 1 metric sabin equals 10.76 ft sabins. Or, 1 sabin = 0.093 metric sabin.

When calculating a room's total absorption, all the materials in the room, according to their area, will contribute to the total absorption:

$$\Sigma A = S_1\alpha_1 + S_2\alpha_2 + S_3\alpha_3 + \cdots \tag{12-2}$$

where $S_1, S_2, S_3 \ldots$ = surface areas, sq ft or sq m
$\quad\quad \alpha_1, \alpha_2, \alpha_3 \ldots$ = respective absorption coefficients

Moreover, the average absorption coefficient can be calculated by dividing the total absorption by the total surface area:

$$\alpha_{average} = \frac{\Sigma A}{\Sigma S} \tag{12-3}$$

When absorptive material is placed over a surface, one must take into account the loss of absorption provided by the original surface. The net increase in absorption over an area is the absorption coefficient of the new material minus the coefficient of the original material.

The absorption coefficient of a material varies with frequency. Coefficients are typically published at the six standard frequencies of 125; 250; 500; 1,000; 2,000; and 4,000 Hz. In some cases, a material's absorption is given as a single-number rating known as the noise reduction coefficient (NRC). The NRC is the average of the coefficients for 250; 500; 1,000; and 2,000 Hz (125 and 4,000 Hz are not used). It is important to remember that the NRC is an average value, and also only accounts for absorption at middle frequencies. NRC is therefore most useful for speech applications. When considering wider-band music, individual coefficients, at a wider range of frequencies, should be used.

In some cases, the sound absorption average (SAA) is used to specify absorption. Like the NRC, the SAA is an arithmetic average, but the SAA uses absorption coefficients from twelve $\frac{1}{3}$-octave bands from 200 Hz to 2.5 kHz. These coefficients are averaged to obtain the SAA value. Finally, the ISO 11654 standard specifies a single-number weighted sound absorption coefficient for materials using the ISO 354 testing standard.

Reverberation Chamber Method

The reverberation chamber method can be used to determine the absorption coefficients of absorbing materials; it measures the average value. This chamber is a large room (perhaps 9,000 ft³) with highly reflective walls, floor, and ceiling. The reverberation time of the room is very long, and the longer it is, the more accurate the measurement. A standard sample of the material to be tested, usually measuring 8 × 9 ft, is mounted on the floor and the reverberation time measured. Comparing this time with the known reverberation time of the empty room yields the number of absorption units the sample adds to the room. From this, the absorption attributed to 1 ft² of material is determined, giving the equivalent of the absorption coefficient in sabins (1 m² of material gives absorption units in metric sabins).

The construction of the chamber is important to ensure a large number of modal frequencies and to equalize mode spacing as much as possible. The position of the sound source and the number and position of the measuring microphones must be considered. Large rotating vanes are used to ensure adequate diffusion of sound. Absorption coefficients supplied by manufacturers for use in architectural acoustic calculations may be measured by the reverberation chamber method.

A square-foot open window is the perfect absorber with coefficient equal to 1.0, but some chamber measurements can show absorption coefficients greater than 1. This is because the diffraction of sound from the edges of the standard sample makes the sample appear, acoustically, of greater area than it really is. There is no standard method of making adjustments for this artifact. Some manufacturers publish the actual measured values if greater than unity; others arbitrarily adjust the values down to unity or to 0.99.

Impedance Tube Method

The impedance tube (also called a standing-wave tube or Kundt tube) has been applied to the measurement of the absorption coefficient of materials. For this application, it can quickly and accurately determine coefficient values. It also has the advantage of small size, modest demands in terms of supporting equipment, and it requires only a small sample. This method is primarily used for porous absorbers because it is not suited to those absorbers that depend on area for their effect such as vibrating panels and large slat absorbers.

The construction and operation of the impedance tube are illustrated in Fig. 12-2. The tube usually has a circular cross section with rigid walls. The sample to be tested is cut to fit snugly into the tube. If the sample is intended to be used while mounted on a solid surface, it is placed in contact with the heavy backing plate. If the material is to be used with a space behind it, it is mounted an appropriate distance from the backing plate.

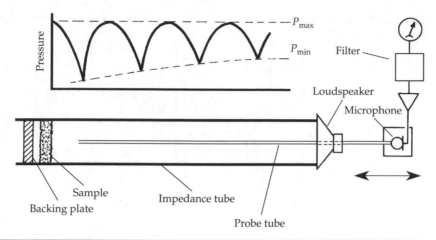

FIGURE 12-2 The standing-wave tube method of measuring the absorption coefficient of absorbing materials at normal incidence.

At the other end of the tube is a small loudspeaker with a hole drilled through its magnet to accommodate a long, slender probe tube coupled to a microphone. Energizing the loudspeaker at a given frequency sets up standing waves due to the interaction of the outgoing wave with the wave reflected from the sample. The form of this standing wave gives important information on the absorption of the material under test.

The sound pressure is maximal at the surface of the sample. As the microphone probe tube is moved away from the sample, the sound pressure falls to the first minimum. Successive, alternating maxima and minima will be detected as the probe tube is further withdrawn. If n is the ratio of the maximum sound pressure to its adjacent minimum, the normal (perpendicular) absorption coefficient α_n is equal to:

$$\alpha_n = \frac{4}{n + (1/n) + 2}$$ (12-4)

This is plotted in Fig. 12-3

The advantages of the standing-wave tube method are offset by the disadvantage that the absorption coefficient so determined is accurate only for normal incidence. In practice, sound impinges on the surface of a material at all angles. Figure 12-4 is a graph for approximately obtaining the random-incidence absorption coefficient from the normal-incidence absorption coefficient as measured by the standing-wave method. The random-incidence coefficients are always higher than normal-incidence coefficients.

Individual reflections affect the timbre of sound. In particular, specific normal reflections, called early sound, are of special interest. Although random-incidence coefficients are of interest in room reverberation calculations, for image control problems normal-incident reflection coefficients are generally required.

Tone-Burst Method

Short pulses of sound can be used to perform anechoic acoustical measurements in ordinary rooms. It takes time for reflections from walls and other surfaces to arrive at the measuring position. With a short-duration pulse, the time gate can be opened only

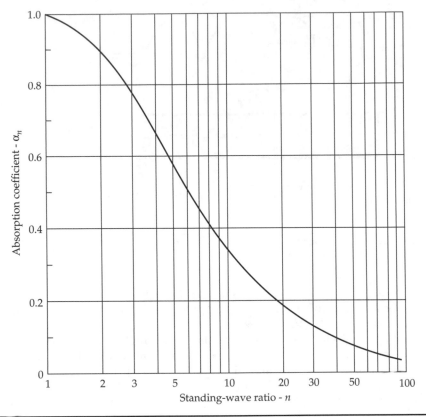

FIGURE 12-3 A graph for interpreting the standing-wave ratio in terms of the absorption coefficient. The standing-wave ratio can be found by dividing any pressure maximum by its adjacent pressure minimum (see Fig. 12-2).

for the desired sound pulse, shutting out the interfering pulses. This tone-burst method can be used to measure the sound-absorption coefficient of a material at any desired angle of incidence.

Such an arrangement is illustrated in principle in Fig. 12-5. The source-microphone system is calibrated at distance x, as shown in Fig. 12-5A. The geometry of Fig. 12-5B is then arranged so that the total path of the pulse reflected from the material to be tested is equal to this same distance x. The strength of the reflected pulse is then compared to that of the unreflected pulse at distance x to determine the absorption coefficient of the sample.

Mounting of Absorbents

The method of mounting the test sample on the reverberation chamber floor is intended to mimic the way the material is used in practice. Table 12-1 lists the standard mountings, both in the ASTM form, and in the older ABPMA form.

The mounting method has a major effect on the absorption characteristics of the material. For example, the absorption of porous materials is much greater with an air-space between the material and the wall. The quarter-wavelength ($\lambda/4$) rule dictates

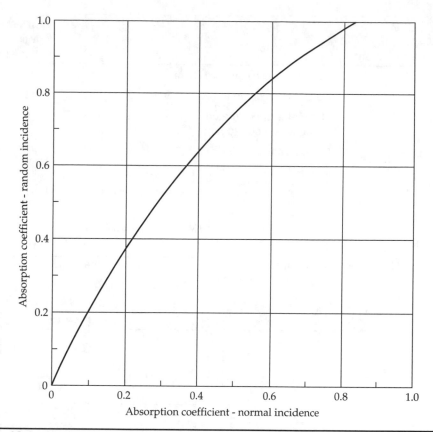

FIGURE 12-4   An approximate relationship between absorption coefficients at normal incidence and those with random incidence.

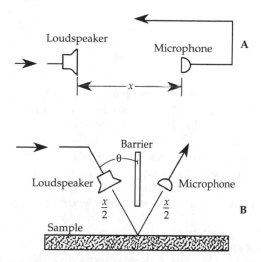

FIGURE 12-5   Determining the absorption coefficients of materials by a tone-burst method. (A) The source-microphone system is calibrated at distance x. (B) The total path length of the pulse reflected from the material under test is equal to distance x.

ASTM Mounting Designation*		ABPMA Mounting Designation†
A	Material directly on hard surface	#4
B	Material cemented to plasterboard	#1
C-20	Material with perforated, expanded or other open facing furred out 20 mm (³/₄")	#5
C-40	Ditto, furred out 40 mm (1¹/₂")	#8
D-20	Material furred out 20 mm (³/₄")	#2
E-400	Material spaced 400 mm (16") from hard surface	#7

*American Society for Testing and Materials (ASTM) designation: E 795-83.
†Mountings formerly listed by ABPMA (Acoustical and Board Products Manufacturers Association).

TABLE 12-1   Mountings Commonly Used in Sound-Absorption Measurements

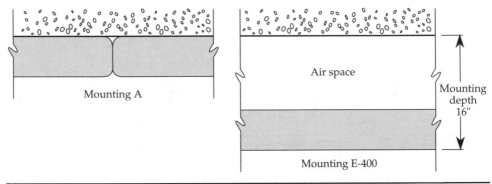

FIGURE 12-6   Commonly used standard mountings associated with listings of absorption coefficients. With mounting A the material is flat against the backing. Mounting E-400 applies to suspended ceilings with lay-in panels. (See Table 12-1.)

that a porous absorber for normal incidence must be at least a quarter-wavelength thick at the frequency of interest. For example, for a frequency of 1 kHz, the minimum absorber thickness should be about 3.4 in. Tables of absorption coefficients should always identify the standard mounting or include a description of the way the material was mounted during the measurements. Otherwise, the coefficients are of little value. Mounting type A with no airspace between the sound absorbing material and the wall is widely used. Another method commonly used is mounting type E-400, which is an approximation to the varying spaces encountered in suspended ceilings, as shown in Fig. 12-6.

Mid/High-Frequency Absorption by Porosity

Materials that have a porous composition, that is, with interstices between their matters, can operate as porous sound absorbers. The key word in this discussion of porous absorbers is interstices, the small cracks or spaces in porous materials. If a sound wave strikes a wad of cotton batting, the sound energy causes the cotton fibers to vibrate. The fiber amplitude will never be as great as the air particle amplitude of the sound wave because of frictional resistance. Restricted as this motion is, some sound energy is changed to frictional heat as fibers are set in motion. The sound penetrates into the interstices of the cotton, losing energy as fibers are vibrated. Cotton and many open-cell foams (such as polyurethane and polyester) are excellent sound absorbers because of their open-cell porosity that allows sound waves to penetrate the material. On the other hand, closed-cell materials (such as polystyrene), such as some used for thermal insulation, do not allow sound to penetrate the material, and are relatively poor absorbers. The better the air flow through a porous material, the better its ability to absorb sound.

Porous absorptive materials most commonly used as sound absorbers are usually fuzzy, fibrous materials in forms such as boards, foams, fabrics, carpets, and cushions. If the fibers are too loosely packed, there will be little energy lost as heat. On the other hand, if fibers are packed too densely, penetration suffers and the air motion cannot generate enough friction to be effective. Between these extremes are many materials that are very good absorbers of sound. These are commonly composed of cellulose or mineral fiber. Their effectiveness depends on the thickness of the material, the airspace, and the density of the material.

The absorption efficiency of materials that depend on the trapping and dissipating of sound energy in tiny pores can be seriously impaired if the surface pores are filled so that penetration is limited. Coarse concrete block, for example, has many such pores and is a fair absorber of sound. Painting may fill the surface pores and greatly reduce sound penetration, and thus absorption. However, if spray painted with nonbridging paint, the absorption may be reduced only modestly. Acoustical tile painted at the factory minimizes the problem of reduced absorption. Under certain conditions, a painted surface can reduce porosity but act as a diaphragm that might actually become a fair absorber on a different principle, that of a damped vibrating diaphragm.

In some rooms, the acoustical treatment may prescribe an overuse of carpeted floors and drapes, which emphasizes a shortcoming of most porous absorbers—that of poor low-frequency absorption. Tiles of cellulose fiber with perforated faces are also deficient in low-frequency absorption. Overly enthusiastic use of porous absorbers causes overabsorption of high frequency sound energy, without addressing a major problem of room acoustics, low-frequency standing waves.

To show the general similarity of the absorption characteristics of sound absorbers depending on porosity for their effectiveness, a comparison is made in Fig. 12-7. The acoustical tile, drapes, and carpet show highest absorption above 500 Hz and relatively low absorption in the low-frequency region dominated by room modes. Coarse concrete blocks show a typical high-frequency absorption peak, but also a more unusual absorption peak around 200 Hz.

Glass Fiber Insulation Materials

Great quantities of glass fiber materials are used in the acoustical treatment of recording studios, control rooms, and public spaces. These glass fibers can consist of both special, high-density materials, and ordinary, low-density building insulation. There is evidence that thicker panels with a rear air cavity are preferred over thin panels, and low density is also preferred over high density. In wood or steel stud single-frame walls, staggered stud walls, and double walls, thermal insulating batts are commonly used.

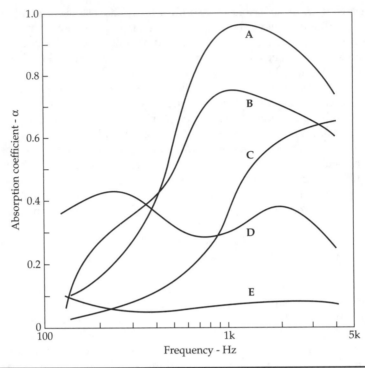

Figure 12-7 Sound absorption coefficients of typical porous materials A, B, and C show a similarity in general shape. Good high-frequency absorption and poor low-frequency absorption characterize porous absorbers. (A) High-grade acoustical tile. (B) Medium weight (14 oz/yd²) velour draped to half area. (C) Heavy carpet on concrete without padding. (D) Coarse concrete blocks, unpainted. (E) Coarse concrete blocks, painted.

This material usually has a density of about 1 lb/ft³. Such material is often identified as R-11, R-19, or other such numbers. These R-prefix designations describe thermal insulating qualities, but are related to thickness. The thickness of R-8 is 2.5 in, R-11 is 3.5 in, and R-19 is 6 in.

Glass fiber insulation commonly comes with a Kraft paper backing. Between walls, this paper has no significant effect, but if building insulation is used as a sound absorber on walls, perhaps behind a fabric facing, the paper becomes significant. Figure 12-8 compares the sound absorption efficiency of R-19 (6-in) and R-11 (3.5-in) glass fiber with the Kraft paper backing exposed and with the glass fiber exposed to the incident sound. When the paper is exposed, it shields the glass fiber from sound above 500 Hz but has little effect below 500 Hz. The net effect is an absorption peak at 250 Hz (R-19) and 500 Hz (R-11), which may be important in room treatment. With insulation exposed, there is essentially perfect absorption above 250 Hz (R-19) or 500 Hz (R-11).

Glass fiber insulation is an excellent and inexpensive absorbing material. When used in panels, a cosmetic and protective cover is usually required, but this is true of denser materials as well. Fabric, expanded metal, metal lath, hardware cloth, or perforated vinyl wall covering can be used as a cover. Absorption coefficients greater than 1.0 can be obtained by glass fiber panels.

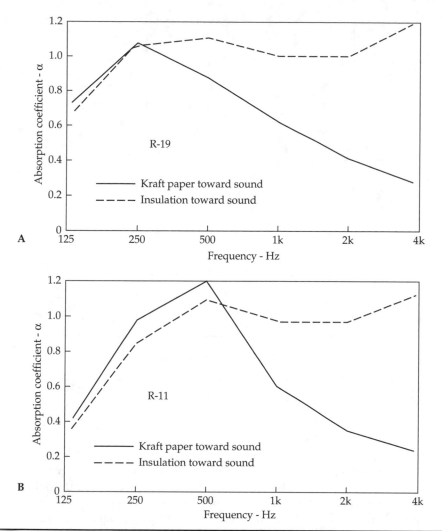

FIGURE 12-8   When ordinary building insulation is used as a wall treatment (perhaps with fabric facing), the position of the Kraft paper backing becomes important. (A) R-19 glass fiber building insulation. (B) R-11 glass fiber building insulation with mounting A.

Glass Fiber: Boards

Semirigid boards of glass fiber can be used in the acoustical treatment of audio rooms. This type of glass fiber is usually of greater density than glass-fiber building insulation. Typical of such materials are Johns-Manville 1000 Series Spin-Glass and Owens-Corning Type 703 Fiberglas, both of 3 lb/ft^3 density. These are available in different thicknesses (e.g., 1 to 4 in) yielding different R values (e.g., 4.3 to 17.4). Other densities are available; for example, Type 701 has a density of 1.5 lb/ft^3 and Type 705 has a density of 6 lb/ft^3. These semirigid boards of glass fiber do not excel cosmetically; hence they are usually covered with fabric. However, they do excel in sound absorption and are widely used for room treatment.

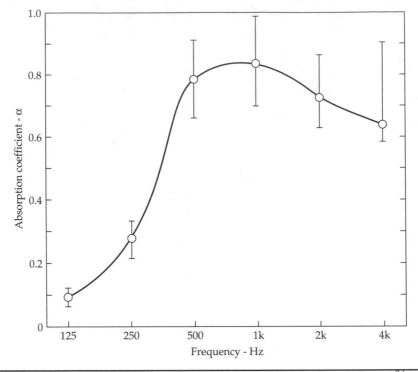

FIGURE 12-9   The average absorption characteristics of eight acoustical tile brands of $^3/_4$-in thickness. The vertical lines show the spread of the data.

Glass Fiber: Acoustical Tile

Manufacturers of acoustical materials offer competitive lines of 12×12 in acoustical tiles. Surface treatments of the tiles include even-spaced holes, random holes, slots, or fissured or other special textures. They are available from local building material suppliers. Such tiles are reputable products for noise and reverberation control provided they are used with full knowledge of their limitations. One of the problems of using acoustical tile in critical situations is that absorption coefficients are not always available for a specific tile. The average of the coefficients for eight cellulose and mineral fiber tiles of $^3/_4$ in thickness is shown in Fig. 12-9. The range of the coefficients is indicated by the vertical lines. The average points could be used for $^3/_4$ in tile for which no coefficients are available. Coefficients 20% lower would be a fair estimate for $^1/_2$ in tiles. When acoustical tiles are used in a drop ceiling, they function primarily as porous absorbers. However, the plenum airspace above the tiles causes them to also perform somewhat as panel absorbers.

Effect of Thickness of Absorbent

It is logical to expect greater sound absorption from thicker porous materials, but this logic holds primarily for lower frequencies. Absorption is greatest when the porous material is placed at a distance of a quarter wavelength ($\lambda/4$) from a hard reflective surface (or odd multiples of this dimension); this is the point where particle velocity is greatest; in practice, this can be difficult to do. Figure 12-10 shows the effect of varying absorbent thickness

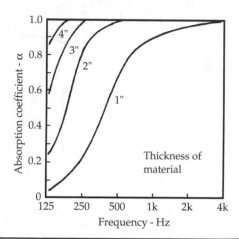

Figure 12-10   The thickness of glass-fiber sound-absorbing material determines the low-frequency absorption (density, 3 lb/ft³). The material is mounted directly on a hard surface.

where the absorbent is mounted directly on a solid surface (mounting type A). There is little difference above 500 Hz in increasing the absorbent thickness from 2 to 4 in, but there is considerable improvement below 500 Hz as thickness is increased. There is also a proportionally greater gain in overall absorption in a 1-in increase of thickness in going from 1 to 2 in than going from 2 to 3 in or 3 to 4 in. A 4-in thickness of glass fiber material of 3-lb/ft³ density has essentially perfect absorption over the 125-Hz to 4-kHz region.

Effect of Airspace behind Absorbent

Effective low-frequency absorption can also be achieved by spacing the porous absorbent out from the wall. A spaced porous absorber can be as effective as a non-spaced absorber of the same thickness. This is an inexpensive way to get improved performance—within limits. Figure 12-11 shows the effect on the absorption coefficient

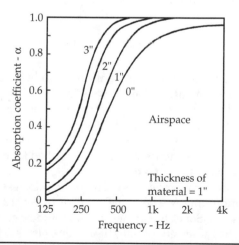

Figure 12-11   The low-frequency absorption of 1-in glass fiber board is improved materially by spacing it out from the solid wall.

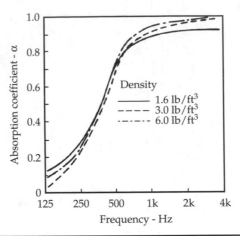

Figure 12-12   The density of glass-fiber absorbing material has relatively little effect on absorption in the range of 1.5 to 6 lb/ft³. The material is mounted directly on solid wall.

of furring 1-in glass-fiber wallboard out from a solid wall. Spacing 1-in material out 3 in makes its absorption approach that of the 2-in material of Fig. 12-10 mounted directly on the wall.

Effect of Density of Absorbent

Glass fiber and other materials come in various densities ranging from soft thermal insulation batts to semirigid and rigid boards. All of these have their proper place in acoustical treatment of spaces. Generally sound is able to penetrate the interstices of a high-density, harder surface material as well as a soft type. Figure 12-12 shows that there is relatively little difference in absorption coefficient as the density is varied over a range of almost 4 to 1. For very low densities, the fibers are so widely spaced that absorption suffers. For extremely dense boards, the surface reflection is high and sound penetration low.

Open-Cell Foams

Flexible polyurethane and polyester foams are widely used in noise quieting of automobiles, machinery, aircraft, and in various industrial applications. Foams also find application as sound absorbers in architectural applications, including recording studios and home listening rooms. Figure 12-13 is a photograph of one form of Sonex, a foam product contoured to simulate the wedges used in anechoic rooms. These are shaped in male and female molds and come in meshed pairs. This material can be glued or stapled to the surface to be treated.

The sound absorption coefficients of Sonex for thicknesses of 2, 3, and 4 in are shown in Fig. 12-14 for mounting A. The 2-in glass fiber of Fig. 12-10 is considerably superior acoustically to the 2-in Sonex but some factors should be considered in this comparison: (1) The Type 703 has a density of 3 lb/ft³ while Sonex is 2 lb/ft³. (2) In 2-in Sonex, the wedge height and the average thickness is far less, while the 703 thickness prevails throughout. (3) Comparing the two products is, in a sense, specious because the higher

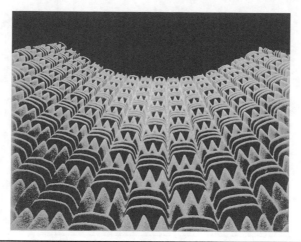

Figure 12-13   Sonex contoured acoustical foam simulating anechoic wedges. This is an open-cell type of foam.

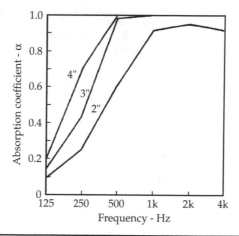

Figure 12-14   Absorption coefficients of Sonex contoured foam of various thicknesses.

cost of Sonex is justified in the minds of many by appearance and ease of mounting rather than straight acoustical considerations.

Drapes as Sound Absorbers

Drapes are a porous type of sound absorber because air can flow through the fabric. This penetration allows absorption to occur. Variables affecting absorbency include type and weight of material, degree of fold, and distance from the wall. The heavier the fabric, the greater the sound absorption. Heavy velour drapes can provide good absorption, while lightweight drapes provide practically no absorption. Figure 12-15 compares the absorption of 10, 14, and 18 oz/yd^2 velour hung straight and presumably at some distance from the wall. The greater absorption in going from 14 to 18 oz/yd^2 than in going from 10 to 14 oz/yd^2 is difficult to explain. The effect is concentrated in the 500- to 1,000-Hz region.

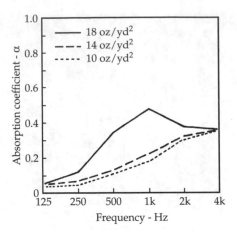

FIGURE 12-15   The sound absorption of velour hung straight for three different weights of fabric. (*Beranek*)

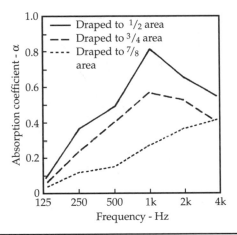

FIGURE 12-16   The effect of draping on the sound absorption of drapes. "Draped to ½ area" means that folds are introduced until the resulting drape area is half that of the straight fabric. (*Mankovsky*)

The higher the degree of folding, the greater the absorption. This is mainly because folds increase the amount of surface area presented to sound. The effect is shown in Fig. 12-16. The "draped to $7/_8$ area" means that the entire $8/_8$ area is drawn in only slightly ($1/_8$) from the flat condition. As can be seen, the deeper the drape fold, the greater the absorption.

The distance a drape is hung from a reflecting surface can have a great effect on its absorption efficiency. In Fig. 12-17A, a drape or other porous material is hung parallel to a solid wall, and the distance *d* between the two is varied. The frequency of the sound impinging on the porous material is held constant at 1 kHz. If the sound absorption provided by the porous material is measured, we find that it varies greatly as the distance *d* from the wall is changed. Examination reveals that the wavelength of the sound is related to maxima and minima of absorption. The wavelength of sound λ is the speed of sound divided by frequency, which in the case of 1,000 Hz, is 1,130/1,000 = 1.13 ft or

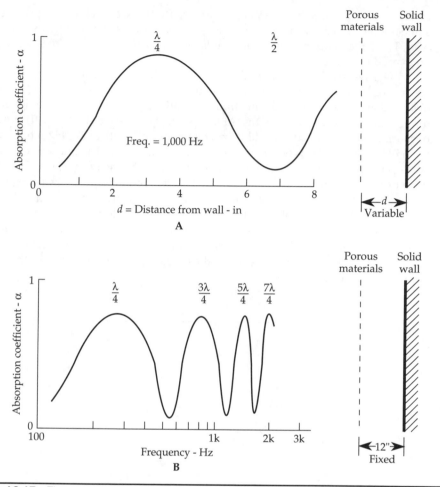

FIGURE 12-17   The sound absorption of porous material such as a drape varies with the distance from a hard wall. (A) The maximum absorption is achieved when the drape is one-quarter wavelength from the wall, the minimum at a half wavelength. (B) The sound absorption of porous material hung at a fixed distance from a wall will show maxima at spacings of a quarter wavelength and odd multiples of a quarter wavelength.

about 13.6 in. A quarter wavelength is 3.4 in, and a half wavelength is 6.8 in. There are absorption peaks at one-quarter wavelength, and if we carry it further than indicated in Fig. 12-17A, peaks are present at each odd multiple of quarter wavelengths. Absorption minima occur at even multiples of quarter wavelengths.

This effect is explained by reflections of the sound from the solid wall. At the wall surface, pressure will be highest, but air particle velocity will be zero because the sound waves cannot move the wall. However, at a quarter wavelength from the wall, pressure is zero, and air particle velocity is maximum. By placing the porous material, such as a drape, a quarter wavelength from the wall, it will have maximum absorbing effect at that corresponding frequency because the particle velocity is maximum at the drape and the greatest frictional losses occur. The same effect occurs at odd multiples of $\lambda/4$, such as $3\lambda/4$, $5\lambda/4$, $7\lambda/4$, and so on. At a half-wavelength distance, particle velocity is at a minimum, hence absorption is minimum. In practice, because drapes usually have some degree of fold, the quarter-wavelength point occurs at different distances from the wall. Thus, the frequency peaks and nulls tend to be broadened, with less specific effect on overall response.

In Fig. 12-17B, the spacing of the drape from the wall is held constant at 12 in as the absorption is measured at different frequencies. The same variation of absorption is observed: maximum when spacing from the wall is at odd quarter wavelengths and minimum at even quarter wavelengths. At this particular spacing of 12 in (1 ft), a wavelength of spacing occurs at $1,130/1 = 1,130$ Hz, a quarter wavelength at 276 Hz, a half wavelength at 565 Hz, and so on. When referring to quarter wavelengths, sine waves are inferred. Absorption measurements are invariably made with bands of random noise. Hence we must expect the variations of Fig. 12-17B to be averaged by the use of such bands.

Figure 12-18 shows reverberation chamber measurements of the absorption of 19 oz/yd² velour. The solid plot is presumably for a drape well removed from any walls. The other plots, very close together, are for the same material spaced about 4 and 8 in from the wall. The 4-in distance is one wavelength at 3,444 Hz; the 8-in distance is at 1,722 Hz. The odd multiples of both the 4- and 8-in quarter wavelengths are seen on the upper part of Fig. 12-18.

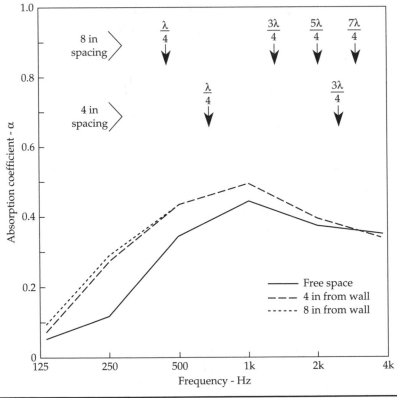

Figure 12-18 Measurements of sound-absorption coefficients of a velvet material (19 oz/yd²) in free space, and about 4 in and 8 in from a solid wall. Also indicated are the points at which absorption will increase due to wall reflection. (*Mankovsky*)

The absorption of the velour is greater when spaced from the wall, and the effect is greatest in the 250- to 1,000-Hz region. At 125 Hz, the 10- and 20-cm spacing adds practically nothing to the drape absorption because at 125 Hz, the quarter wavelength spacing is 2.26 ft.

Carpet as Sound Absorber

Carpet commonly dominates the acoustical picture in many types of spaces. It is the one amenity the owner often specifies in advance and the reason is more often comfort and appearance than acoustic. Carpet and its underlay can provide significant absorption at mid and high frequencies. Suppose that carpet is placed in a recording studio with a floor area of 1,000 ft^2. Further, suppose that reverberation time is specified to be about 0.5 second, which requires 1,060 sabins of absorption in this room. At higher audio frequencies, a heavy carpet and pad with an absorption coefficient of around 0.6 gives 600 sabins of absorption at 4 kHz or 57% of the required absorption for the entire room before the absorption needs of walls and ceiling are even considered. The acoustical design is quite limited before it is started.

There is another, more serious problem. This high absorbance of carpet is only at higher audio frequencies. Carpet having an absorption coefficient of 0.60 at 4 kHz may offer only 0.05 at 125 Hz. In other words, the 1,000 ft^2 of carpet introduces 600 sabins at 4 kHz but only 50 sabins at 125 Hz. This is a major problem encountered in many acoustical treatments. The unbalanced absorption of carpet can be compensated in other ways, principally with resonant-type, low-frequency absorbers.

To compound the problem of unbalanced absorption of carpet, dependable absorption coefficients are hard to come by. An assortment of types of carpet and variables in underlay add to the uncertainty. Experience and judgment must be used when deciding which absorption coefficients should be employed, particularly in wall-to-wall carpeting.

Effect of Carpet Type on Absorbance

There are considerable variations in sound absorption between types of carpet. Figure 12-19 shows the difference between a heavy Wilton carpet and a velvet carpet with and without a latex backing. The latex backing seems to increase absorption materially above 500 Hz and to decrease it a modest amount below 500 Hz.

Effect of Carpet Underlay on Absorbance

Foam rubber, sponge rubber, felts, polyurethane, or combinations are often used as the padding under a carpet. Foam rubber is made by whipping a latex water dispersion, adding a gelling agent, and pouring into molds. The result is always open-celled. Sponge rubber, on the other hand, formed by chemically generated gas bubbles, can yield either open or closed cells. Open cells provide the interstices required for good sound absorption while closed cells do not.

The influence of underlay on carpet absorption is significant. Figure 12-20 shows reverberation-chamber measurements of absorption coefficients for a single Axminster type of carpet with different underlay conditions. Graphs A and C show the effect of hair felt of 80 and 40 oz/yd^2 weight. Graph B shows an intermediate combination of hair felt and foam. While these three graphs differ considerably, they all

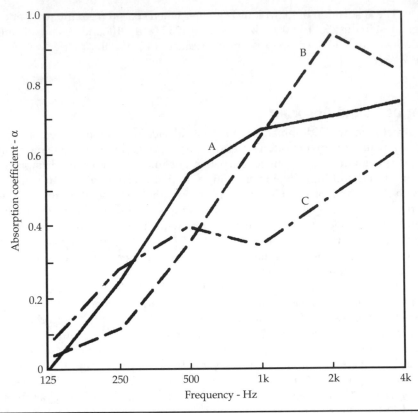

Figure 12-19 A comparison of the sound-absorption characteristics of three different types of carpet. (A) Wilton, pile height 0.29 in, 92.6 oz/yd². (B) Velvet, latex backed, pile height 0.25 in, 76.2 oz/yd². (C) The same velvet without latex backing, 37.3 oz/yd², all with 40-oz hair felt underlay. (*Harris*)

stand in contrast to graph D for the carpet laid directly on bare concrete. We conclude that the padding underneath the carpet contributes markedly to overall carpet absorption.

Carpet Absorption Coefficients

The absorption coefficients plotted in Figs. 12-19 and 12-20 are taken from Harris's 1957 paper, an exhaustive study of carpet characteristics then available. The coefficients in Fig. 12-21 and included in the appendix are plotted for comparison with Figs. 12-19 and 12-20. Carpets vary widely, which can account for some of the great variability with which any designer of acoustical systems is confronted.

Sound Absorption by People

The people comprising a concert-hall audience account for a significant portion of the sound absorption of the room—perhaps up to 75% for a full house. It can also make an acoustical difference whether one or several people are in a small monitoring room. The

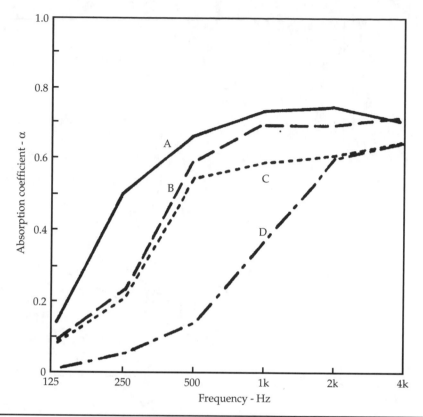

FIGURE 12-20 Sound-absorption characteristics of the same Axminster carpet with different underlay. (A) 80-oz hair felt. (B) Hair felt and foam. (C) 40-oz hair felt. (D) No underlay, on bare concrete. (*Harris*)

problem is how to rate human absorption and how to use it in calculations. One method uses the surface area over which an audience is seated. Or, we may simply consider the number of people present. In any case, the absorption units (sabins) due to people must be considered at each frequency and then added to the sabins of the carpet, drapes, and other absorbers in the room at each frequency. Table 12-2 lists the absorption per person of informally dressed college students in a classroom along with a range of absorption for more formally dressed people in an auditorium environment.

For 1 kHz and higher, the absorption offered by college students in informal attire in the furnishings of a classroom falls at the lower edge of the range of a more average audience. The low-frequency absorption of the students, however, is considerably lower than that of the more formally dressed people. The rule of thumb employed by some acousticians simply attributes 5 sabins at 500 Hz, per seated person.

Sound propagated across rows of people, as in an auditorium or concert hall, is subjected to an unusual type of attenuation. In addition to the normal decrease in sound with distance from the stage, there is an additional dip of up to 15 to 20 dB around 150 Hz and spreading over the 100- to 400-Hz region. In fact, this is not strictly an audience effect because it prevails even when the seats are empty. A similar dip in sound-pressure level affects important first reflections from the side walls. This apparently results from interference. The angle of incidence also plays a role. When an

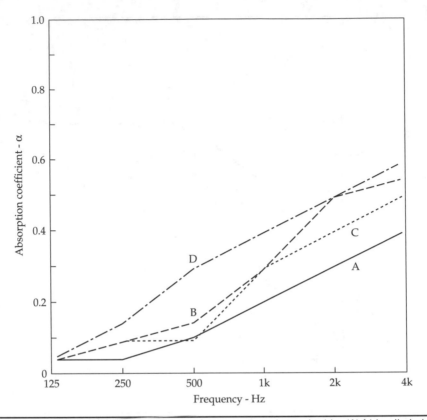

FIGURE 12-21 Carpet absorption coefficients from a commonly used table. (A) $\frac{1}{8}$-in pile height. (B) $\frac{1}{4}$-in pile height. (C) $\frac{3}{16}$-in combined pile and foam. (D) $\frac{5}{16}$-in combined pile and foam. Compare these graphs with those of Figs. 12-18 and 12-19.

	Frequency (Hz)					
	125	**250**	**500**	**1,000**	**2,000**	**4,000**
College students informally dressed seated in tablet arm chairs	NA	2.5	2.9	5.0	5.2	5.0
Audience seated, depending on spacing and upholstery of seats	2.5–4.0	3.5–5.0	4.0–5.5	4.5–6.5	5.0–7.0	4.5–7.0

TABLE 12-2 Sound Absorption by People (Sabins per Person)

audience is seated on a relatively flat floor, the angle of sound incidence is low and there is greater absorption. With higher angle of incidence (e.g., in stadium seating), there is less absorption.

Sound Absorption in Air

For frequencies of about 2 kHz and above and for large auditoriums, the absorption of sound by the air in the space becomes important. Air absorption could account for 20% to

25% of the total absorption in the space, a significant factor. Air absorption can be esti-mated from:

$$A_{air} = mV \qquad\qquad (12\text{-}5)$$

where m = air attenuation coefficient, sabins/ft^3 or sabins/m^3
$\quad\quad V$ = volume of room, ft^3 or m^3

The value of the air attenuation coefficient m varies with humidity. With humidity between 40% and 60%, the values of m at 2, 4, and 8 kHz are: 0.003, 0.008, and 0.025 sabins/ft^3 and 0.009, 0.025, and 0.080 sabins/m^3, respectively.

For example, a church seating 2,000 people has a volume of 500,000 ft^3. At 2 kHz and 50% relative humidity, the air absorption is 0.003 sabins/ft^3. This yields 1,500 sabins of absorption at 2 kHz.

Panel (Diaphragmatic) Absorbers

The absorption of sound at lower audible frequencies can be effectively achieved by resonant (or reactive) absorbers. Glass fiber and acoustical tiles are common forms of porous absorbers in which the sound energy is dissipated as heat in the interstices of the fibers. However, the absorption of glass fiber and other fibrous absorbers at low audio frequencies is quite poor. To absorb well, the thickness of the porous material must be comparable to the wavelength of the sound. At 100 Hz, the wavelength is 11.3 ft, and using any porous absorber approaching this thickness would be impractical. For this reason, resonant absorbers are often used to obtain absorption at low frequencies.

A mass suspended from a spring will vibrate at its natural frequency. Panels designed with an air cavity behind them act similarly. The mass of the panel and the springiness of the air in the cavity are together resonant at some particular frequency. Sound is absorbed as the panel is flexed because of the damping caused by frictional heat losses of the material within the panel. (Similarly, a mass on a spring will stop oscillating because of damping.) Absorption provided by panel absorbers is usually relatively modest because the resonant motion also radiates some sound energy. Panels made of limp materials with high damping provide greater absorption.

Damping increases as the velocity of the panel increases, and velocity is highest at the resonant frequency. There the absorption of sound is maximal at the frequency at which the structure is resonant. As noted, the enclosed and sealed air cavity behind the panel acts as spring; the greater the depth of the airspace, the less stiff the spring. Like-wise a smaller airspace acts as a stiff spring. The frequency of resonance for a flat, unperforated panel can be estimated from:

$$f_0 = \frac{170}{\sqrt{(m)(d)}} \qquad\qquad (12\text{-}6)$$

where f_0 = frequency of resonance, Hz
$\quad\quad m$ = surface density of panel, lb/ft^2 or kg/m^2
$\quad\quad d$ = depth of airspace, in or m

Note: In metric units, change 170 to 60.

For example, consider a piece of $\frac{1}{4}$-in plywood spaced out from the wall on 2 × 4 studs on edge, which gives close to a $3\frac{3}{4}$-in airspace behind. The surface density of $\frac{1}{4}$-in plywood, 0.74 lb/ft^2, can be measured or found in references. Substituting, we obtain a resonant frequency of about 102 Hz.

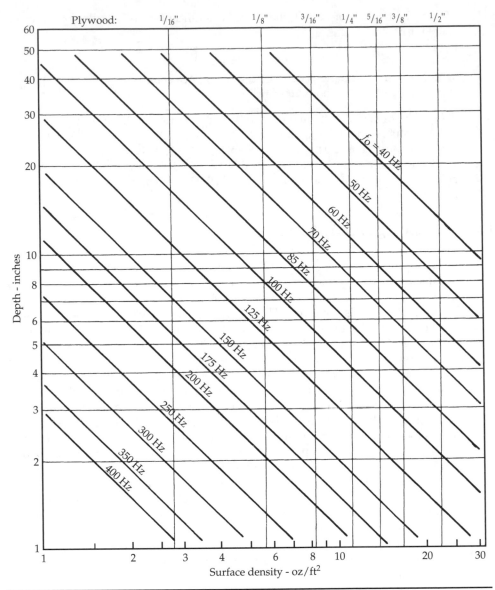

FIGURE 12-22   Design chart for resonant panel absorbers. (See also Fig. 12-34.)

Figure 12-22 shows panel resonant frequencies as a function of cavity depth in inches, and surface density in ounce per square foot. Knowing the thickness of the plywood and the depth of the space behind the plywood, the frequency of resonance can be read off the diagonal lines. Equation (12-6) applies to membranes and diaphragms of materials other than plywood such as Masonite, fiberboard, or even Kraft paper. For other than plywood, the surface density must be determined. The surface density is easily found by weighing a piece of the material of known area. The surface area of a panel absorber should be at least 5 ft^2.

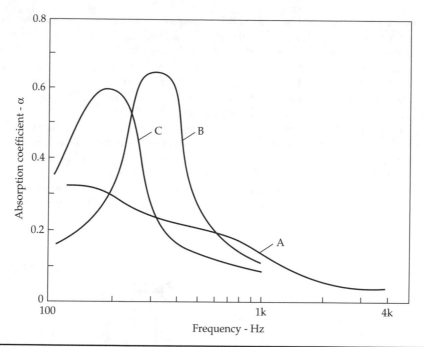

FIGURE 12-23   Absorption measurements of three panel absorbers. (A) $^3/_{16}$-in plywood with 2-in airspace. (B) $^1/_{16}$-in plywood with 1-in mineral wool and $^1/_4$-in airspace. (C) The same as (B) but for $^1/_8$-in panel.

How accurate are Eq. (12-6) and Fig. 12-22? Actual measurements on three plywood membrane absorbers are shown in Fig. 12-23. Graph A shows the simple case of $^3/_{16}$-in plywood panels on 2-in battens. We estimate that this structure is resonant at about 175 Hz. The peak coefficient is about 0.3 which is about as high as one can expect for such structures. Graph B is for $^1/_{16}$-in plywood with a 1-in glass fiber blanket and $^1/_4$-in airspace behind it. Graph C is the same except for a $^1/_8$-in panel. Note that the glass fiber filler has about doubled the peak absorption. The glass fiber has also shifted the peaks to a paint about 50 Hz lower. Such calculations of the frequency of peak absorption at resonance are not perfect, but they are a good first approximation of sufficient accuracy for most purposes.

Absorption is increased when the cavity is filled with a porous material such as glass fiber insulation. This is because the material increases damping. The material may be stuffed loosely inside the cavity or attached to the back of the panel. The panel is most effective when placed at a pressure maximum for the desired absorption frequency; this might be on an end wall, a midpoint, or a corner of a room. Panels are relatively ineffective when placed at pressure minima.

Some music rooms owe their acoustical excellence to the low-frequency absorption offered by extensive paneled walls. Plywood or tongue-and-groove flooring or sub flooring vibrates as a diaphragm and contributes to low-frequency absorption. Drywall construction on walls and the ceiling does the same thing. All such components of absorption must be included in the acoustical design of a room, large or small.

Drywall or gypsum board plays an important role in the construction of homes, studios, control rooms, and other spaces. Drywall absorbs sound by a flexural, diaphragmatic action, working as a resonant system. Drywall is particularly important in the absorption of low-frequency sound. Usually, such low-frequency absorption is welcome, but in larger spaces designed for music, drywall surfaces can absorb so much low-frequency sound as to prevent the achievement of the desired reverberate conditions. Drywall of $\frac{1}{2}$-in thickness on studs spaced at 16 in offers an absorption coefficient of 0.29 at 125 Hz and even higher at 63 Hz (which would be of interest in music recording studios). Drywall absorption in small audio rooms must be recognized and its low-frequency absorption included in calculations. This is sometimes difficult because different thicknesses of drywall are used, and the frequency of peak absorption varies according to thickness and airspace. A $\frac{1}{2}$-in drywall has a surface mass of 2.1 lb/ft^2 and double $\frac{5}{8}$-in drywall has a surface mass of 5.3 lb/ft^2. With an airspace of $3\frac{3}{4}$ in, the $\frac{1}{2}$-in panel resonates at 60.6 Hz, and the double $\frac{5}{8}$-in panel resonates at 38.1 Hz.

We have noted that porous materials commonly show their greatest absorption in the high-frequency region. Vibrating panel arrangements show their best absorption in the

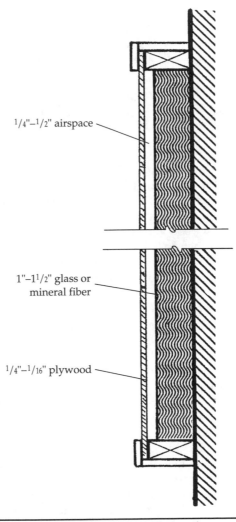

$\frac{1}{4}"–\frac{1}{2}"$ airspace

$1"–1\frac{1}{2}"$ glass or mineral fiber

$\frac{1}{4}"–\frac{1}{16}"$ plywood

Figure 12-24 Typical resonant panel absorber with wall mounting.

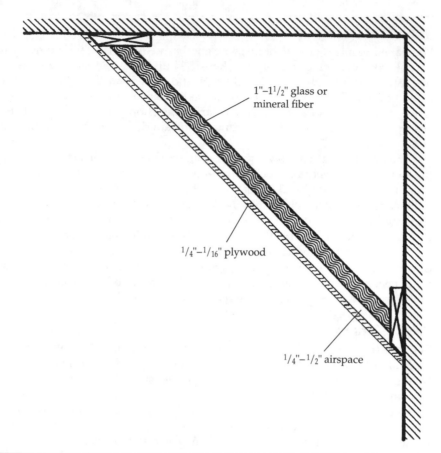

1"–1^1/$_2$" glass or
mineral fiber

^{1}/$_4$"–^{1}/$_{16}$" plywood

^{1}/$_4$"–^{1}/$_2$" airspace

FIGURE 12-25   Typical resonant panel absorber with either vertical or horizontal corner mounting.

low frequencies. In treating small listening rooms and studios we find that structures giving good low-frequency absorption are invaluable in controlling room modes.

Panel sound absorbers are quite simple to build. An example of a panel absorber to be mounted on a flat wall or ceiling surface is shown in Fig. 12-24. A ^{1}/$_4$- or ^{1}/$_{16}$-in plywood panel is fastened to a wooden framework to give the desired spacing from the wall. A glass or mineral fiber blanket of 1 to 1^1/$_2$ in is glued to the wall surface. An airspace of ^{1}/$_4$ or ^{1}/$_2$ in should be maintained between the absorbent and the rear surface of the plywood panel.

A corner panel absorber is shown in Fig. 12-25. For computations, an average depth is used. Depths greater and smaller than the average simply mean that the peak of absorbance is broader than that of an absorber with uniform depth. Spacing the absorber ^{1}/$_4$ to ^{1}/$_2$ in from the rear of the plywood panel is simple if a mineral fiber board such as Tectum is used. Using a flexible blanket of glass fiber requires support by hardware cloth, open-weave fabric, or expanded metal. For applications in which reflected mid/high-frequency sound from the panel absorber might create problems, a facing of glass fiber board would not interfere with the low-frequency absorbing action if it was spaced to avoid damping of the vibration of the plywood panel. All room modes terminate in the corners of a room. A corner panel absorber could be used to control such modes.

Polycylindrical Absorbers

Flat paneling in a room might do some good acoustically, but wrapping a plywood or hardboard skin around some semi-cylindrical bulkheads can provide some very attractive features. With polycylindrical elements (polys), it is acoustically possible to achieve a good diffuse field along with liveness and brilliance, factors that tend to oppose each other in rooms with flat surfaces. The larger the chord dimension, the better the bass absorption. Above 500 Hz there is little significant difference between the polys of different sizes.

The overall length of polys is rather immaterial, ranging in actual installations from the length of a sheet of plywood to the entire length, width, or height of a room. However, it is advisable to break up the cavity behind the poly skin with randomly spaced bulkheads. The polys of Fig. 12-26 incorporate such bulkheads.

Polys perform differently, depending on whether they are empty or filled with absorbent material. Figures 12-26C and D show the increase in bass absorption resulting from filling the cavities with absorbent. If needed, this increased bass absorption can be easily achieved by simply filling the polys with glass fiber. If the bass absorption is not needed, the polys can be used empty. This adjustability can be an important asset in the acoustical design of listening rooms and studios.

Poly Construction

The construction of polycylindrical absorbers is reasonably simple. A framework for vertical polys is shown in Fig. 12-27, mounted above a structure intended for a low-frequency slat absorber. The variable chord dimensions are apparent, as are the random placement of bulkheads so that cavities will be of various volumes, resulting in different natural cavity frequencies. It is desirable that each cavity be essentially airtight, isolated from adjoining cavities by well-fitted bulkheads and framework. Irregularities in the wall can be sealed with a nonhardening acoustical sealant. The bulkheads of each poly are cut to the same radius on a bandsaw. Sponge rubber weather stripping with an adhesive on one side is stuck to the edge of each bulkhead to ensure a tight seal against the plywood or hardboard cover. If such precautions are not taken, rattles and coupling between cavities can result.

The polys of Fig. 12-28 use $1/8$-in tempered Masonite as the poly skin. A few hints can simplify the job of stretching this skin. In Fig. 12-28, slots of a width to snugly fit the Masonite are cut along the entire length of strips 1 and 2 with a radial saw. Assume that poly A is already mounted and held in place by strip 1, which is nailed or screwed to the wall. Working from left to right, the next job is to mount poly B. First the left edge of Masonite sheet B is inserted in the remaining slot of strip 1. The right edge of Masonite sheet B is then inserted in the left slot of strip 2. If all measurements and cuts have been accurately made, swinging strip 2 against the wall should make a tight seal over the bulkheads 3 and weather stripping 4. Securing strip 2 to the wall completes poly B. Poly C is mounted in a similar fashion and so on to the end of the series of polys. In some designs, the axes of symmetry of the polys on the side wall are made perpendicular to those on the rear wall. If polys were used on the ceiling, their axes should be perpendicular to both the others.

It is practical and acceptable to construct each poly as an entirely independent structure rather than build them on the wall. Such independent polys can be spaced at will.

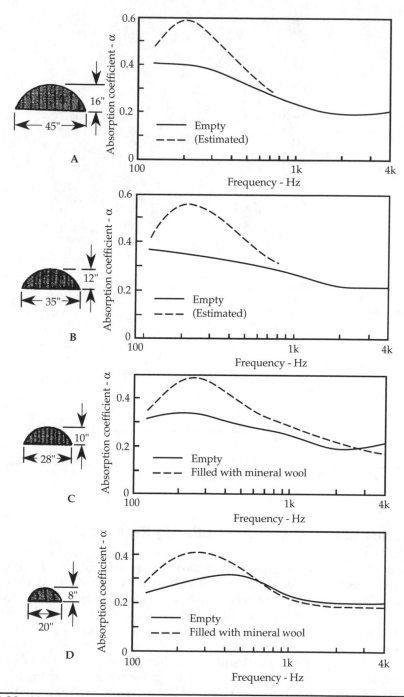

Figure 12-26 Measured absorption of polycylindrical absorbers of various chord and height dimensions. (A and B) Only empty poly data are available; the broken lines are estimated absorption when filled with mineral wool. (C and D) Both empty polys and polys filled with mineral wool. (*Mankovsky*)

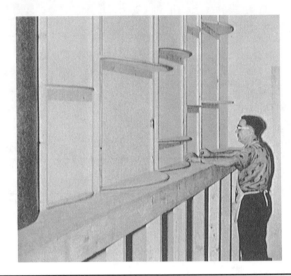

FIGURE 12-27   The construction of polys in a motion picture sound mixing studio. Foam rubber anti-rattle strips are applied on the edge of each bulkhead. Also note the random spacing of the bulkheads. (*Moody Institute of Science*)

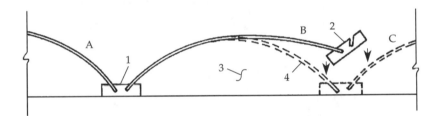

FIGURE 12-28   The method of stretching the plywood or hardboard skin over the poly bulkheads shown in Fig. 12-27.

Bass Traps: Low-Frequency Absorption by Resonance

The term bass trap describes many kinds of low-frequency sound absorbers, including panel absorbers, but perhaps the term should be reserved for a particular type of reactive cavity absorber. Sound absorption at the lowest octave or two of the audible spectrum is often difficult to achieve. The bass trap is commonly used in recording studio control rooms to reduce standing waves at these bass frequencies. A true bass trap is shown in Fig. 12-29. It is a box or cavity of critical depth and with a mouth opening of size to suit particular purposes. It is a tuned cavity with a depth of a quarter wavelength (point of maximum particle velocity) at the design frequency at which maximum absorption is desired.

The concept of wall reflection and quarter-wavelength depths (graphically portrayed for drapes in Fig. 12-17) also applies to bass traps. The sound pressure at the bottom of the cavity is maximal at the quarter-wavelength design frequency. The air particle velocity is zero at the bottom. At the mouth, the pressure is zero and the particle velocity is maximal, which results in two phenomena. First, a glass fiber semirigid board across the opening offers great friction to the rapidly vibrating air particles, resulting in maximum absorption at this frequency. In addition, the zero pressure at the

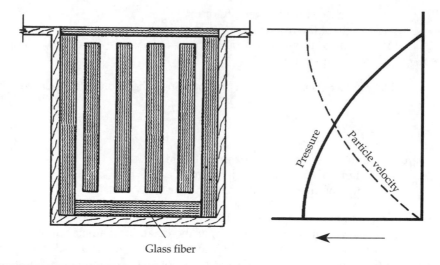

Glass fiber

FIGURE 12-29   The operation of a bass trap depends on reflections of sound from the bottom of the trap. The pressure for the frequency at which the depth is a quarter wavelength is maximal at the bottom and the particle velocity is zero at the bottom. At the mouth, the pressure is zero (or very low) and the particle velocity is maximal. Absorbent placed where the particle velocity is maximal will absorb sound very effectively. The same action occurs at odd multiples of the quarter wavelength.

opening constitutes a vacuum that acts like a sound sink. The bass trap's effect, then, is greater than its opening area would suggest.

The bass-trap effect, like the drape spaced from a reflective wall, occurs not only at a quarter-wavelength depth, but also at odd multiples of a quarter wavelengths. Large trap depths are required for very low bass frequencies. For example, a quarter wavelength for 40 Hz is 7 ft. Unused spaces above control room ceilings and between inner walls and outer shells are often used for trap space. The famous Hidley bass trap design is an example of this type of trap.

Helmholtz (Volume) Resonators

The Helmholtz type of resonator is widely used to achieve absorption at lower audio frequencies. The operation of resonator absorbers can be easily demonstrated. Blowing across the mouth of any bottle produces a tone at its natural frequency of resonance. The air in the cavity is springy, and the mass of the air in the neck of the bottle reacts with this springiness to form a resonating system, much as a weight on a spring vibrating at its natural period. Absorption is maximal at the frequency of resonance and decreases at nearby frequencies. The resonant frequency of a Helmholtz resonator with square opening is given by:

$$f_0 = \left(\frac{c}{2\pi}\right)\sqrt{\frac{S}{V(1 + 2\Delta l)}}$$

(12-7)

where c = speed of sound in air, 1130 ft/sec or 343 m/sec
        S = cross-sectional area of the resonator opening, ft² or m²

V = volume of the resonator, ft³ or m³

l = length of the resonator opening, ft or m

$2\Delta l$ = resonator mouth correction factor = $0.9a$, where a is the edge length of the square opening

For Helmholtz resonators with circular openings, the resonant frequency is given by:

$$f_0 = \frac{(30.5R)}{\sqrt{V(l+1.6R)}}$$

(12-8)

where R = radius of the circular opening, ft or m

V = volume of the resonator, ft³ or m³

l = length of the resonator opening, ft or m

Note: In metric units, change 30.5 to 100.

Change the volume of the air cavity, or the length or diameter of the neck, and the frequency of resonance changes. The width of this absorption band depends on the friction of the system. A glass bottle offers little friction to the vibrating air and would have a very narrow absorption band. Adding a bit of gauze across the mouth of the bottle or stuffing a wisp of cotton into the neck, the amplitude of vibration is reduced and the width of the absorption band is increased. For maximum effectiveness, Helmholtz absorbers should be placed in areas of high modal sound pressure for the tuned frequency.

To demonstrate the effectiveness of a continuously swept narrow-band technique of measuring absorption coefficients, Riverbank Acoustical Laboratories measured the absorption of Coca-Cola bottles. A tight array of 1,152 empty 10-oz bottles was arranged in a standard 8×9 ft space on the concrete floor of a reverberation chamber. A single, well-isolated bottle had an absorption of 5.9 sabins at its resonance frequency of 185 Hz, but with a bandwidth (between −3dB points) of only 0.67 Hz. Absorption of 5.9 sabins is about what a person would absorb at 1 kHz, or what 5.9 ft² of glass fiber (2 in thick, 3 lb/ft³ density) would absorb at midband. The sharpness of this absorption characteristic corresponds to a very high Q (quality factor) of $185/0.67 = 276$.

The sound impinging on a Helmholtz resonator that is not absorbed is reradiated. As the sound is reradiated from the resonator opening, it tends to be radiated in a hemisphere. This means that unabsorbed energy is diffused, and diffusion of sound is very desirable in a studio or listening room.

In Helmholtz resonators, we have acoustical artifacts that far antedate Helmholtz himself. Bronze jars have been found in ancient Greek and Roman open-air theaters. Large jars may have been used to absorb sound at lower frequencies. Groupings of smaller jars may have supplied sound absorption at higher frequencies. In medieval times, resonators were used in a number of churches in Sweden and Denmark. Pots like those of Fig. 12-30 were embedded in the walls, presumably to reduce low-frequency modes. Ashes have been found in some of the pots, perhaps introduced to lower the Q of the ceramic pot and thus broaden the frequency of its effectiveness.

Helmholtz resonators are often used in the form of acoustical blocks. These blocks are formed of concrete with an open slot facing the closed cavity. A two-cell unit would have two slots. In some cases, a metal divider is placed inside each cavity, or a porous absorber is placed in the slot inside the cavity. Figure 12-31 shows idealized square bottles with tubular necks forming a perforated face resonator. Stacking these bottles enhances the resonator action. Consider a box of length L, width W, and depth H with

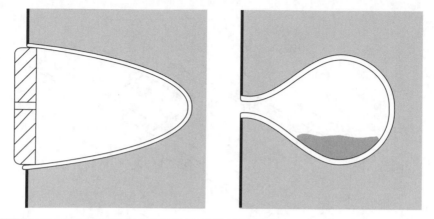

FIGURE 12-30 Pots embedded in the walls of medieval churches in Sweden and Denmark served as Helmholtz resonators, absorbing sound. Ashes, found in some of the pots, may have served as a dissipative agent. (*Brüel*)

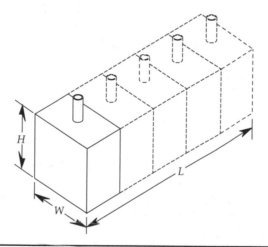

FIGURE 12-31 Development of perforated-face Helmholtz resonator from a single rectangular-bottle resonator.

a lid of thickness equal to the length of necks of the bottles. In this lid, holes are drilled having the same diameter as the holes in the neck. The partitions between each segment can be removed without greatly affecting the Helmholtz action.

Figure 12-32 illustrates a square bottle with an elongated slit neck. These can also be stacked in multiple rows. This design is similar to a slot-type resonator. The separating walls in the air cavity can also be eliminated without destroying the resonator action. A word of caution is in order, however. Subdividing the airspace can improve the action of perforated face or slit resonators but only because this reduces spurious, unwanted modes of vibration being set up within the air cavity.

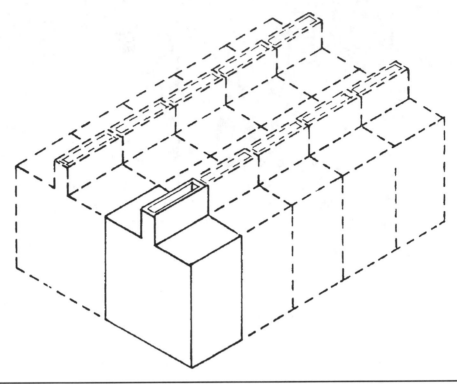

Figure 12-32 Design of a slot-type Helmholtz resonator from a single rectangular-bottle resonator.

Acoustical concrete masonry units such as the Cinderblox and its many derivatives have been in use since 1917. SoundBlox and SoundCell units, manufactured by Proudfoot, supply load-bearing ability and the mass required for sound isolation, and also enhanced low-frequency absorption through Helmholtz resonators formed by slots and cavities in the blocks.

Perforated Panel Absorbers

Perforated panels, for example, using hardboard, plywood, aluminum, or steel, spaced from the wall constitute a resonant type of sound absorber. Each hole acts as the neck of a Helmholtz resonator, and the share of the cavity behind belonging to that hole is comparable to the cavity of a Helmholtz resonator. In fact, we can view this structure as a host of coupled resonators. If sound arrives perpendicular to the face of the perforated panel, all the tiny resonators are in phase. For sound waves striking the perforated board at an angle, the absorption efficiency is somewhat decreased. This loss can be minimized by sectionalizing the cavity behind the perforated face with an egg crate type of divider of wood or corrugated paper.

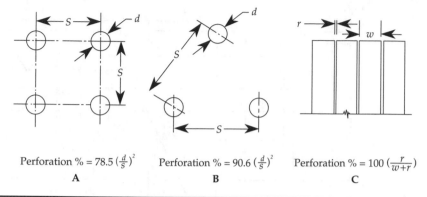

$$\text{Perforation \%} = 78.5 \left(\tfrac{d}{S}\right)^2 \qquad \text{Perforation \%} = 90.6 \left(\tfrac{d}{S}\right)^2 \qquad \text{Perforation \%} = 100 \left(\tfrac{r}{w+r}\right)$$

      A   **B**   **C**

FIGURE 12-33 Formulas for calculating perforation percentage for perforated panel resonators, including slat absorbers. (A and B) The perforation percentage for two types of circular-hole configurations. (C) The perforation percentage for slat absorbers.

The frequency of resonance of perforated panel absorbers with circular holes backed by a subdivided airspace is given approximately by:

$$f_0 = 200 \sqrt{\frac{p}{(d)(t)}} \qquad\qquad (12\text{-}9)$$

where f_0 = frequency of resonance, Hz
     p = perforation percentage
       = hole area divided by panel area $\times$ 100 (see Fig. 12-33A and B)
     t = effective hole length, in, with correction factor applied
       = (panel thickness) + (0.8) (hole diameter), in
     d = depth of airspace, in

There is some confusion in the literature concerning p, the perforation percentage. Some writers use the decimal ratio of hole area to panel area, rather than the percentage of hole area to panel area, introducing an uncertainty factor of 100. This perforation percentage is easily calculated by reference to Fig. 12-33. The perforation percentage for two types of circular-hole configurations is shown in Fig. 12-33A and B, and the percentage for slat absorbers (described below) is shown in Fig. 12-33C.

The frequency of resonance for perforated panels with circular holes presented in Eq. (12-9) is presented in graphical form in Fig. 12-34 for a panel thickness of $^3/_{16}$ in. Common pegboard with holes with diameter $^3/_{16}$ in, spaced on 1-in centers with the square configuration of Fig. 12-33 has 2.75% of the area in holes. If this pegboard is spaced out from the wall by 2 $\times$ 4 studs on edge, the system resonates at about 420 Hz and the peak absorption appears near this frequency.

In commonly available perforated materials, such as pegboard, the holes are so numerous that resonances at only higher frequencies can be obtained with practical airspaces. To obtain low-frequency absorption, the holes can be drilled by hand. Drilling $^7/_{32}$-in holes on 6-in centers gives a perforation percentage of about 1.0%. Of course, with zero perforation percentage, the panel behaves as a solid panel absorber.

Figure 12-35 shows the effect of varying hole area from 0.18% to 8.7% in a structure of otherwise fixed dimensions. The plywood is $^5/_{32}$-in thick perforated with

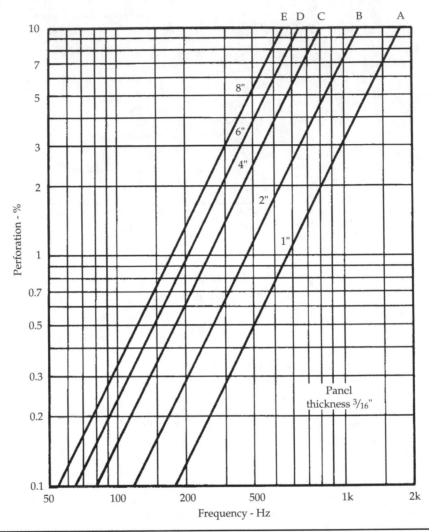

FIGURE 12-34   A graphic presentation of Eq. (12-9) relating percent perforation of perforated panels, the depth of airspace, and the frequency of resonance. The graphs are for a panel of $^3/_{16}$-in thickness. (See also Fig. 12-22.) (A) For 1-in furring lumber. The lines are drawn to correspond to furring lumber which is furnished; the line for 8 in is actually $7^3/_4$-in airspace. (B) For 2-in furring lumber. (C) For 4-in furring lumber. (D) For 6-in furring lumber. (E) For 8-in furring lumber.

$^3/_{16}$-in holes, except for the 8.7% case in which the hole diameter is about $^3/_4$ in. The perforated plywood sheet is spaced 4 in from the wall and the cavity is half filled with glass fiber and half is airspace.

Figure 12-36 is identical to Fig. 12-35, except that the perforated plywood is spaced 8 in and glass fiber of 4-in thickness is mounted in the cavity. The effect of these changes is a substantial broadening of the absorption curve.

It would be unusual to employ such perforated panel absorbers without acoustic resistance in the cavity in the form of glass fiber batts or boards. Without such resistance,

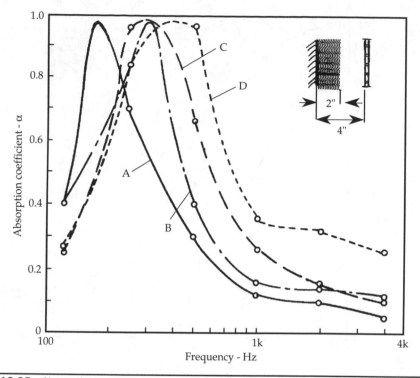

FIGURE 12-35 Absorption measurements of perforated panel absorbers of 4-in airspace, half filled with mineral wool and for panel thickness of $5/32$-in. (A) Perforation 0.18%. (B) Perforation 0.79%. (C) Perforation 1.4%. (D) Perforation 8.7%. The presence of the mineral wool shifts the frequency of resonance considerably from the theoretical values of Eq. (12-9) and Fig. 12-34. (*Data from Mankovsky*)

the absorption bandwidth is very sharp. However, one possible use of such sharply tuned absorbers would be to control specific troublesome room modes or isolated groups of modes with otherwise minimum effect on the signal and overall room acoustics. When a perforated panel is placed over a porous absorber, the panel adds low-frequency absorption compared to the porous material alone. However, the panel may also reduce the high-frequency absorption of the porous material.

Table 12-3 includes the calculated frequency of resonance of 48 different combinations of airspace depth, hole diameter, panel thickness, and hole spacing. This listing should assist in approximating the desired condition.

Slat Absorbers

Another type of resonant absorber utilizes closely spaced slats over an air cavity. The mass of the air in the slots between the slats reacts with the springiness of air in the cavity to form a resonant system, again comparable to a Helmholtz resonator. The glass fiber board usually introduced behind the slots acts as a resistance, broadening the peak of absorption. The narrower the slots and the deeper the cavity, the lower the frequency of maximum absorption. The percent perforation of a slat absorber is shown in Fig. 12-33C.

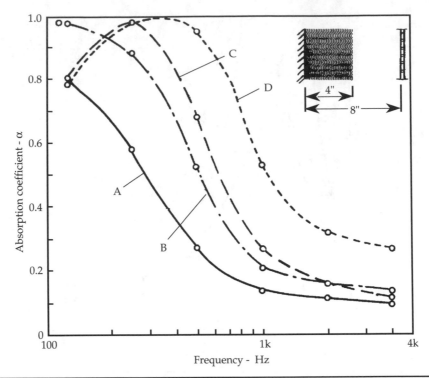

Figure 12-36 Absorption measurements on the same perforated panel absorbers of Fig. 12-35, except that the airspace is increased to 8 in, half of which is taken up with mineral fiber. Panel thickness is $^5/_{32}$ in. (A) Perforation 0.18%. (B) Perforation 0.79%. (C) Perforation 1.4%. (D) Perforation 8.7%. (*Data from Mankovsky*)

Depth of Airspace	Hole Diameter	Panel Thickness	% Perforation	Hole Spacing	Frequency of Resonance
$3^5/_8$	$^1/_8$	$^1/_8$	0.25%	2.22	110 Hz
			0.50	1.57	157
			0.75	1.28	192
			1.00	1.11	221
			1.25	0.991	248
			1.50	0.905	271
			2.00	0.783	313
			3.00	0.640	384
$3^5/_8$	$^1/_8$	$^1/_4$	0.25%	2.22	89 Hz
			0.50	1.57	126
			0.75	1.28	154
			1.00	1.11	178

Table 12-3 Helmholtz Low-Frequency Absorber Perforated-Face Type (All Dimensions in Inches)

Depth of Airspace	Hole Diameter	Panel Thickness	% Perforation	Hole Spacing	Frequency of Resonance
			1.25	0.991	199
			1.50	0.905	217
			2.00	0.783	251
			3.00	0.640	308
$3^5/_8$	$^1/_4$	$^1/_4$	0.25%	4.43	89 Hz
			0.50	3.13	126
			0.75	2.56	154
			1.00	2.22	178
			1.25	1.98	199
			1.50	1.81	217
			2.00	1.57	251
			3.00	1.28	308
$5^5/_8$	$^1/_8$	$^1/_8$	0.25%	2.22	89 Hz
			0.50	1.57	126
			0.75	1.28	154
			1.00	1.11	178
			1.25	0.991	199
			1.50	0.905	218
			2.00	0.783	251
			3.00	0.640	308
$5^5/_8$	$^1/_8$	$^1/_4$	0.25%	2.22	74 Hz
			0.50	1.57	105
			0.75	1.28	128
			1.00	1.11	148
			1.25	0.991	165
			1.50	0.905	181
			2.00	0.783	209
			3.00	0.640	256
$5^5/_8$	$^1/_4$	$^1/_4$	0.25%	4.43	63 Hz
			0.50	3.13	89
			0.75	2.56	109
			1.00	2.22	126
			1.25	1.98	141
			1.50	1.81	154
			2.00	1.57	178
			3.00	1.28	218

TABLE 12-3 (*Continued*)

The resonance frequency of the slat absorber can be estimated:

$$f_0 = 216\sqrt{\frac{p}{(d)(D)}} \qquad (12\text{-}10)$$

where f_0 = frequency of resonance, Hz
  p = perforation percentage (see Fig. 12-33C)
  D = depth of airspace, in
  d = thickness of slat, in

Placement of Materials

Placing sound-absorbing materials in distributed patches, as opposed to a continuous area, effectively contributes to greater absorption. (Such placement of absorbers can also improve diffusion.) If several types of absorbers are used, it is desirable to place some of each type on ends, sides, and ceiling so that all three axial modes (longitudinal, transverse, and vertical) will come under their influence. In rectangular rooms, it is most effective to place absorbing material near corners and along edges of room surfaces. In speech studios, some absorbent effective at higher audio frequencies should be applied at head height on the walls. In fact, material applied to the lower portions of high walls can be as much as twice as effective as the same material placed elsewhere. Untreated surfaces should never face each other.

Reverberation Time of Helmholtz Resonators

There exists the possibility that acoustically resonant devices, such as Helmholtz absorbers, can "ring" with a "reverberation time" of their very own, thus causing timbral changes to audio signals. It is true that any resonant system, electronic or acoustical, has a certain time constant associated with it. The Q-factor (quality factor) describes the sharpness of tuning, as shown in Fig. 12-37. Once the tuning curve has been obtained

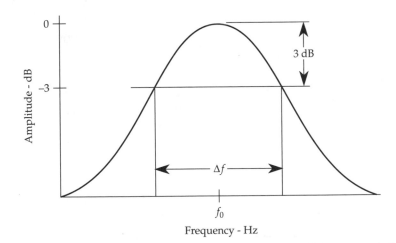

FIGURE 12-37   After the tuning curve of a Helmholtz type resonant absorber has been determined, its Q factor can be found from the expression, $f_0/\Delta f$. The "reverberation time" of such absorbers is very short for Q factors normally encountered.

Q	Reverberation Time (sec)
100	2.2
5	0.11
1	0.022

*f_0 = 100 Hz

TABLE 12-4   Sound Decay in Resonant Absorbers*

experimentally, the width of the tuning curve at the −3 dB points gives Δf. The Q of the system is then $Q = f_0 / \Delta f$, where f_0 is the frequency to which the system is tuned. Measurements on a number of perforated and slat Helmholtz absorbers gave Q factors around 1 or 2 but some as high as 5. Table 12-4 shows how the decay rate of resonant absorbers of several Q factors relates to reverberation time.

With resonant absorber Q factors of 100, problems could be encountered in a room having a reverberation time of perhaps 0.5 sec as the absorbers regenerated sound for several seconds. However, Helmholtz absorbers with such Q factors would be very special devices, perhaps made of ceramic. Absorbers made of wood with glass fiber to broaden the absorption curve have low Q factors so their sound dies away much faster than the studio or listening room itself.

Reducing Room Modes with Absorbers

Because of the relatively narrow absorption bandwidth possible with resonator absorbers, they are ideal in reducing room modes. For example, using ETF (energy, time, frequency) analysis the low-frequency modal structure of a room can be revealed, as shown in Figs. 12-38 and 12-39. A mode that could cause audible distortion produces a reverberant tail at 47 Hz, as seen on the extreme left of Fig. 12-38. Steps can be taken to reduce its energy so that it will behave as the other modes of the room.

The solution is a highly tuned absorber placed in the room at a point of high pressure of the 47-Hz mode. High-pressure points of the 47-Hz mode are located by energizing the room with a 47-Hz sine wave with a loudspeaker and measuring the response with a sound-level meter. A location that is both convenient and effective will probably be found in a corner. Figure 12-39 shows the low-frequency modal structure of the same room after the introduction of a tuner Helmholtz resonator absorber.

The resonator, shown in Fig. 12-40, can be constructed from a concrete-forming tube available in hardware stores. Laminated wood covers are tightly fitted into both ends of the tube. The length of a PVC pipe forming the duct opening can be varied to tune the resonator to a specific frequency. An absorbent would partially fill the resonator.

Increasing Reverberation Time

Low-Q Helmholtz resonators are capable of shortening reverberation time by increasing absorption. High-Q resonators can increase reverberation time through storage of energy as described by Gilford. To achieve the high Q factors necessary, plywood, particleboard, Masonite, and other such materials must be abandoned and materials such as ceramics, plaster, and concrete used in resonator construction. By proper tuning of the resonators, the increase in reverberation time can be placed where needed in regard to frequency.

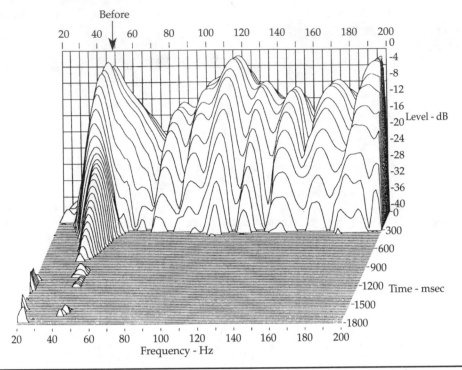

Figure 12-38 Low-frequency modal structure of the sound field of a small room *before* introduction of the tuned Helmholtz resonator absorber.

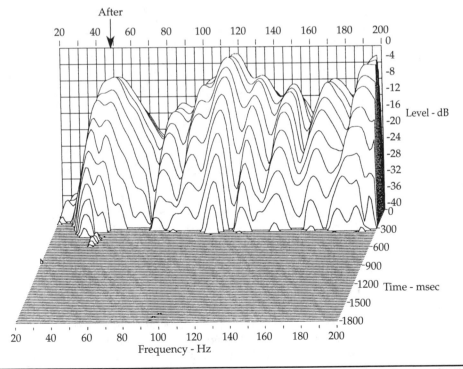

Figure 12-39 Low-frequency modal structure of the sound field of the same small room *after* the introduction of the tuned Helmholtz resonator absorber.

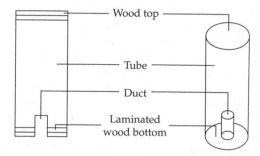

FIGURE 12-40   Details of Helmholtz resonator design.

Modules

The British Broadcasting Corporation (BBC) proposed a modular approach to the acousti-cal treatment of their small voice studios. It has been applied in several hundred such studios with satisfactory acoustical results. In this design, the walls are covered with standard-sized modules, say 2 × 3 ft, having a maximum depth of perhaps 8 in. These can be framed on the walls to give a flush surface appearing very much like an ordinary room, or they can be made into boxes with grill cloth covers mounted on the walls in regular pat-terns. All modules can be made to appear identical, but the similarity is only skin deep.

There are commonly three, or perhaps four different types of modules, each mak-ing its own distinctive acoustical contribution. Figure 12-41 shows the radically differ-ent absorption characteristics obtained by merely changing the covers of the standard

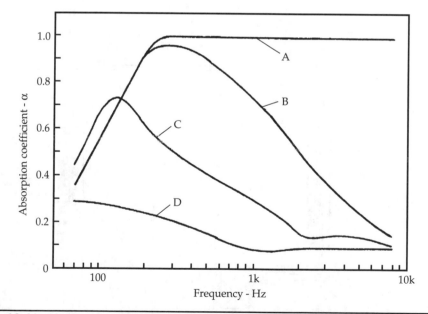

FIGURE 12-41   Modular absorber having a 7-in airspace and 1-in semirigid glass fiber board of 9 to 10 lb/ft³ density behind the perforated cover. (A) No perforated cover at all, or at least more than 25% perforation. (B) 5% perforated cover. (C) 0.5% perforated cover. (D) ³⁄₄-in plywood cover, essentially to neutralize the module. (*Data from Brown*)

module. This is for a module measuring 2 × 3 ft having a 7-in airspace and a 1-in semi-rigid glass fiber board of 3 lb/ft³ density inside. The wideband absorber has a highly perforated cover (25% or more perforation percentage) or no cover at all, yielding essentially complete absorption down to about 200 Hz. Even better low-frequency absorption is achieved by breaking up the airspace with egg-crate-type dividers of corrugated paper to discourage unwanted resonance modes. A cover ¹/₄-in thick with a 5% perforation percentage will peak in the 300- to 400-Hz range. A true bass absorber is obtained with a low-perforation cover (0.5% perforation). If essentially neutral modules are desired, they can be covered with plywood with thickness of ³/₈ or ¹/₄ in, which would give relatively low absorption with a peak around 70 Hz. Using these three or four modules as acoustical building blocks, the desired effect can be designed into a studio by specifying the number and distribution of each of the basic types.

Figure 12-42 shows an adaptation from BBC practice where the wall is used as the bottom of the module box. In this case, the module size is 2 × 4 ft. The modules are fastened to the 2 × 2-in mounting strips that in turn are fastened to the wall. A studio wall 10 ft high and 23 or 24 ft long might use 20 modules of distributed types, four modules high and five long. It is good practice to have acoustically dissimilar modules opposing each other on opposite walls. BBC experience has shown that careful distribution of the different types of modules results in adequate diffusion.

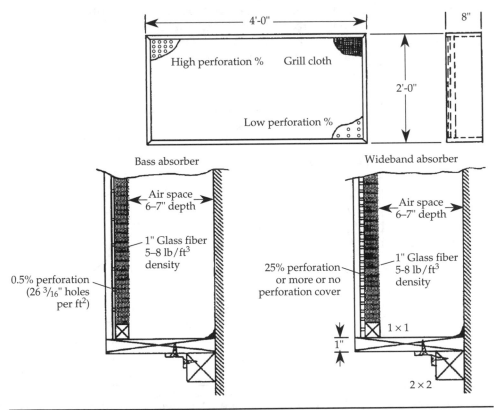

Figure 12-42   Plan for a practical module absorber utilizing the wall as the bottom of the module. (Left) Bass absorber. (Right) Wideband absorber.

CHAPTER 13

Modal Resonances

$\mathbf{S}$ound in enclosed spaces behaves fundamentally differently from sound in a free field. In a free field, sound leaves the source and travels outward unimpeded. It is easy to determine the sound-pressure level at any distance from the source. In most enclosed spaces, sound from the source is reflected from boundary surfaces such as walls, floor, and ceiling. The resulting sound-pressure level at any point in the enclosure is a combination of the direct sound and the reflected sound. In particular, modal resonances are established; the sound-pressure level will be different at different places in the enclosed space and will also vary according to frequency. These resonant frequencies and the variations they create are a function of the dimensions of the enclosed space.

Most rooms can be considered as enclosed spaces; they are essentially containers of air. As such, they exhibit modal resonances, preferred frequencies where energy is concentrated. Likewise, these resonances create frequencies where energy is decreased. These complex energy distributions exist in three-dimensional space throughout the room. Many musical instruments use pipes or tubes in their construction; these also operate as enclosed spaces, and exhibit modal resonance. Changing the parameters of the modal resonances changes the musical pitch.

We live most of the hours of our lives in enclosed spaces; thus our acoustical lives are governed by the effects of these large resonators. In any case, room resonances profoundly affect the sounds we hear whenever we are indoors.

Early Experiments and Examples

Hermann Von Helmholtz performed early acoustical experiments with resonators. His resonators were a series of metal spheres of graded sizes, each fitted with a neck, appearing somewhat like the round-bottom flask found in a chemistry laboratory. In addition to the neck there was another small opening to which he applied his ear. The resonators of different sizes resonated at different frequencies, and by pointing the neck toward the sound under investigation he could estimate the energy at each frequency by the loudness of the sound of the different resonators.

There were numerous applications of this principle long before the time of Helmholtz. A thousand years ago, resonators were embedded in church walls in Sweden and Denmark with the mouths flush with the wall surface, apparently for sound absorption. The interior walls of some modern buildings use concrete blocks with slits, leading to enclosed resonant cavities inside. Energy absorbed from sound in the room causes the resonators to vibrate at a characteristic frequency. Part of the energy is absorbed, part reradiated. The energy reradiated is sent in every direction, contributing to the diffusion of sound in the room. The resonator principle, old as it is, continually appears in new applications.

Resonance in a Pipe

The two ends of the closed pipe of Fig. 13-1A can be likened to two opposing walls of a room. The pipe gives us a simple, one-dimensional example; what happens between opposite walls of a rectangular room can be examined without being bothered by the reflections from the other four surfaces. The pipe is a resonator capable of vibrating at its characteristic frequencies when excited in some way. We will consider its behavior with sound with wavelength much larger than the diameter of the pipe, so that sound

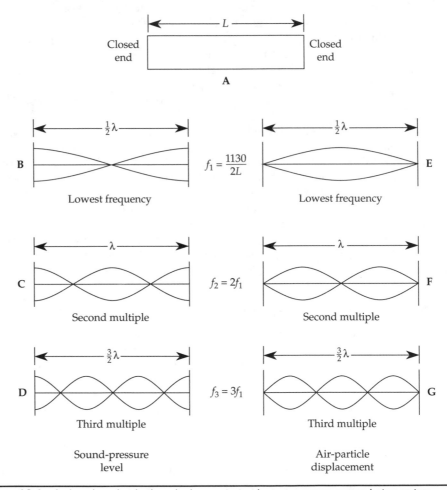

Figure 13-1   A pipe closed at both ends demonstrates how resonance occurs between two opposing walls of a room. The distance between the walls determines the characteristic frequency of resonance and its harmonics. (A) Closed pipe. (B) Plot of sound-pressure level of f_1. (C) Plot of sound-pressure level of f_2. (D) Plot of sound-pressure level of f_3. (E) Plot of air-particle displacement of f_1. (F) Plot of air-particle displacement of f_2. (G) Plot of air-particle displacement of f_3.

travels the length of the pipe. Air inside an organ pipe can be set to vibrating by blowing a stream of air across a lip at the edge of the pipe. More precisely, a small loudspeaker can be placed inside the pipe. A sine wave signal is fed to the loudspeaker and varied in frequency. A small listening hole drilled in the pipe in the end opposite the loudspeaker makes it possible to hear the tones radiated by the loudspeaker. As the frequency is increased, nothing unusual is noted until the frequency radiated from the loudspeaker coincides with the natural frequency of the pipe. At this frequency f_1 modest energy from the loudspeaker is strongly reinforced, and a relatively loud sound is heard at the listening hole. We note that the wavelength of this frequency is twice the length of the pipe. As the frequency is increased, the loudness is again low until a frequency of $2f_1$ is reached, at which point another strong reinforcement is noted. Such resonant peaks can also be detected at $3f_1$, $4f_1$ . . . , and at all multiples of f_1.

Now let us assume that means of measuring and recording the sound pressure all along the pipe are available. Figure 13-1B, C, and D shows how the sound pressure varies along the length of the pipe for three different excitation frequencies. A sound wave traveling to the right is reflected from the right end, returning as a sound wave out of polarity with itself (delayed by $\frac{1}{2}$ period), traveling to the left. The left-going waves react with the right-going waves to create, by superposition, a standing wave at the natural modal frequency of the pipe or one of its multiples. This standing wave has areas of cancellation (nodes) and reinforcement (antinodes) between the reflecting surfaces. We note that the resonant frequencies f_r of a closed pipe are determined by the length of the pipe. In particular,

$$f_r = \frac{nc}{2L} \tag{13-1}$$

where n = an integer ≥ 1
  L = length of the pipe, ft or m
  c = speed of sound = 1,130 ft/sec or 344 m/sec

For example, if the closed pipe is 5-ft long, the fundamental resonant frequency f_1 is $1130/10 = 113$ Hz. The second resonant frequency f_2 is $(2)(1130)/10 = 226$ Hz, and so on.

When sounding the f_1 frequency, as shown in Fig. 13-1B, measuring probes inserted through tiny holes along the pipe would detect zero pressure at the center of the pipe, high pressure near the closed ends of the pipe, and so on. Similar nodes (minima) and antinodes (maxima) can be observed at $2f_1$ and $3f_1$, as shown in Fig. 13-1C and D. Likewise, other resonant frequencies would exist at higher-order multiples of f_1.

In the same way, the dimensions of a room determine its characteristic frequencies much as though there were a north-south pipe, an east-west pipe, and a vertical pipe, the pipes corresponding to the length, width, and height of the room, respectively. In other words, rooms behave the same as closed pipes, but in three dimensions. In addition, rooms can exhibit other kinds of modes that reflect from more than two surfaces.

We also note that the air-particle displacement is exactly defined in any standing wave, as shown in Fig. 13-1E, F, and G. In particular, at any point, the particle displacement is exactly out of phase with the sound-pressure level; there is a displacement node at every pressure antinode, and a displacement antinode at every pressure node. For example, particle displacement is always zero at both ends of a closed pipe. For the lowest mode (Fig. 13-1E) particle displacement is maximum at the center point; this center position from the end of the pipe corresponds to $\lambda/4$ for this mode. Similarly, for any mode, although the physical distance differs, the first position of maximum particle displacement (and minimum particle velocity) corresponds to $\lambda/4$. This explains why in a closed pipe, and in a room, porous absorbers are most effective when placed at a distance of $\lambda/4$ from a boundary.

Any room with reflective walls will illustrate the effect of resonance and the resulting reinforcement of sound at certain frequencies related to the dimensions of the room. Exciting the air at frequencies far removed from these characteristic frequencies results in weaker sounds.

Imagine yourself inside the pipe in Fig. 13-1D. At that resonant frequency, you would hear the sound pressure rise and fall as you traversed its length. A person in a room is, in a sense, inside a large organ pipe. However, there is one important difference; it is now a three rather than an essentially one-dimensional system like the pipe. When the walls of a room are reflective, there is a characteristic modal frequency of resonance associated with the length, another with the width, and still another with the height of the room. In the case of a cubical room, all three modal frequencies coincide to give a mighty reinforcement at the basic characteristic modal frequency and multiples of it.

Reflections Indoors

Anyone can appreciate the difference between sound conditions indoors and outdoors. Outdoors in an open space, the only reflecting plane may be the ground. If that happens to be covered with a foot of snow, which is an excellent absorber of sound, it may be difficult to carry on a conversation with someone 20 ft away. Indoors, sound energy is contained, resulting in a louder sound with a given effort. In an auditorium, a speaker can be heard and understood by hundreds of people with no reinforcement but that of reflecting surfaces.

Consider sound reflections from a single wall. In Fig. 13-2, a point source of sound, S, is a given distance from a rigid wall. The spherical wave fronts (solid lines, traveling

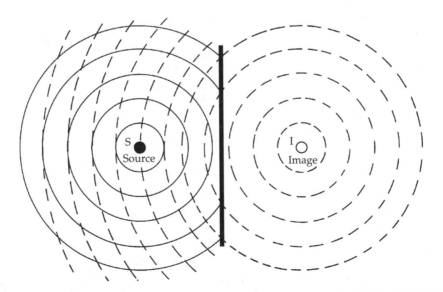

FIGURE 13-2   Sound radiated by a point S is reflected by the rigid wall. The reflected wave can be considered as coming from I, an image of S.

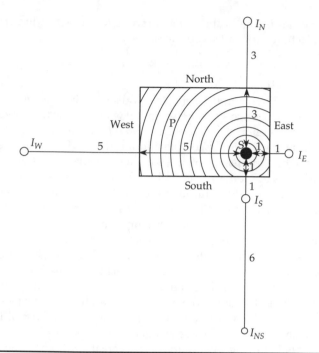

FIGURE 13-3 In an enclosure, the source S has six primary images, one in each of the six surfaces of the enclosure. An infinite number of other images are generated by multiple reflections, and reflections from multiple surfaces. Sound intensity at point P is made up of the direct sound from S plus the contributions from all images.

to the right) are reflected from this surface (broken lines). The reflections from the surface traveling to the left act exactly as though they were radiating from another identical point source, an image source I, an equal distance from the reflecting surface but on the opposite side. This is the simplest case of one source, one image, and a reflecting surface, all in free space.

Now let's consider the isolated reflecting surface of Fig. 13-2 as the east wall of the rectangular room, as shown in Fig. 13-3. Source S still has its image in the east wall of the room, now labeled I_E. The source also has other images. I_N is the image of S in the north wall reflecting surface, I_W is the image of S in the west wall, and I_S is the image of S in the south wall. Use your imagination to visualize I_F, the image in the floor, and I_C, the image in the ceiling. All of these six images, like S, are sending sound energy back into the room. Because of absorption at the reflecting surfaces, their contributions will be somewhat weaker at a given point P in the room, but they all make their contribution.

There are also images from sound that reflects from more than one surface. For example, the I_{NS} image is a result of sound that leaves S, reflects from the north wall, then reflects from the south wall. Similarly, the other surfaces produce the same multiple reflections. And, the process continues, for example, reflecting from north to south to north again, ad infinitum. Finally, other sounds may reflect at many other angles, for example, bouncing from each of the north, west, south, and east walls.

After many surface reflections, images are so weak that they can be neglected for the sake of simplicity. In any case, the sound field at a point P in a room is built up from the direct sound from the source S plus the vector sum of the contributions of all the images of S. That is, the sound at P comprises the direct sound from S plus single or multiple reflections from all six surfaces.

Two-Wall Resonance

Figure 13-4 shows two parallel, reflective walls of infinite extent. When a loudspeaker radiating noise excites the space between the walls, this wall-air-wall system exhibits a resonance at a frequency of $f_1 = 1,130/2L$ or $565/L$, when L = the distance in feet between the two walls and 1,130 the speed of sound in ft/sec. The fundamental frequency f_1 is considered a natural frequency of the space between the reflective walls, and it is accompanied by a series of modes each of which also exhibits resonance. Thus, similar resonances occur at $2f_1$, $3f_1$, $4f_1$ . . . through the spectrum. Various names have been applied to such resonances, including standing waves, room resonances, eigentones, permissible frequencies, natural frequencies, or simply modes. In adding two more mutually perpendicular pairs of walls, to form a rectangular enclosure, we also add two more resonance systems, each with its own fundamental and modal series.

Actually, the situation is far more complicated. Thus far, only axial modes have been discussed, of which each rectangular room has three, plus a modal series for each. An axial mode reflects from two opposite and parallel wall surfaces, tangential modes reflect from four wall surfaces, and oblique modes reflect from all six surfaces. If axial modes are referenced at a 0-dB level, tangential modes are plotted at −3 dB, and oblique modes at −6 dB. In practice, wall surfaces will greatly influence the actual energy in any particular mode. These three types of modes are shown in Fig. 13-5.

What these diagrams offer in terms of clarity, they lack in rigor. In these diagrams, rays of sound are pictured as obeying the law: The angle of incidence equals the angle

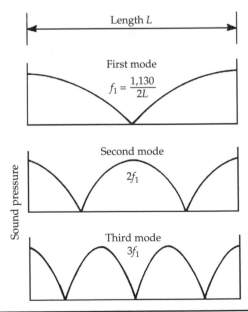

FIGURE 13-4   The space between two parallel, reflective walls can be considered a resonant system with a frequency of resonance of $f_1 = 1,130/2L$. This system is also resonant at integral multiples of f_1.

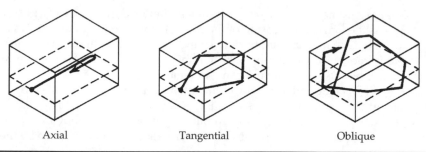

Axial   Tangential   Oblique

FIGURE 13-5 Visualization of axial, tangential, and oblique room modes using the ray concept.

of reflection. For higher audio frequencies, this ray concept is quite useful. However, when the size of the enclosure becomes comparable to the wavelength of the sound, problems arise and the ray approach collapses. For example, a studio 30-ft long is only 1.3 wavelengths long at 50 Hz. Rays lose all meaning in such a case. A wave approach is employed to study the behavior of sound of longer wavelengths.

Frequency Regions

The audible spectrum is very wide when viewed in terms of wavelength. At 16 Hz, considered the low-frequency limit of the average human ear, the wavelength is 1,130/16 = 70.6 ft. At the upper limit of hearing, say 20 kHz, the wavelength is only 1,130/20,000 = 0.056 ft or about 0.7 in. The behavior of sound is greatly affected by the wavelength of the sound in comparison to the size of objects encountered. In a room, sound of 0.7-in wavelength is scattered (diffused) significantly by a wall irregularity of a few inches. The effect of the same irregularity on sound of 70-ft wavelength would be insignificantly small. The heart of the acoustical problem is that no single analytical approach can cover sound of such a wide range of wavelengths.

In considering the acoustics of small rooms, the audible spectrum can be arbitrarily divided into four regions: A, B, C, and D, as shown in Fig. 13-6. Room size determines how the audible spectrum must be divided for acoustical analysis. Very small rooms, with too few modal resonances spaced too far apart, are characterized by domination of a great stretch of the audible spectrum by modal resonances.

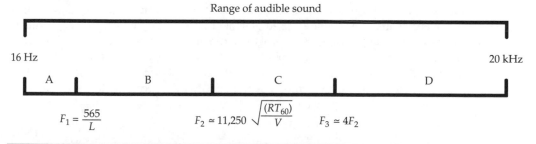

Range of audible sound

16 Hz   20 kHz

A   B   C   D

$$F_1 = \frac{565}{L} \qquad F_2 \approx 11{,}250 \sqrt{\frac{(RT_{60})}{V}} \qquad F_3 \approx 4F_2$$

FIGURE 13-6 When dealing with the acoustics of enclosed spaces, it is convenient to consider the audible frequency range as composed of four regions: A, B, C, and D delineated by frequencies F_1, F_2, and F_3. There is no modal boost for sound in region A. In region B, room modes dominate. Region C is a transition zone in which diffraction and diffusion dominate. In region D, specular reflections and ray acoustics prevail.

Region A is the very-low-frequency region below a frequency of 1130/2L or 565/L, where L is the longest dimension of the room. Below the frequency of this lowest axial mode, there is no resonant support for sound in the room. This does not mean that such very-low-frequency sound cannot exist in the room, only that it is not boosted by room resonances because there are none in that region.

Region B is that region in which the dimensions of the room are comparable to the wavelength of sound being considered. It is bounded on the low-frequency end by the lowest axial mode, 565/L. The upper boundary is not definite but an approximation is given by what has been called the cutoff or crossover frequency given by the equation:

$$F_2 = 11,250\sqrt{\frac{RT_{60}}{V}}$$ (13-2)

where F_2 = cutoff or crossover frequency, Hz
RT_{60} = reverberation time of the room, sec
V = volume of the room, ft³

Region C is a transition region between region B, in which wave acoustics must be used, and region D in which ray acoustics are valid. It is bounded on the low-frequency end approximately by the cutoff frequency F_2 and on the high end approximately by $F_3 = 4F_2$. This region is more difficult to analyze, dominated by wavelengths often too long for ray acoustics and too short for wave acoustics.

Region D describes the spectral area above F_3 that covers higher audible frequencies with short wavelengths; geometric acoustics apply. Specular reflections (angle of incidence equals angle of reflection) and the sound ray approach to acoustics prevail. In this region statistical approaches are generally possible.

In summary, as an example, consider a room measuring 23.3 × 16 × 10 ft. Volume is 3,728 ft³, and reverberation time is 0.5 second. Region A is below 565/23.3 = 24.2 Hz. There is no resonant boost for sound. Region B is between 24.2 and 130 Hz. The wave acoustical approach of modal resonances is used to predict response. Region C is between 130 Hz and (4)(130) = 520 Hz. This is a transitional region. Region D is above 520 Hz. The modal density is very high, statistical conditions generally prevail, and geometrical acoustics can be used.

Room-Mode Equation

A solution of the wave equation can be used to calculate the room mode frequencies for a rectangular enclosure. The geometry used is shown in Fig. 13-7, which fits the familiar, mutually perpendicular x, y, z coordinates of three-dimensional space to a room. For convenience, the longest dimension L (length of the room) is placed on the x axis, the next longest dimension W (width) is placed on the y axis, and the smallest dimension H (height) on the z axis. The goal is to calculate the permissible frequencies corresponding to the modes of a rectangular enclosure. We employ the room-mode equation first stated by Rayleigh:

$$\text{Frequency} = \frac{c}{2}\sqrt{\frac{p^2}{L^2} + \frac{q^2}{W^2} + \frac{r^2}{H^2}}$$ (13-3)

where L, W, H = room length, width, and height, ft or m
p, q, r = integers 0, 1, 2, 3, . . .
c = speed of sound, 1,130 ft/sec or 344 m/sec

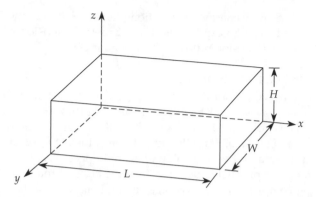

FIGURE 13-7 Orientation of rectangular room of length *L*, width *W*, and height *H* with respect to the *x*, *y*, and *z* coordinates for calculating room-mode frequencies.

This equation gives the frequency of every axial, tangential, and oblique mode of a rectangular room. The integers p, q, and r are the only variables once L, W, and H are set for a given room. Room modes are possible only when p, q, and r are whole numbers (or zero) because this is the condition that creates a standing wave. There are many combinations of integers when fundamentals (associated with 1), second modes (associated with 2), and third modes (associated with 3), and so on, are introduced.

These integers serve to identify the mode as axial, tangential, or oblique, and also identify the frequency of a given mode. An axial mode has two zero terms such as (1, 0, 0) or (0, 0, 3); a tangential mode has one zero term such as (1, 1, 0) or (0, 3, 3); an oblique mode has no terms such as (1, 1, 1) or (3, 3, 3). Furthermore, the mode number indicates the frequency multiple. For example, the axial modes (0, 2, 0) and (0, 0, 2) are a multiple of the (0, 1, 0) and (0, 0, 1) modes, respectively; their frequency is twice as high. Similarly, higher integer values describe higher multiples of the fundamental mode. Tangential and oblique modes are described in the same way.

If $p = 1$, $q = 0$, and $r = 0$ (the 1, 0, 0 mode), the width and height terms drop out and the equation becomes:

$$\text{Frequency} = \frac{c}{2}\sqrt{\frac{p^2}{L^2}} = \frac{c}{2L} = \frac{1{,}130}{2L} = \frac{565}{L} \qquad (13\text{-}4)$$

This is the axial mode corresponding to the length of the room. We observe that it is identical to the mode equation for a closed pipe. The width axial mode (0, 1, 0) and the height axial mode (0, 0, 1) are calculated similarly by substituting the appropriate dimensions.

The relative dimensions of a room determine the modal response, and the suitability of the dimensions. If a rectangular room has two or three dimensions that are equal or a multiple of each other, the modal frequencies will coincide, resulting in peaks at those frequencies and dips at other frequencies. For example, if a room has the dimensions of $10 \times 20 \times 30$ ft (height, width, length), all three axes of the room will have modal frequencies at 56.5, 113, 169.5, 226, 282.5, 339 Hz, and other frequencies. These coincident frequencies, called degenerate frequencies, will result in poor room frequency response and unequal energy distribution. As we will see, modal response can be greatly improved through careful selection of room dimension ratios. In this case, we see that the room dimension ratio of 1:2:3 is a poor choice.

Mode Calculations—An Example

The utility of the room-mode equation [Eq. (13-3)] is best demonstrated by an example. The dimensions of a room are: length $L = 12.46$ ft, width $W = 11.42$ ft, and average height $H = 7.90$ ft. (The ceiling actually slopes along the length of the room with a height of 7.13 ft on one end and 8.67 ft on the other.) The values of L, W, and H are inserted into the mode equation along with combinations of integers p, q, and r. Note that the room is not a rectangular parallelepiped, which is the basis of the room-mode equation. For the following computation, the average ceiling height is used; in practice, this would introduce some error in the results.

Table 13-1 lists some modes for this example room in ascending frequency. It lists p, q, and r and the resulting modal frequency for each combination. Furthermore, each frequency is identified as axial, tangential, or oblique by the number of zeros in that particular p, q, r combination. The lowest natural room frequency is 45.3 Hz, which is the 1, 0, 0 axial mode, associated with the longest dimension, the length L, of the room below which there is no modal boost of the sound. Mode 7, the 2, 0, 0 mode, which is the second axial mode associated with the length L yields a frequency of 90.7 Hz. In the same manner, mode 18, the 3, 0, 0 mode, is the third axial mode associated with the length L, and mode 34 is the fourth mode.

Axial modes and their spacings have been emphasized in studio design. However, there is more to room acoustics than axial modes. As Table 13-1 shows, there are many tangential and oblique modes between the axial modal frequencies that have an effect, though weaker.

Room modes are certainly more than theoretical curiosities. They play a dominant role in the low-frequency response of a room where modes are more isolated in frequency and thus more audible. It is easy to demonstrate their profound effect. If a modal frequency (e.g., a 136-Hz sine wave in this example room) is played through a loudspeaker, and the listener walks through the room, the loud antinodes and soft nodes will be clearly audible at different locations along the length of the room. Depending on the mode, the rise and fall of the tone's loudness may be heard across the room's length, width, or height, or a combination of those. Alternatively, if a listener is seated in one location, and a sine-wave generator is used to sweep across a low-frequency range, great variations in level will be heard at different frequencies because of room modes.

Consider what would happen if a sound source was placed at a node; it would only weakly sound that frequency. Or, consider a mixing engineer seated at an antinode; the loudness of certain frequencies would be accentuated. Room modes can greatly affect a room's frequency response, depending on the physical positioning of sound sources and receivers in the room.

Experimental Verification

The modal frequencies listed in Table 13-1 largely determine the response of this particular room for the frequency range specified. To evaluate their relative effects, a swept sine-wave transmission experiment was performed. In effect, this measures the frequency response of the room. Knowing that all room modes terminate in the corners of a room, a loudspeaker (JBL 2135) was placed in one low tricorner and a measuring

Mode Number	Integers p, q, r	Mode Frequency (Hz)	Axial	Tangential	Oblique
1	100	45.3	x		
2	010	49.5	x		
3	110	67.1		x	
4	001	71.5	x		
5	101	84.7		x	
6	011	87.0		x	
7	200	90.7	x		
8	201	90.7		x	
9	111	98.1			x
10	020	98.9	x		
11	210	103.3		x	
12	120	108.8		x	
13	021	122.1		x	
14	012	122.1		x	
15	211	125.6			x
16	121	130.2			x
17	220	134.2		x	
18	300	136.0	x		
19	002	143.0	x		
20	310	144.8		x	
21	030	148.4	x		
22	221	152.1			x
23	301	153.7		x	
24	112	158.0			x
25	311	161.5			x
26	031	164.8		x	
27	320	168.2		x	
28	202	169.4		x	
29	131	170.9			x
30	022	173.9		x	
31	230	173.9		x	
32	212	176.4			x
33	122	179.7			x

TABLE 13-1   Mode Calculations (Room Dimensions: 12.46 × 11.42 × 7.90 ft) (*Continued*)

Mode Number	Integers p, q, r	Mode Frequency (Hz)	Axial	Tangential	Oblique
34	400	181.4	x		
35	321	182.8			x
36	231	188.1			x
37	222	196.2			x
38	040	197.9	x		
39	302	197.9		x	
40	330	201.3		x	
41	312	203.5			x
42	032	206.1		x	
43	132	211.1			x
44	003	214.6	x		
45	103	219.3		x	
46	013	220.2		x	
47	322	220.8			x
48	113	224.8			x
49	232	225.2			x
50	203	232.9		x	
51	430	234.4		x	
52	023	236.3		x	
53	213	238.1		x	
54	340	240.2		x	
55	123	240.6			x
56	332	247.0			x
57	223	253.1			x
58	303	254.0		x	
59	033	260.9		x	
60	323	272.6			x
61	233	276.2			x
62	403	281.0		x	
63	004	286.1	x		
64	043	291.1		x	
65	304	316.8		x	
66	034	322.3		x	

TABLE 13-1 Mode Calculations (Room Dimensions: 12.46 × 11.42 × 7.90 ft) (*Continued*)

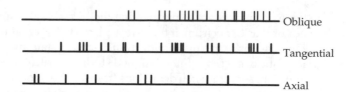

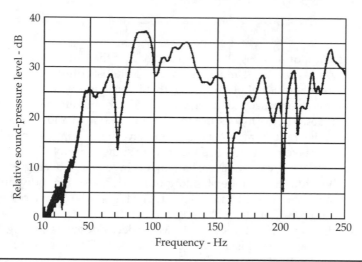

Figure 13-8   Swept sine-wave transmission run in the test room of Table 13-1. The location of each axial, tangential, and oblique modal frequency is indicated.

microphone in the diagonal high tricorner of the room. The loudspeaker was then energized by a slowly swept sine-wave signal. The room response to this signal from the microphone was recorded. This resulted in a linear sweep from 50 to 250 Hz. The resulting trace is shown in Fig. 13-8.

Attempts have been made to identify the effects of each mode in reverberation rooms with all six surfaces hard and reflective. In such cases, the prominent modes stand out as sharp spikes. The practical room in which the recording of Fig. 13-8 was made is not a reverberation chamber. Instead of concrete, the walls are of frame construction covered with gypsum board (drywall); carpet over plywood comprises the floor, and doors almost cover one wall. There is a large window, pictures on the walls, and some furniture. It is evident that this is a fairly absorbent room. The reverberation time is 0.33 second at 125 Hz. This room is much closer acoustically to studios and control rooms than to reverberation chambers, and this is why it was chosen.

There are 11 axial modes, 26 tangential modes, and 21 oblique modes in this 45.3- to 254.0-Hz response, and the transmission trace shown in Fig. 13-8 is the composite effect of all 58 modes. Using only Fig. 13-8, an attempt to correlate the peaks and valleys of the experimentally obtained room response to specific axial, tangential, and oblique modes would be disappointing. The response can be attributed to modes and the interaction of modes; for example, modes close together would be expected to reinforce room response if in phase, and cancel if out of phase. In practice, as we will see, a more extensive analysis is needed to completely understand modal room response.

The three major dips are so narrow that they would take little energy from a distributed speech or music spectrum. If they are neglected, the remaining fluctuations are more modest. Fluctuations of this magnitude in such steady-state swept-sine transmission tests are characteristic of even carefully designed studios, control rooms, and listening rooms. The ear commonly accepts such deviations from a flat response. The modal structure of a space always gives rise to these kinds of fluctuations. However, there is more to acoustical performance than the steady-state response considered so far.

Mode Decay

The steady-state response of Fig. 13-8 tells only part of the story. The ear is sensitive to transient effects, and speech and music are made up almost entirely of transients. Reverberation decay is one transient phenomenon that is easily measured. When broadband sound such as speech or music excites the modes of a room, our interest centers on the decay of the modes. The 58 modes in the 45.3- to 254.0-Hz band of Fig. 13-8 are the microstructure of the reverberation of the room. Reverberation is commonly measured in octave bands. Octave bandwidths of interest are shown in Table 13-2.

Each reverberatory decay by octave bands thus involves an average of the decay of many modes, but it takes an understanding of the decay of the individual modes to explain the decay of the octave band of sound. The higher the octave center frequency, the greater the number of modes included.

All modes do not decay at the same rate. Mode decay depends, among other things, on the way absorbing material is distributed in the room. Carpet on the floor of the test room has no effect on the (1, 0, 0) or (0, 1, 0) axial modes involving only walls. Tangential and oblique modes, which involve more surfaces, would be expected to die away faster than axial modes that involve only two surfaces. On the other hand, absorption is greater for the axial modes in which the sound impinges on the surfaces at right angles than for the low angle of incidence common for tangential and oblique modes.

The reverberation decays in the test room using sine-wave excitation are shown in Fig. 13-9. Single prominent modes decaying alone yield smooth, logarithmic traces. The dual slope decay at 240 Hz is interesting because the low value of the early slope (0.31 second) is probably dominated by a single mode involving much absorption, later giving way to other modes that encounter much less absorption.

Identifying these modes from Table 13-1 is difficult, although you might expect mode 44 to die away slowly and the group of three near 220 Hz (45, 46, 47) to be highly damped. It is common to force nearby modes into oscillation, which then decay at their natural frequency.

Limits		Modes		
Octave	**(−3-dB Points)**	**Axial**	**Tangential**	**Oblique**
63 Hz	45–89 Hz	3	3	0
125 Hz	89–177 Hz	5	13	7
250 Hz	177–254 Hz	3*	10*	14*

*Partial octave.

TABLE 13-2 Modes in Octave Bands

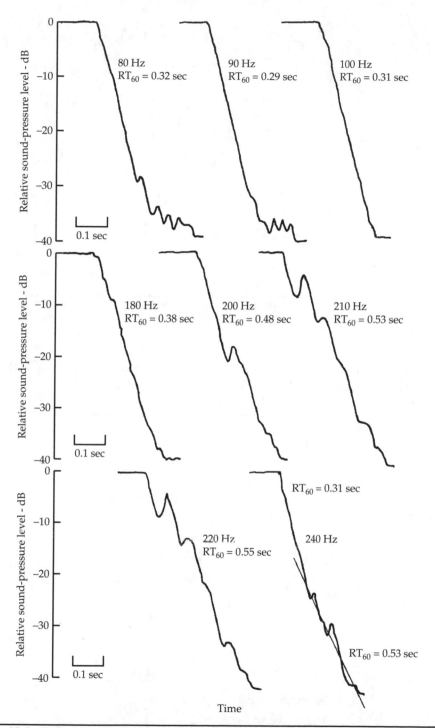

Figure 13-9 Pure-tone reverberation-decay recordings made in a test room. Single modes decaying alone yield smooth, logarithmic traces. Beats between neighboring modes cause irregular decays. The two-slope pattern at 240 Hz reveals the smooth decay of a single prominent mode for the first 20 dB, after which one or more lightly absorbed modes dominate.

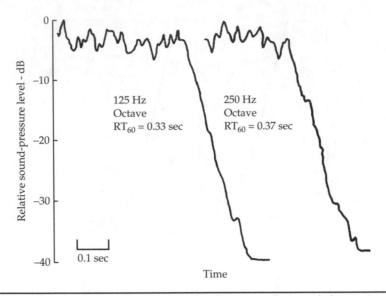

Figure 13-10 Octave-band pink-noise reverberation-decay recordings made in a test room. Decay of the 125- and 250-Hz octave bands indicates the average decay of all modes in those octaves.

Figure 13-10 shows measurements using octave bands of pink noise. The decay of the 125-Hz octave band of noise (0.33 second) and the 250-Hz octave band (0.37 second) averages many modes and should be considered more or less the "true" values for this frequency region. However, normally, many decays for each band would be taken to provide statistical significance. The measured reverberation times for this room are summarized in Table 13-3.

Frequency (Hz)	Average Reverberation Time (sec)
80	0.32
90	0.29
100	0.31
180	0.38
200	0.48
210	0.53
220	0.55
240	0.31 and 0.53 (double slope)
125-Hz octave noise	0.33
250-Hz octave noise	0.37

Table 13-3 Measured Reverberation Time of Test Room

Mode Bandwidth

Normal modes determine room resonances. Taken separately, each normal mode exhibits a resonance curve such as shown in Fig. 13-11. Bandwidth is defined as the half-power (−3-dB) points on either side of a resonant peak. Each mode has a bandwidth determined by the expression:

$$\text{Bandwidth} = f_2 - f_1 = \frac{2.2}{\text{RT}_{60}} \tag{13-5}$$

where RT_{60} = reverberation time, sec

Bandwidth is inversely proportional to reverberation time. The deader the room, the broader the bandwidth of the resonance curve. In electrical circuits, the sharpness of the tuning curve depends on the amount of resistance in the circuit; the greater the resistance, the broader the tuning curve. In room acoustics, the reverberation time depends on absorption (resistance). The analogy is fitting (the more absorption, the shorter the reverberation time, and the wider the mode resonance). For convenient reference, Table 13-4 lists a few values of bandwidth in relation to reverberation time.

Figure 13-12 shows an expanded version of the 40- to 100-Hz portion of Fig. 13-8. The axial modes from Table 13-1 falling within this range are plotted with bandwidth of 6 Hz at the −3-dB points. The axial modes are peaked at zero reference level. The tangential modes have only one-half the energy of the axial modes, so their peaks are plotted 3 dB (10 log 0.5) below the axial modes. The oblique modes have only one-fourth the energy of the axial modes, so the single oblique mode at 98.1 Hz that falls within this range is plotted 6 dB (10 log 0.25) below the axial mode peaks.

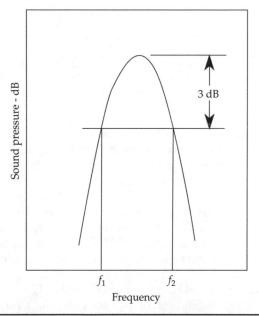

Figure 13-11 Each room mode has a finite bandwidth. Bandwidth is customarily measured at −3-dB points. The more absorbent the room, the greater the bandwidth.

Reverberation Time (sec)	Mode Bandwidth (Hz)
0.2	11
0.3	7
0.4	5.5
0.5	4.4
0.8	2.7
1.0	2.2

TABLE 13-4   Mode Bandwidth

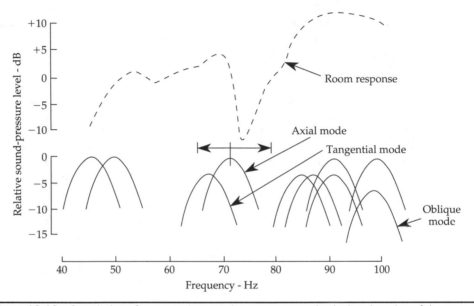

FIGURE 13-12   Correlation of measured swept-sine response and calculated modes of the test room. The portion of the room response from 40 to 100 Hz from Fig. 13-8 is shown. Axial, tangential, and oblique modes are plotted.

The mode bandwidth of a typical room is about 5 Hz. This means that adjacent modes tend to overlap for rooms with short reverberation time, which is desirable. As the skirts of the resonance curves of adjacent modes overlap (e.g., as indicated by the axial and tangential modes labeled in Fig. 13-12), energizing one mode by driving the room at its resonant frequency will also tend to force the other mode into excitation. When the first excitation frequency is removed, the energy stored in the other mode decays at its own frequency. The two will beat with each other during the decay. Very uniform decays, such as the 80, 90, and 100-Hz modes in Fig. 13-9, are probably single modes sufficiently removed from neighbors so as to act independently without irregularities caused by beats.

The overall response of the test room (labeled in Fig. 13-12) is most certainly made up of the collective contributions of the various modes tabulated in Table 13-1. Can the response be accounted for by the collective contributions of the axial and tangential modes, and the single oblique mode in this frequency region? It seems reasonable to account for the 12-dB peak in the room response between 80 and 100 Hz by the combined effect of the two axial, three tangential, and one oblique mode in that frequency range. The falloff below 50 Hz is undoubtedly due to loudspeaker response. This leaves the 12-dB dip at 74 Hz yet to be accounted for.

The axial mode at 71.5 Hz is the vertical mode of the test room with a sloping ceiling. The frequency corresponding to the average height is 71.5 Hz, but the one corresponding to the height at the low end of the ceiling is 79.3 Hz, and for the high end it is 65.2 Hz. The uncertainty of the frequency of this mode is indicated by double arrow above the 71.5-Hz mode (marked in Fig. 13-12). If this uncertain axial mode were shifted to a slightly lower frequency, the 12-dB dip in response could be better explained. It would seem that a dip in response should appear near 60 Hz, but none was found.

Although this test room experimentally verifies many aspects of room-mode theory, more importantly, it demonstrates several practical lessons. The test room is not a rectangular parallelepiped, which is the basis of the mode equation. In addition, when combining the effects of individual modes to obtain an overall response, phase must be taken into account. These components must be combined vectorially with both magnitude and phase fully considered. Finally, and importantly, each mode is not a fixed entity, but one whose magnitude varies from zero to maximum, depending on the location in the room. Individual pressure plots vary across the room, and it is not possible to determine overall magnitude at a point in the room by simply summing all the modes. For example, at any point, a mode could be zero or maximum. Room response should be considered as combined modal responses distributed over the physical dimensions of the room, as discussed below.

Mode Pressure Plots

The modal pattern of a room creates a very complex sound field. To illustrate this, several sketches of sound pressure distributions are shown. The one-dimension organ pipe of Fig. 13-1 can be compared to the (1, 0, 0) axial mode of Fig. 13-13 for a three-dimensional room. The sound pressure is higher (1.0) near the ends and zero along the center of the

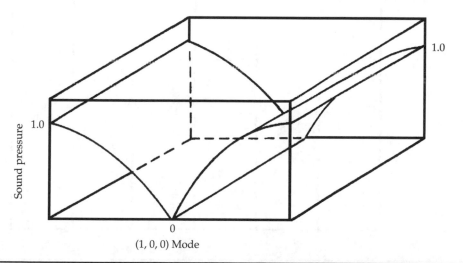

(1, 0, 0) Mode

FIGURE 13-13 A representation of the sound-pressure distribution of the (1, 0, 0) axial mode of a room. The sound pressure is zero in the vertical plane at the center of the room and maximum at the ends of the room. This is comparable to f_1 of the closed pipe of Fig. 13-1.

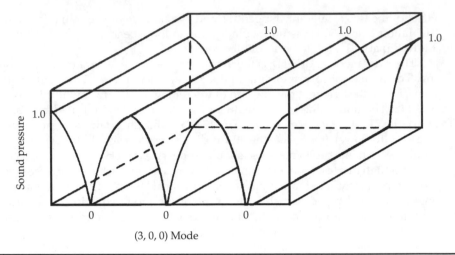

(3, 0, 0) Mode

Figure 13-14   A representation of the sound-pressure distribution of the (3, 0, 0) axial mode of a room.

room. Figure 13-14 shows sound-pressure distribution when only the (3, 0, 0) axial mode is energized. The sound pressure nodes and antinodes in this case are straight lines as shown in Fig. 13-15.

Three-dimensional sketches of sound-pressure distribution throughout a room become difficult, but Fig. 13-16 is an attempt for the (2, 1, 0) tangential mode. We see sound-pressure maxima in each corner of the room with two more maxima at the center edges. This is topographically portrayed in Fig. 13-17 in which the pressure contour

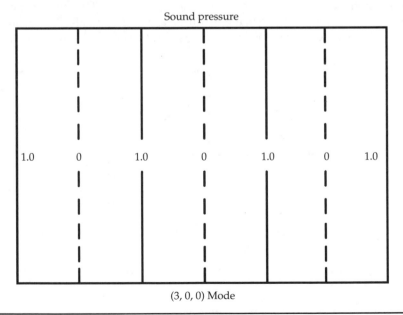

(3, 0, 0) Mode

Figure 13-15   Sound pressure contours on a section through a rectangular room for the (3, 0, 0) axial mode.

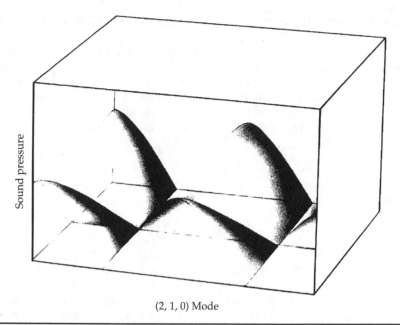

(2, 1, 0) Mode

FIGURE 13-16 Three-dimensional representation of the sound-pressure distribution in a rectangular room for a (2, 1, 0) tangential mode. (*Bruël & Kjaer Instruments, Inc.*)

Sound pressure

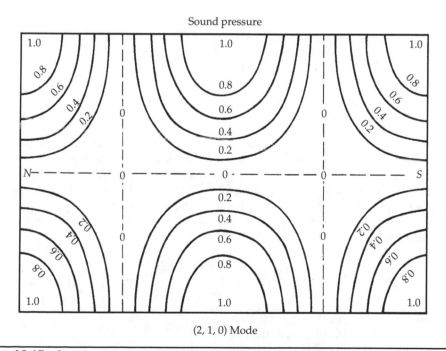

(2, 1, 0) Mode

FIGURE 13-17 Sound-pressure contours of the rectangular room of Fig. 13-16 for a (2, 1, 0) tangential mode.

lines are drawn. The broken lines crisscrossing the room between the maxima of sound mark the zero pressure regions.

Imagine how complicated the sound pressure pattern would be if all the modes were concurrently or sequentially excited by voice or music energy moving up and down the spectrum while constantly shifting in intensity. The plot of Fig. 13-17 shows sound-pressure maxima in the corners of the room. These maxima always appear in room corners for all modes. To excite all modes, place the sound source in a corner. Conversely, if you wish to measure all modes, a corner is the place to locate the microphone.

Mode Density

Mode density increases with frequency. For example, in Table 13-1, in the one-octave spread between 45 and 90 Hz, only 8 modal frequencies are counted. In the next highest octave there are 25 modes. Even in this very limited low-frequency range below 200 Hz, we see modal density increasing with frequency. Figure 13-18 shows that at higher frequencies the rate of increase dramatically rises. Above about 300 Hz or so, the mode spacing is so small that the room response smoothes markedly with frequency. The following equation can be used to determine the approximate number of modes in a given bandwidth at a given center frequency:

$$\Delta N = \left[\frac{4\pi V f^2}{c^3} + \frac{\pi S f}{2c^2} + \frac{L}{8c} \right] \Delta f \tag{13-6}$$

where ΔN = number of modes
Δf = bandwidth, Hz
f = center frequency, Hz
$V = (l_x l_y l_z)$ = room volume, ft^3
$S = 2(l_x l_y + l_y l_z + l_x l_z)$ = room surface area, ft^2
$L = 4(l_x + l_y + l_z)$ = lengths of all edges in room, ft
c = speed of sound = 1,130 ft/sec

This equation shows the modal density at a given bandwidth increases with frequency. Similarly, modal density increases as room size increases.

Room modes are most important in smaller rooms where modal density is low; that particularly occurs at low frequencies. In large rooms, the modal density is relatively high except at extremely low frequencies, and room modes play a less important role. The cutoff frequency between a small room and a large room can be estimated with the Schroeder equation:

$$f_c = 11,885 \sqrt{\frac{RT_{60}}{V}} \tag{13-7}$$

where f_c = cutoff frequency, Hz
RT_{60} = reverberation time, sec
V = room volume, ft^3 or m^3

Note: In metric units, change 11,885 to 2,000.

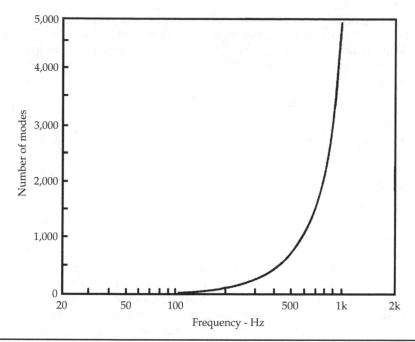

FIGURE 13-18   The number of modal frequencies increases with frequency.

Mode Spacing and Timbral Defects

Timbral defects are frequency-response anomalies in an audio signal and are potentially audible. They can influence the quality of sound for any acoustically sensitive room, ranging from a listening room to a concert hall. The task is to determine which, if any, of the hundreds of modal frequencies in a room are likely to cause audible defects.

Timbral defects can contribute to poor characteristics for recording or other critical work. A musical pitch falling between widely separated modes may be abnormally weak and may die away faster than other notes. In a sense, it is as though that particular note were sounded outdoors while the other notes were simultaneously sounded indoors.

The spacing of the modal frequencies is a critically important factor. In the D region of Fig. 13-6, the modal frequencies of a small room are so close together that they tend to merge helpfully and harmlessly. In the B and C regions, below about 300 Hz, their separation is greater and it is in this region that problems can arise. To avoid timbral defects, modes should not be isolated from one another, but neither should they coincide with each other.

How much separation is allowed between modal frequencies before timbral defects arise? Gilford states his opinion that an axial mode separated more than 20 Hz from the next axial mode will tend to be isolated acoustically. It will tend not to be excited through coupling due to overlapping skirts but will tend to act independently. In this isolated state it can respond to a component of the signal near its own frequency and give this component its proportional resonant boost, thus potentially creating timbral defects.

Gilford's main concern was how wide axial-mode spacing can cause frequency-response deviations resulting from independent and uncoupled modal action. Another criterion for mode spacing has been suggested by Bonello. He analyzes separations to avoid degeneracy (coincident) effects. In this type of analysis, some separation is preferred. Also, he considers all three types of modes, not axial modes alone. He states that it is desirable to have all modal frequencies in a critical band at least 5% of their frequency apart. For example, one modal frequency at 20 Hz and another at 21 Hz would be barely acceptable. However, a similar 1-Hz spacing would not be acceptable at 40 Hz (5% of 40 Hz is 2 Hz).

Zero spacings between modal frequencies are a common source of frequency-response deviations. Zero spacing means that two modal frequencies are coincident (called a degeneracy), and such degeneracies tend to overemphasize signal components at that frequency. Modes should not be widely separated from one another, but neither should they coincide with each other.

Audibility of Timbral Defects

Any ear can be offended by timbral defects caused by isolated or boosted modes, but even a trained ear needs some instrumental assistance in identifying and evaluating such response deviations. In a BBC Research Department study, observers listened to persons speaking into a microphone in the studio under investigation, the voices being reproduced in another room over a high-quality system. Observers' judgments were aided by selectively amplifying a narrow frequency band (10 Hz) to a level about 25 dB above the rest of the spectrum. The output was mixed in small proportions with the original signal to the loudspeaker, the proportions being adjusted until they were barely perceptible as a contribution to the whole output. Any frequency-response deviations were then made clearly audible when the selective amplifier was tuned to the appropriate frequency.

In most studios tested this way, only one or two obvious timbral defects were found in each. Figure 13-19 is a plot of audible timbral defects in 61 male voices observed over a period of two years. Most fall between 100 and 175 Hz. Female voice response variations occur most frequently between 200 and 300 Hz.

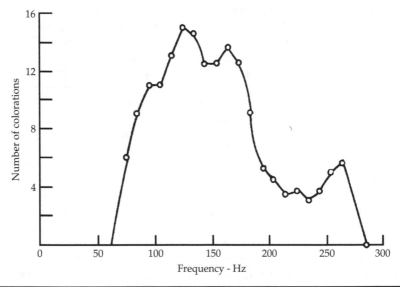

FIGURE 13-19 A plot of timbral defects in 61 male voices observed over a period of two years in BBC studios. Most occur in the 100- to 175-Hz region. Most response variations in female voices occur in the 200- to 300-Hz region. (*Gilford*)

Optimal Room Shape

Can a room be proportioned to achieve optimal modal distribution throughout the room? This opens up a field in which there are strong opinions—some of them supported by quite convincing experiments—and some just as strong without much support.

Room geometry greatly affects acoustical performance. The popularity of rectangular rooms is largely due to economy of construction, but the geometry has acoustical advantages. The axial, tangential, and oblique modes can be calculated with little effort and their distribution studied. For a first approximation, a good approach is to consider only the more dominant axial modes. Degenerate modes can be identified and other room faults revealed.

The relative proportioning of length, width, and height of an acoustically sensitive room is most important. When plans are being drawn for such a room, one should start with basic room proportions. Cubical rooms are anathema because their modal distribution is poor.

The literature contains early quasi-scientific guesses, and later statistical analyses of room proportions that give good modal distribution. Bolt gives a range of room proportions producing smooth room characteristics at low frequencies in small rectangular rooms. Volkmann's 2:3:5 proportion is sometimes used. Boner suggested the $1 : \sqrt[3]{2} : \sqrt[3]{4}$ (or 1:1.26:1.59) ratio as optimum. Sepmeyer suggests several favorable ratios. Louden lists 125 dimension ratios arranged in descending order of room acoustical quality.

Table 13-5 summarizes the best rectangular-room proportions suggested by these researchers. To compare these with the favorable area suggested by Bolt, they are plotted in Fig. 13-20. Most of the ratios fall on or very close to the "Bolt area." This would suggest that a ratio falling in the Bolt area will yield acceptable low-frequency room quality as far as distribution of axial-mode frequencies is concerned. However, particularly with small rooms, where satisfactory modal response is most difficult, any proposed ratio must be tested. The frequency range of validity for the Bolt specification varies with room volume; for example, in an 8,000-ft³ room, the validity range is about 20 to 80 Hz.

One cannot tell by looking at a room's dimensional ratio whether it is desirable or not; analysis is required. Assuming a room height of 10 ft, and choosing the other two

Author		Height	Width	Length	In Bolt's Range?
Sepmeyer	A	1.00	1.14	1.39	No
	B	1.00	1.28	1.54	Yes
	C	1.00	1.60	2.33	Yes
Louden (3 best ratios)	D	1.00	1.4	1.9	Yes
	E	1.00	1.3	1.9	No
	F	1.00	1.5	2.5	Yes
Volkmann (2:3:5)	G	1.00	1.5	2.5	Yes
Boner ($1 : \sqrt[3]{2} : \sqrt[3]{4}$)	H	1.00	1.26	1.59	Yes

TABLE 13-5   Rectangular Room Dimension Ratios for Favorable Mode Distribution

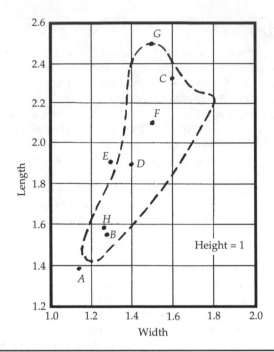

Figure 13-20   A chart of favorable room dimensional ratios to achieve uniform distribution of modal frequencies of a room. The broken line encloses the "Bolt area." The frequency range of validity for the Bolt specification varies with room volume. The letters refer to Table 13-5.

dimensions as favorable ratios, an axial mode analysis can be made. The resulting modes for room dimension ratios *A* through *H* (from Table 13-5) are plotted in Fig. 13-21.

All of these are relatively small rooms and therefore suffer the same problem of having axial-mode separation that is greater than desired. The more uniform the spacing, the better. Degeneracies, or mode coincidences, are a potential problem and they are identified in Fig. 13-21 by the "2" or "3" superscripts above them to indicate the number of coincident resonances. Modes very close together, even though not actually coincident, can also present problems. With these rules to follow, which of distributions of Fig. 13-21 are preferred?

In this example, we might reject *G* because of two triple coincidences spaced apart from neighboring modes. We might also eliminate *F* because of three double coincidences and some spacings. We can neglect the effect of the double coincidences near 280 Hz in *C* and *D* because frequency-response deviations are rarely experienced above 200 Hz. Aside from these reservations, there is little to differentiate between the remainder. Each has flaws, but each would probably serve well, alerted as we are to potential problems. This simple analysis of mode distribution only considers axial modes, knowing that the weaker tangential and oblique modes can only help by filling in between the more widely spaced axial modes. This may alter our opinion of a particular preferred ratio.

If a new room is being constructed, you may have the freedom to move a wall or raise or lower the ceiling to improve distribution. When choosing room dimensions, a primary goal is to avoid coincidences of axial modes. For example, if a cubical space were analyzed, the three fundamentals and all harmonics would coincide. This produces a triple coincidence at each modal frequency and gaps in between. Unquestionably,

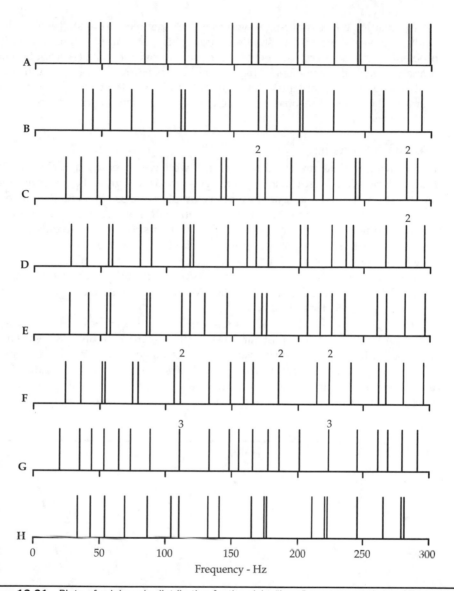

Figure 13-21 Plots of axial mode distribution for the eight "best" room proportions of Table 13-5. The small numbers indicate the number of modes coincident at those particular frequencies. A room height of 10 ft is assumed.

sound in such a cubical space would be acoustically very poor. Rather, room modes should be as evenly spaced in frequency as possible.

For example, a room measuring 19-ft, 5-in long, 14-ft, 2-in wide, and 8-ft high will have 22 axial modes between 29.1 and 316.7 Hz. If evenly spaced, the spacing would be about 13 Hz, but spacings vary from 3.2 to 29.1 Hz. However, there are no coincidences; the closest pair are 3.2 Hz apart. While this cannot be represented as the best proportioning possible, this room, properly treated, will yield good sound quality. The proper starting point is proper room proportions.

In adapting an existing space, you may lack the freedom to shift walls. However, a study of the axial modes can still be very helpful. For example, if such a study reveals problems and space permits, a new wall might improve the modal distribution markedly. If analysis reveals a coincidence that is well separated from neighboring modes, one is alerted to potential problems with an understanding of the cause. As one solution, for example, a Helmholtz resonator may be tuned to the offending coincidence frequency to control its effect.

The Bonello Criterion

In reviewing various preferred room ratios, we have observed that choosing the right ratio of dimensions is crucial in obtaining good modal response. For example, the ratio of any two dimensions should not be a whole number or close to a whole number. Bonello suggests a method for determining the acoustical desirability of the proportions of rectangular rooms. He divides the low end of the audible spectrum into bands $\frac{1}{3}$-octave wide and considers the number of modes in each band below 200 Hz. The $\frac{1}{3}$-octave bands are chosen because they approximate the critical bands of the human ear.

To meet Bonello's criterion, each $\frac{1}{3}$ octave should have more modes than the preceding one, or at least the same number. Modal coincidences are not tolerated unless there are at least 5 modes in that band.

How does a 15.4 × 12.8 × 10 ft room qualify by this criterion? Figure 13-22 shows that it passes this test. The plot climbs steadily upward with no downward anomalies. The horizontal section at 40 Hz is allowed. This suggests good modal response.

Although many authors have suggested various room dimension ratios for optimal modal response, it is important to remember that there is no ideal ratio of dimensions. Furthermore, the quest for a perfect ratio is an impossible one. In real rooms, the

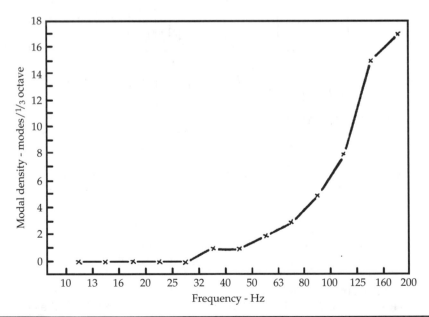

FIGURE 13-22   A plot showing the number of modes in $\frac{1}{3}$-octave bands for a 15.4 × 12.8 × 10 ft room. The plot climbs steadily upward with no downward anomalies; hence the room meets the Bonello's criterion.

structural integrity of the room is not uniform at low frequencies; with a given sound source location, the various modes are not excited equally; and a seated listener can only hear a few of the modes. Modal response is a genuine problem, but predicting the response with certainty using generalized assumptions is very difficult. In other words, working from general guidelines and recommendations, the modal response of each room must be considered on a case-by-case basis.

Splaying Room Surfaces

Splaying (canting) one or two walls of an acoustically sensitive room does not eliminate modal problems, although it might shift them slightly in the room and produce nominally better diffusion. In new construction, splayed walls cost no more, but they may be quite expensive in adapting an existing space. Flutter echoes can be controlled by splaying one of two opposing walls. The amount of splaying is usually between 1 ft in 20 ft and 1 ft in 10 ft. In some recording studio control-room designs, the front surfaces of the room are splayed so that early reflections from the monitor loudspeakers are directed away from the mixing position; this type of design is sometimes called a reflection-free zone.

Nonrectangular Rooms

The acoustical benefit derived from the use of nonrectangular shapes in audio rooms is controversial. As Gilford noted, slanting the walls to avoid parallel surfaces does not remove timbral defects; it only makes them more difficult to predict. Trapezoidal-shaped spaces, commonly used as the outer shell of recording studio control rooms, guarantee asymmetrical low-frequency sound fields even though it is generally conceded that bilateral symmetry with the control position is desirable.

Computer studies based on the finite element approach reveal what happens to a low-frequency sound field in a nonrectangular room. The results of a study conducted by van Nieuwland and Weber are shown in Figs. 13-23 through 13-26, comparing a

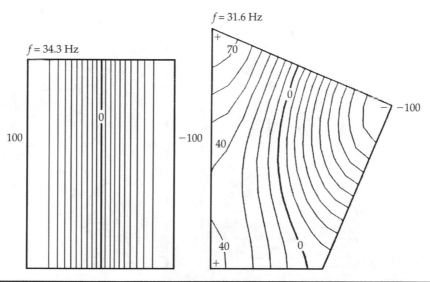

Figure 13-23 Comparison of the modal patterns for a 5 × 7 m two-dimensional rectangular room and a nonrectangular room of the same area. The sound field of the (1, 0, 0) mode is distorted in the nonrectangular room and the frequency of the mode is shifted slightly.

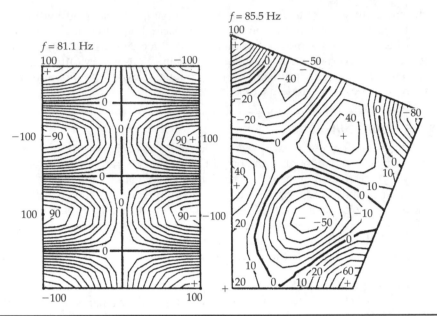

Figure 13-24   The same two-dimensional rectangular room and nonrectangular room of Fig. 13-23 comparing the (1, 3, 0) mode. The sound field is distorted and the frequency is shifted.

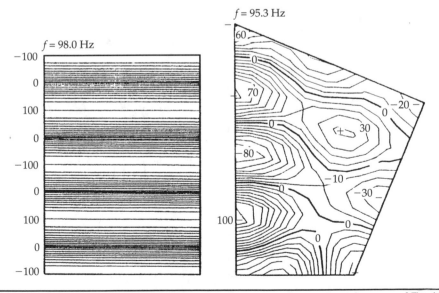

Figure 13-25   The same two-dimensional rectangular room and nonrectangular room of Fig. 13-23 comparing the (0, 4, 0) mode. The sound field is distorted and the frequency is shifted.

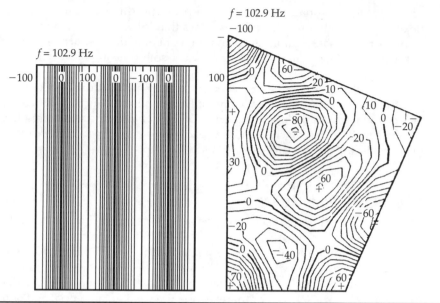

FIGURE 13-26 The same two-dimensional rectangular room and nonrectangular room of Fig. 13-23 comparing the (3, 0, 0) mode. The sound field is distorted and the frequency is shifted.

rectangular geometry to a nonrectangular one. Highly contorted sound fields are shown for the four modes in the nonrectangular geometry. A shift in frequency of the standing wave from that of the rectangular geometry of the same area is indicated: −8.6%, −5.4%, −2.8%, and +1% in the four cases illustrated. This would support the common statement that splaying of walls helps slightly in breaking up degeneracies, but shifts of 5% or more are needed to avoid the effects of degeneracies. The proportions of a rectangular room can be selected to eliminate, or at least greatly reduce degeneracies, while in the case of the nonrectangular room, a prior examination of degeneracies is difficult. Making the sound field asymmetrical by splaying walls introduces unpredictability in the design. If the decision is made to splay walls in a room, say 5%, a reasonable approximation would be to analyze the equivalent rectangular room having the same volume.

Controlling Problem Modes

As noted, a Helmholtz resonator is one possible solution for controlling room modes. If the goal is to bring a mode or closely spaced group of modes under control, the placement of the resonator is important. Suppose that the (2, 1, 0) mode of Fig. 13-17 is causing timbral changes in voice quality and it is necessary to introduce narrow-band absorption at the (2, 1, 0) frequency. For maximum effectiveness, Helmholtz resonators should be placed in areas of high modal sound pressure for the tuned frequency. If the Helmholtz resonator were placed at a pressure node (zero pressure), it would have little effect. Placed at one of the antinodes (pressure peaks) it would have high interaction with the (2, 1, 0) mode. Therefore, any corner would be acceptable, as would the pressure peaks on the walls.

Building a resonator with very sharp tuning (high Q) can be demanding. The flexing of wooden boxes introduces losses that lower the Q. To attain a truly high-Q resonator with sharp tuning, the cavity must be made of concrete, ceramic, or other hard, nonyielding material, but fitted with some means of varying the resonance frequency.

Sound absorption at the lowest octave or two of the audible spectrum is often difficult to achieve. Bass traps are commonly used in recording studio control rooms to reduce room modes at bass frequencies. Large trap depths are required for absorption of very low bass frequencies. Unused spaces above control room ceilings and between inner walls and outer shells are often used for trap space.

Simplified Axial Mode Analysis

To summarize axial-mode analysis, the method is applied to an example room. The dimensions of the room are $28 \times 16 \times 10$ ft. The 28-ft length resonates at $565/28 = 20.2$ Hz, the two side walls 16 ft apart resonate at $565/16 = 35.3$ Hz, and the floor-ceiling combination resonates at $565/10 = 56.5$ Hz. These three axial resonances and the series of multiples for each are plotted in Fig. 13-27. There are 27 axial resonance frequencies below 300 Hz. For this exercise, the weaker tangential and oblique modes are neglected.

Because many timbral defects are traceable to axial modes, their spacings will be examined in detail. Table 13-6 illustrates a convenient form for this simplified analysis of axial modes. The resonance frequencies from the L, W, and H columns are arranged in ascending order in the fourth column. This makes it easy to examine the critical factor of axial mode spacing.

The L-f_7 resonance at 141.3 Hz coincides with the W-f_4 resonance also at 141.3 Hz. This means that these two axial modes can act together to create a potential response deviations at that frequency. This coincidence is also separated 20 Hz from neighboring modes. At 282.5 Hz we see a triple degeneracy of L-f_{14}, W-f_8, and H-f_5 modes which, together, would seem to be a troublesome source of timbral defects. They are also separated from neighbors by 20 Hz. Adjustment of dimensions of a proposed room would be a logical approach. In an existing room, a Helmholtz resonator, sharply tuned and properly located, offers a possible solution.

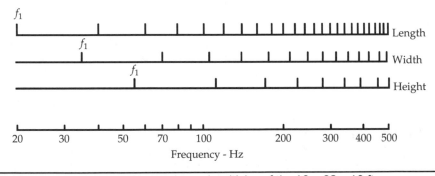

FIGURE 13-27 The axial mode frequencies and multiples of the $16 \times 28 \times 10$ ft room.

Room Dimensions = 28.0 × 16.0 × 10.0 ft				
Axial Mode Resonances (Hz)				
Length L = 28.0 ft $f_1 = 565/L$ (Hz)	Width W = 16.0 ft $f_1 = 565/W$ (Hz)	Height H = 10.0 ft $f_1 = 565/H$ (Hz)	Arranged in Ascending Order	Axial Mode Spacing (Hz)
f_1	20.2	35.3	56.5	20.2	15.1
f_2	40.4	70.6	113.0	35.3	5.1
f_3	60.5	105.9	169.5	40.4	16.1
f_4	80.7	141.3	226.0	56.5	4.0
f_5	100.9	176.6	282.5	60.5	10.1
f_6	121.1	211.9	339.0	70.6	10.1
f_7	141.3	247.2		80.7	20.2
f_8	161.4	282.5		100.9	5.0
f_9	181.6	317.8		105.9	7.1
f_{10}	201.8			113.0	8.1
f_{11}	222.0			121.1	20.2
f_{12}	242.1			141.3 ⎫	0
f_{13}	262.3			141.3 ⎭	20.1
f_{14}	282.5			161.4	8.1
f_{15}	302.7			169.5	7.1
				176.6	5.0
				181.6	20.2
				201.8	10.1
				211.9	10.1
				222.0	4.0
				226.0	16.1
				242.1	5.1
				247.2	15.1
				262.3	20.2
				282.5 ⎫	0
				282.5 ⎬	0
				282.5 ⎭	20.2
				302.7	

TABLE 13-6   Axial Mode Analysis Form

Summary

- Acoustical resonances, generally known as normal modes or standing waves, exist naturally in enclosures. They yield unequal energy distribution throughout the enclosure, particularly at low frequencies in most rooms.

- Small rooms whose dimensions are comparable to the wavelength of audible sound may have the problem of excessive separation between modes.

- As frequency is increased, the number of modes greatly increases. In most rooms, above 300 Hz, average mode spacing becomes so small that room response tends to become smoother.

- The frequency spacing of room modes is determined by the relative dimensions of the enclosure. Through proper selection of room proportions, modes can be placed to avoid both coincidences as well as large separations between modes.

- Axial modes are comprised of two waves traveling in opposite directions, moving parallel to one axis, and striking only two walls. Axial modes make the most prominent contribution to the acoustical characteristics of a space. Because there are three axes to a rectangular room, there are three fundamental axial frequencies, each with its own series of modes.

- Tangential modes are formed by four traveling waves that reflect from four walls and move parallel to two walls. Tangential modes have only half the energy of axial modes, yet their effect on room acoustics can be significant. Each tangential mode has its series of modes.

- Oblique modes involve eight traveling waves reflecting from all six walls of an enclosure. Oblique modes, having only one-fourth the energy of axial modes, are less prominent than the other two.

- Axial, tangential, and oblique modes decay at different rates. To be effective in absorbing a given mode, absorbing material must be located on surfaces where the modal pressure is high. For example, carpet on the floor has no effect on horizontal axial modes. Tangential and oblique modes are associated with more surface reflections than axial modes; thus more treatment locations are possible.

- Predicting modal response with certainty is very difficult. Working from general guidelines and recommendations, the modal response of each room must be considered on a case-by-case basis.

CHAPTER 14

Schroeder Diffusers

After remarkable insight and much experimentation, Manfred R. Schroeder developed a particularly efficient type of diffuser. This quadratic residue diffuser (QRD) is a construction that uses a series of wells of constant width, separated by thin dividers. The relative depth of each well is calculated using one of several number sequences that optimize diffusion in a plane transverse to the alignment of the wells. The sequence can be repeated through multiple periods to extend the size of the diffuser. As with any reflection phase-grating diffuser, the maximum frequency for diffusion is determined by the well width, and the minimum frequency for diffusion is determined by the well depth.

Experimentation

Using number theory, Schroeder hypothesized that a surface with grooves arranged in a certain way would efficiently diffuse sound to a degree unattainable by many other methods. In particular, he discovered that maximum-length sequences can be used to create pseudo-random noise by application of certain sequences of +1 and −1. The power spectrum (from the Fourier transform) of such noise is essentially flat. A wide and flat power spectrum is related to reflection coefficients and angles, and this indicates that diffusion can be obtained by applying the +1 and −1 in a maximum-length sequence. The −1 suggested a reflection from the bottom of a groove in a wall with a depth of a quarter wavelength. The +1 reflection is a reflection from the wall itself without any groove.

Using 3-cm microwaves, Schroeder tested a piece of sheet metal bent into the shape of Fig. 14-1. This shape followed the binary maximum-length sequence with period length 15: − + + − + − + + + + − − − + −. Notably, well depths are one-quarter wavelength.

The resulting reflection pattern, shown in Fig. 14-2A, indicated excellent diffusion. Reflected sound is diffused over a wide angle. In contrast, a metal sheet with well depth of one-half wavelength yielded strong specular reflection, but little diffusion of energy, as shown in Fig. 14-2B.

When a narrow strip of metal covered just one of the (−) grooves, the reflection pattern showed essentially specular reflection of most of the energy back toward the source. In other words, covering one of the slots greatly degraded the favorable diffusion properties. The particular sequence of well depths, as predicted by theory, is crucial in providing efficient diffusion.

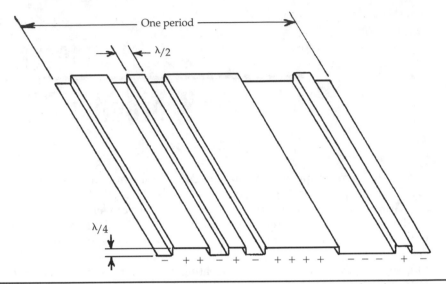

Figure 14-1   An experimental metal sheet folded to conform to a maximum-length sequence used by Schroeder to check the diffusion of 3-cm radio waves. The well depths are all one-quarter wavelength.

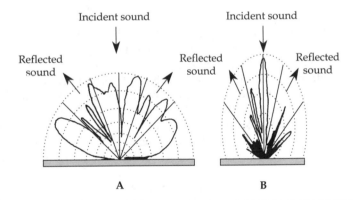

Figure 14-2   Diffusion is profoundly affected by well depth. (A) When the diffuser well depths of Fig. 14-1 are one-quarter wavelength, a very favorable diffusion pattern is obtained. (B) When the wells are one-half wavelength, almost pure specular reflection results, the same as for a flat sheet of metal. (*Schroeder*)

Reflection Phase-Grating Diffusers

Schroeder diffusers, otherwise known as reflection phase-grating diffusers, provide good performance. The diffuse reflection pattern of Fig. 14-2A is superior to many other sound diffusers. Adjustment of room proportions, splaying of walls, the use of semi-spherical, polycylindrical, triangular, cubical, and rectangular geometrical protrusions, and the distribution of absorbing material are helpful in promoting diffusion but are typically not as efficient as the Schroeder diffuser.

Diffraction phase-grating diffusers do have limitations. Because of the one-quarter wavelength groove depth requirement of the binary maximum-length diffuser, the sound-diffusing properties of the surface depend on the wavelength of the incident sound. Experience has indicated that reasonable diffusion results over a band of plus or minus one-half octave of the frequency around which the diffuser is designed. For example, consider a maximum-length sequence diffuser with a sequence length of 15. A design frequency of 1 kHz gives a one-half wavelength groove depth of 7.8 in, and a one-quarter wavelength groove depth of 3.9 in. A single period of this diffuser would be about 5 ft wide and effective from about 700 to about 1,400 Hz. Many such units would be required to provide diffusion over a reasonable portion of the audible band. Even so, diffraction phase-grating diffusers provide good results.

Quadratic-Residue Diffusers

Schroeder reasoned that an incident sound wave falling on a reflection phase grating would diffuse sound almost uniformly in all directions. The phase shifts (or time shifts) can be obtained by an array of wells of depths determined by a quadratic-residue sequence. The maximum well depth is determined by the longest wavelength to be diffused. The well width is about a half wavelength at the shortest wavelength to be scattered. The depths of the sequence of wells are determined by the statement:

$$\text{Well depth proportionality factor} = n^2 \text{ modulo } p \qquad (14\text{-}1)$$

where n = integer ≥ 0
       p = prime number

A prime number is a positive integer that is divisible without a remainder by only 1 and the prime number itself. Examples of prime numbers are 5, 7, 11, 13, and so on. The modulo refers to the residue or remainder. For example, inserting $n = 5$ and $p = 11$ into the above equation gives 25 modulo 11. The modulo 11 means that 11 is subtracted from 25 until the significant residue is left. In other words, 11 is subtracted from 25 twice and the residue 3 is the well depth factor. As another example, for $n = 8$, $p = 11$, the result is (64 modulo 11) = 9. In a similar fashion, the well depth factors for all the wells in a QRD panel are calculated.

A value is selected for prime number p, and the well depth factor for each corresponding integer n value is obtained. In Fig. 14-3, quadratic-residue sequences are listed for the prime numbers 5, 7, 11, 13, 17, 19, and 23, a separate column for each. To check the above example for $n = 5$ and $p = 11$, enter the p column marked 11, run down to $n = 5$ and find 3, which checks the previous computation. The numbers in each column of Fig. 14-3 are proportional to well depths of different quadratic-residue diffusers. At the bottom of each column of Fig. 14-3 is a sketch of a quadratic residue diffuser profile with well depths proportional to the numbers in the sequence. The broken lines indicate thin dividers between the wells. As an example, Fig. 14-4 shows a model for a quadratic-residue reflection phase-grating diffuser for $p = 17$. In this example, the sequence period is repeated twice. Also, note the symmetry of the sequence.

This and rigid separators between the wells, usually metallic, are commonly used to maintain the acoustical integrity of each well. Without separators the effectiveness of the diffuser is decreased. The stepped phase shifts for sound arriving at angles other than the perpendicular tend to be confused in the absence of dividers.

Quadratic-residue sequences

n	p 5	7	11	13	17	19	23
0	0	0	0	0	0	0	0
1	1	1	1	1	1	1	1
2	4	4	4	4	4	4	4
3	4	2	9	9	9	9	9
4	1	2	5	3	16	16	16
5	0	4	3	12	8	6	2
6		1	3	10	2	17	13
7		0	5	10	15	11	3
8			9	12	13	7	18
9			4	3	13	5	12
10			1	9	15	5	8
11			0	4	2	7	6
12				1	8	11	6
13				0	16	17	8
14					9	6	12
15					4	16	18
16					1	9	3
17					0	4	13
18						1	2
19						0	16
20							9
21							4
22							1
23							0

Well depth or proportionality = n^2 modulo p
n = integer
p = prime number

FIGURE 14-3 Quadratic-residue sequences for prime numbers from 5 to 23. In the diffuser profile at the foot of each column, the depths of the wells are proportional to the sequence of numbers above.

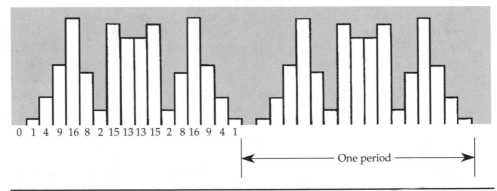

0 1 4 9 16 8 2 15 13 13 15 2 8 16 9 4 1

◄——————— One period ———————►

FIGURE 14-4 A typical quadratic-residue diffuser based on the prime number 17 column of Fig. 14-3. The depths of the wells are proportional to the sequence of numbers in the prime 17 column. Two periods are shown illustrating how adjacent periods are fitted together.

Primitive-Root Diffusers

Primitive-root diffusers use a number theory sequence that is different from the quadratic-residue sequence:

$$\text{Well depth proportionality factor} = g^2 \text{ modulo } p \qquad (14\text{-}2)$$

where g = least primitive root of p
p = prime number

Figure 14-5 shows primitive-root sequences for six different combinations of g and p. The sketches at the bottom of each column show that primitive-root diffusers are not symmetrical like those of the quadratic-residue diffusers. In most cases this is

Primitive-root sequences

n	$p = 5$ $g = 2$	$p = 7$ $g = 3$	$p = 11$ $g = 2$	$p = 13$ $g = 2$	$p = 17$ $g = 3$	$p = 19$ $g = 2$
1	2	3	2	2	3	2
2	4	2	4	4	9	4
3	3	6	8	3	10	8
4	1	4	5	3	13	16
5		5	10	6	5	13
6		1	9	12	15	7
7			7	10	11	14
8			3	9	16	9
9			6	5	14	18
10			1	10	8	17
11				7	7	15
12				1	4	11
13					12	3
14					2	6
15					6	12
16					1	5
17						10
18						1

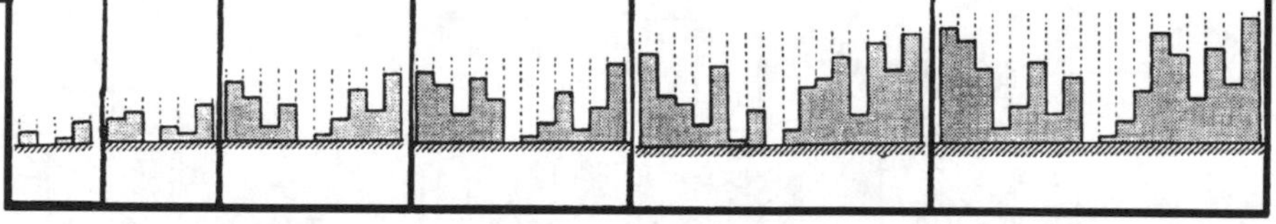

Well depth or proportionality = g^n modulo p
p = prime number
g = least primitive root of p

Figure 14-5 Primitive-root sequences for six combinations of prime number and least primitive roots. Sound diffuser profiles at the foot of each column have depths proportional to the sequence of numbers above. Note that these diffusers are not symmetrical as with quadratic-residue diffusers.

a disadvantage, but in some cases it is an advantage. There is an acoustical limitation with the primitive-root diffusers in that the specular mode is not suppressed as well as it is in the quadratic-residue diffuser. Commercial development has largely utilized quadratic-residue sequences.

Performance of Diffraction-Grating Diffusers

In designing an audio space, the acoustician has three building blocks: absorption, reflection, and diffusion. In many cases in the room to be treated, it is common to find too much reflection but too little diffusion. The effects on incident sound of the three physical principles of absorption, reflection, and diffusion are compared in Fig. 14-6. Figure 14-6A shows sound impinging on the surface of a sound absorber. Energy is largely absorbed, but a small fraction is reflected. The temporal response shows a greatly attenuated reflection from the surface of the absorber.

The same sound wave falling on a hard, reflective surface, as shown in Fig. 14-6B, yields a reflection of almost the same intensity as the sound falling on the surface itself, just reduced slightly by losses at the reflective surface. The polar plot shows the energy concentrated about the angle of reflection. The width of the response in the polar plot is a function of wavelength and the size of the reflecting surface.

A sound wave falling upon a diffuser, such as a quadratic residue, as shown in Fig. 14-6C, is diffracted throughout the hemidisc. The diffused energy falls off

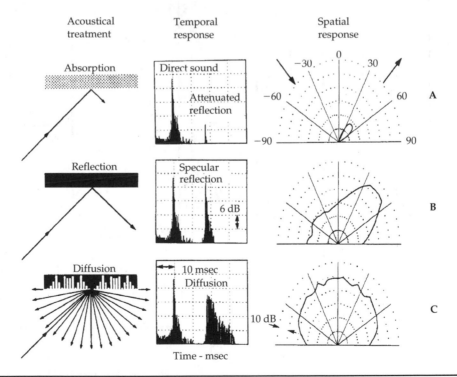

Figure 14-6 A comparison of the physical characteristics of three treatments, showing temporal response and spatial response. (A) Absorption. (B) Reflection. (C) Diffusion. (*D'Antonio, RPG Diffusor Systems, Inc.*)

exponentially. The polar plot shows energy spread more or less equally throughout 180°, but somewhat reduced at grazing angles.

The uniformity of the angular distribution from a diffraction-grating diffuser of scattered energy through a wide frequency range is shown in Fig. 14-7. The left column shows polar distribution of octave bands of energy centered from 250 to 8,000 Hz, a span of five octaves. The right column shows the effect for sound incident at 45° for the same frequencies. The results for all angles of incident sound for the lowest frequency are dependent on the diffuser's well depth. The upper frequency is directly proportional to the number of wells per period and inversely proportional to the well width. The diffusion of a flat panel is shown by a light line for comparison. The uniformity of spatial diffusion is determined by the length of the diffuser period. Good broadbandwidth, wide-angle diffusion requires a large period with a large number of deep, narrow wells. For example, a diffuser might have 43 wells with width of only 1.1 in, and a maximum well depth of 16 in.

These polar plots are smoothed by averaging over octave bands. Comparable polar plots, for single frequencies based on a far-field diffraction theory, show a host of tightly packed lobes that have little practical significance. Near-field Kirchoff diffraction theory shows less lobing.

Figure 14-8 compares the reflective return from a flat panel with that from a quadratic-residue diffuser. The large peak to the left is the direct sound. The second large peak is the specular reflection from a flat panel. Note that the energy of this sharp specular panel reflection is only a few dB below that of the incident sound. The diffused energy from the quadratic residue diffuser is significantly spread out in time. Most importantly, and revealed in the polar diagrams of Fig. 14-7, the grating diffracts sound throughout 180°, not just in the specular direction like a flat panel.

Reflection Phase-Grating Diffuser Applications

The acoustical treatment of large and small spaces is well served by phase-grating diffusers. A large space is one whose normal modal frequencies are so closely spaced as to avoid low-frequency resonance problems. This includes music halls, auditoriums, and many churches. The sound quality of a music hall is greatly influenced by the reflections from side walls. Side walls provide necessary lateral reflections, and diffraction-grating diffusers placed along the center of a hall at ceiling level can diffuse sound from the stage laterally to people in the seats. Troublesome specular reflections can be controlled by strategic placement of diffusers.

In churches, there is often acoustical design conflict between the intelligibility of the spoken word and conditions for enjoyment of music. The rear wall is often the source of reflections that create disturbing echoes. To make this wall absorbent is often detrimental to music conditions. Making the rear wall diffusive, however, minimizes the echo problem while at the same time conserving useful music and speech energy. Music directors are often faced with the problem of singers or instrumental players not hearing each other well. Surrounding the music group with reflection phase-grating diffusers both conserves music energy and spreads it around to achieve ensemble between musicians.

Difficult small-room acoustics are also helped by diffraction-grating diffusing elements. The need to cant walls and distribute absorption material to achieve sound diffusion is relaxed by the proper use of grating diffusers. For example, through proper

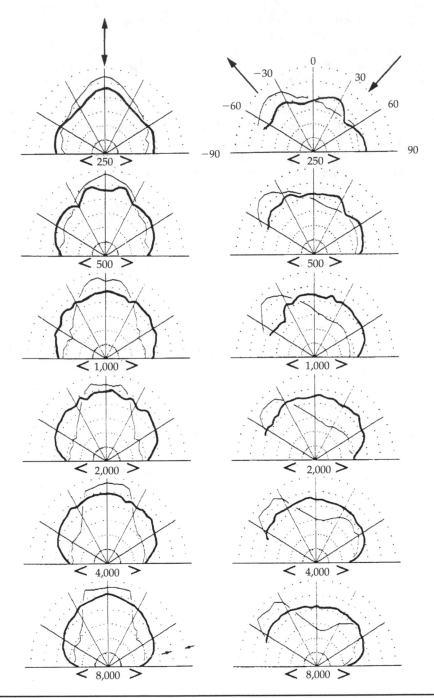

FIGURE 14-7 Polar plots of a commercial quadratic-residue diffuser, smoothed by averaging over octave bands. The angular distribution of the energy is shown at six frequencies and two angles of incidence. The diffusion of a flat panel is shown by a light line. (*D'Antonio, RPG Diffusor Systems, Inc.*)

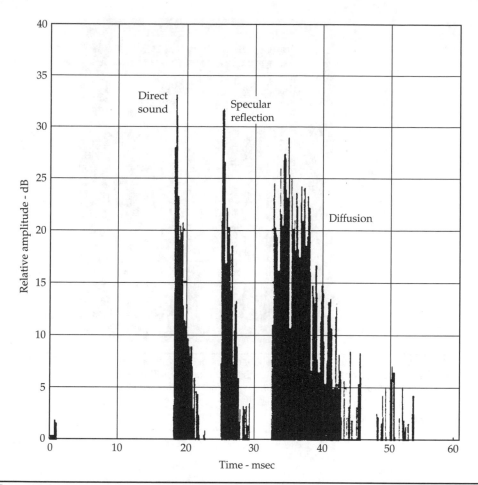

Figure 14-8   An energy-time plot comparing a specular reflection from a flat panel with the energy diffused from a quadratic residue diffuser. The peak energy from a diffusing surface is somewhat lower than that of the flat panel but it is spread out in time. (*D'Antonio, RPG Diffusor Systems, Inc.*)

design, it is possible to obtain acceptable voice recordings from small announce booths because diffusing elements create the sound of a larger room.

Numerous commercial diffuser products incorporating quadratic residue theory are available from RPG Diffusor Systems, Inc. Figure 14-9 shows a QRD-1911 above and two QRD-4311 models below. In the model numbers, the "19" indicates that it is built on prime 19 and the "11" specifies well widths of 1.1 in. (The sequence of numbers in the prime 19 column of Fig. 14-3 specify the proportionality factors for well depths of the diffuser.) The QRD-4311 in the lower portion of Fig. 14-9 is based on prime 43 with well widths also of 1.1 in. (For practical reasons, the columns of Fig. 14-3 stop at prime 23; primes between 23 and 43 are 29, 31, 37, and 41.)

This particular cluster of quadratic-residue diffusers offers excellent diffusion in the horizontal hemidisc, as shown in Fig. 14-10A. The specular reflection from the face of

FIGURE 14-9   A cluster of commercial quadratic-residue diffusers with a single QRD-1911 unit mounted above, and two QRD-4311 units mounted below. The hemidisc of diffusion for the lower unit is horizontal, that of the upper unit is vertical. (*D'Antonio, RPG Diffusor System, Inc.*)

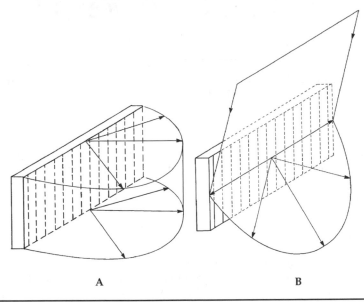

A   B

FIGURE 14-10   Sound is diffused over a hemidisc geometry. (A) A one-dimensional quadratic-residue diffuser scatters sound in a hemidisc. (B) This hemidisc may be one specularly directed by orientation of the source with respect to the diffuser.

the diffuser is shown in Fig. 14-10B. The vertical wells of the QRD-4311 scatter horizontally and the horizontal wells of the QRD-1911 scatter vertically. Together they produce a virtual hemisphere of diffusion.

Figure 14-11A shows the QRD-734 diffuser; this is a 2 × 4 ft model suitable for use in suspended ceiling T-frames as well as other applications. Figure 14-11B shows the Abffusor, which combines broadband absorption and diffusion in the same unit. Figure 14-11C shows the Triffusor which incorporates reflective, absorptive, and diffusive faces. A group of these, set into a wall, offers wide options in the acoustics of a space by rotating the individual units. These three commercial products are all manufactured by RPG Diffusor Systems.

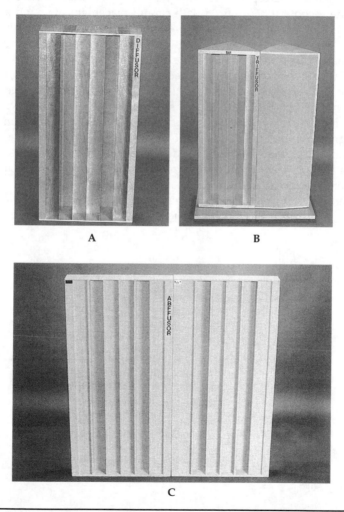

Figure 14-11 Three proprietary sound diffusing systems. (A) A broad bandwidth and wide-angle QRD-734 unit. (B) Triffusor unit having one side absorptive, one diffusive, and one reflective for acoustical variability. (C) Abffusor diffuser/absorber unit with broad bandwidth. (*D'Antonio, RPG Diffusor Systems, Inc.*)

Flutter Echo

If two opposing reflective surfaces of a room are parallel, there exists the possibility of periodic or near-periodic echoes called flutter echoes. This applies to either horizontal or vertical modes. Such successive, repetitive reflections, equally spaced in time, can be audibly disturbing, and can degrade the intelligibility of speech and the quality of music. When the time between echoes is short, they can produce a perception of a pitch or a timbral change in music. If the time between echoes is greater than about 30 to 50 msec, the periodicity is audible as a distinct flutter. This time period is within the Haas fusion zone, and might otherwise be inaudible, but the periodicity of the echo makes it more audible. The lack of ornamentation in modern architecture, and hence poor diffusion, can result in a greater possibility of flutter echoes. Flutter echoes are often associated with specific relative placement of sound source and listener.

Flutter echo can be reduced with careful placement of sound absorbing material. Two opposing, parallel surfaces of a room should never be highly reflective. Flutter echo can also be reduced by splaying walls by 5° to 10°. However, splaying is impractical in many cases, and increasing absorption may degrade the acoustical quality of a space. Diffuser wall treatments can reduce reflections by scattering rather than absorbing. The Flutterfree, as shown in Fig. 14-12, is a commercial example of an architectural hardwood molding, which reduces specular reflection and provides diffusion. The molding, measuring 4 in wide by 4 or 8 ft long, works as a one-dimensional reflection phase-grating diffuser because of the wells routed into its surface. The depths of the wells follow a prime 7 quadratic residue sequence. These moldings may be affixed to a wall butted together or spaced, horizontally or vertically. If they are vertical, spectral reflections are controlled in the horizontal plane and vice versa.

These moldings can also be employed as slats for a Helmholtz slat-type low-frequency absorber. While low-frequency sound is being absorbed by the Helmholtz

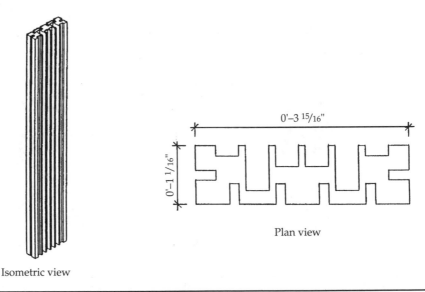

Isometric view

0'–3 15/16"

0'–1 1/16"

Plan view

FIGURE 14-12 The Flutterfree unit is a nonabsorptive flutter echo control molding. It is a quadratic-residue diffuser based on the prime 7. It can also serve as slats on a slat-type Helmholtz low-frequency absorber. (*D'Antonio, RPG Diffusor Systems, Inc.*)

absorber, the surface of each slat performs as a mid-high-frequency sound diffuser. The Flutterfree is manufactured by RPG Diffusor Systems.

Application of Fractals

Certain production limitations have been encountered in the development of reflection phase-grating diffusers. For example, the low-frequency diffusion limit is determined principally by well depth, and the high-frequency limit is determined principally by well width. Manufacturing constraints can place a well-width limit of about 1 in, and a well-depth limit of about 16 in, beyond which the units become diaphragmatic.

To increase the effective bandwidth, the self-similarity principle has been applied in the form of diffusers employing fractals. For example, a fractal design could create a triple-level unit with a diffuser within a diffuser within a diffuser, as shown in the progressive illustration of Fig. 14-13. The Diffractal is a commercial example of this type of unit. Three sizes of quadratic residue diffusers are required to make up the complete Diffractal. The various diffusers operate analogously to the woofer, midrange, and tweeter in a three-way loudspeaker, operating independently to reproduce a wideband response.

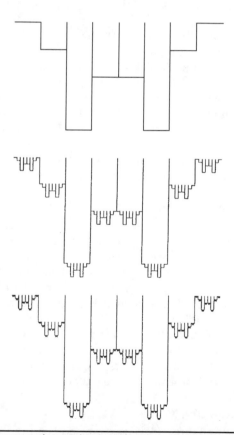

Figure 14-13 Fractal theory can be used to increase the effective bandwidth of diffusers. This essentially allows the design of multiple diffusers within diffusers. In this way, high-frequency fractal sound diffusers can be nested within low-frequency diffusers.

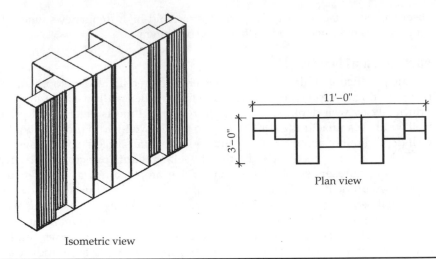

Isometric view

Plan view

11'–0"

3'–0"

FIGURE 14-14 The DFR-82LM Diffractal is a two-way wideband diffuser employing a fractal design. It is composed of a low-freqency unit with a midrange unit embedded at the bottom of each well, creating a diffuser within a diffuser. (*D'Antonio, RPG Diffusor Systems, Inc.*)

Figure 14-14 shows the DFR-82LM Diffractal that is 7 ft, 10 in high and 11 ft wide with a depth of 3 ft. This unit covers the range of 100 to 5,000 Hz. The low-frequency portion is based on a prime 7 quadratic residue sequence. A mid-frequency Diffractal is embedded at the bottom of each well of the larger unit. The frequency range of each section as well as the crossover points of these composite units are calculable.

Figure 14-15 shows the larger DFR-83LMH unit that is 6 ft, 8 in high and 16 ft wide with a depth of 3 ft. This is a three-way unit covering a frequency range of 100 Hz to 17 kHz. The depths of the wells of the low-frequency unit follow a prime 7 quadratic residue sequence. Fractals are set in the wells of fractals that are set in the low-frequency wells. The Diffractal is manufactured by RPG Diffusor Systems.

Diffusion in Three Dimensions

The reflection phase-grating diffusers discussed previously have rows of parallel wells. These can be called one-dimensional units because the sound is scattered in a hemidisc, as shown in Fig. 14-16A. There are occasions in which hemispherical coverage is desired, as shown in Fig. 14-16B. The Omniffusor is an example of a commercial unit that provides hemispherical diffusion. It consists of a symmetrical wooden array of 64 square cells, as shown in Fig. 14-17. The depth of these cells is based on the phase-shifted prime 7 quadratic residue number theory sequence.

Similarly, the FRG Omniffusor consists of an array of 49 square cells based on the two-dimensional phase-shifted quadratic residue number theory. This unit is made of fiberglass-reinforced gypsum. It is light in weight and is therefore useful for application to large surface areas. These units are manufactured by RPG Diffusor Systems.

Diffusing Concrete Blocks

Concrete masonry units are widely used for load-bearing ability and the mass required for sound isolation. The DiffusorBlox system provides load-bearing ability and transmission loss, as well as low-frequency absorption and sound diffusion.

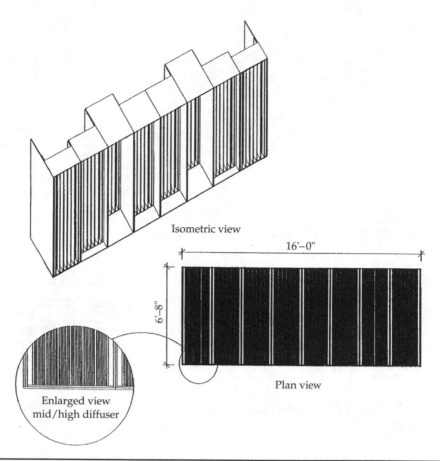

Isometric view

16'–0"

6'–8"

Plan view

Enlarged view
mid/high diffuser

FIGURE 14-15   The larger DFR-83LMH Diffractal is a three-way wideband diffuser employing a fractal design. Fractals are set in the wells, creating a diffuser within a diffuser within a diffuser. (*D'Antonio, RPG Diffusor Systems, Inc.*)

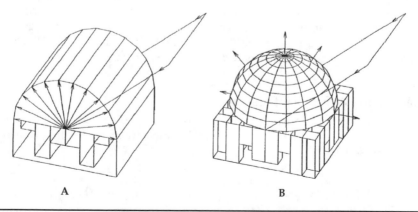

A   B

FIGURE 14-16   Comparison of diffraction patterns. (A) The hemicylindrical form of the one-dimensional quadratic-residue diffuser. (B) The hemispherical form of the two-dimensional diffuser. (*D'Antonio, RPG Diffusor Systems, Inc.*)

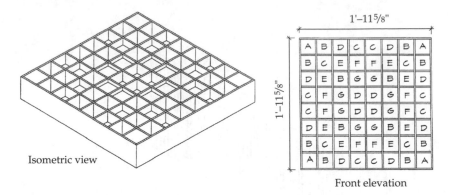

FIGURE 14-17   The Omniffusor is a two-dimensional unit which diffuses sound in both the horizontal and vertical planes for all angles of incidence. (*D'Antonio, RPG Diffusor Systems, Inc.*)

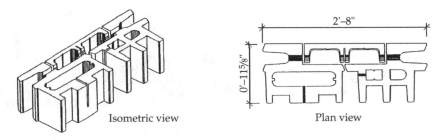

FIGURE 14-18   The DiffusorBlox concrete block offers the transmission loss of a heavy wall, absorption via Helmholtz resonator action, and diffusion through quadratic residue action. The blocks are formed on standard block machines using licensed molds. (*D'Antonio, RPG Diffusor systems, Inc.*)

The DiffusorBlox is made up of three distinct blocks, all of which nominally measure $8 \times 16 \times 12$ in. A typical block is shown in Fig. 14-18. These concrete blocks are characterized by a surface containing a partial sequence of varying well depths, separated by dividers; an internal five-sided cavity that can accept a fiberglass insert; an optional rear half-flange for reinforced construction; and an optional low-frequency absorbing slot. Typical walls constructed of DiffusorBlox are illustrated in Fig. 14-19. DiffusorBlox are manufactured under license by RPG Diffusor Systems.

Measuring Diffusion Efficiency

A measure of the effectiveness of a diffuser can be obtained by comparing the sound intensity in the specular direction with the intensity at ±45° of that direction. This can be expressed as:

$$\text{Diffusion coefficient} = \frac{I(\pm 45°)}{I(\text{specular})} \qquad (14\text{-}3)$$

The diffusion coefficient is 1.0 for the perfect diffuser. This coefficient varies with frequency and is commonly expressed in graphical form. The variation of diffusion coefficient with frequency for several typical units is shown in Fig. 14-20. For comparison,

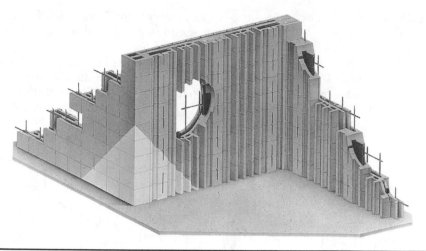

FIGURE 14-19 Typical wall configurations using DiffusorBlox. (*D'Antonio, RPG Diffusor Systems, Inc.*)

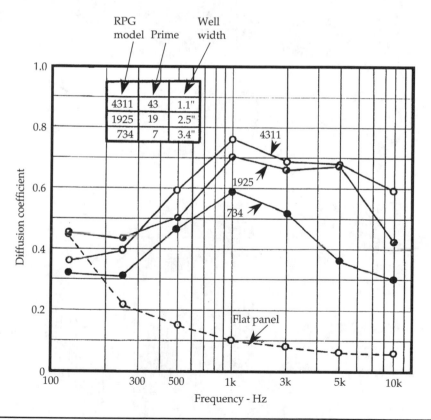

FIGURE 14-20 Comparison of diffusion coefficients with respect to frequency of three commercial diffusion units (QRD-4311, -1025, -734), and a flat panel. (*D'Antonio, RPG Diffusor Systems, Inc.*)

diffusion from a flat panel is included as a broken line. These measurements were all made under reflection-free conditions on sample areas of 64 ft² and using the time-delay-spectrometry technique.

The number of wells and well widths affect the performance of diffuser units. For example, the QRD-4311 diffuser has relatively deep well depths and narrow well widths that are feasible from a manufacturing standpoint. It yields high diffusion coefficients over a wide frequency range, as shown in Fig. 14-20. For comparison, the QRD-1925

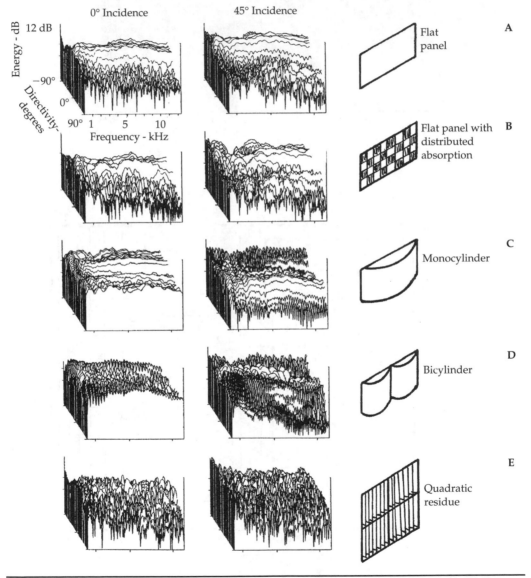

FIGURE 14-21 Energy-frequency-directivity plots comparing diffuser surfaces. (A) Flat panel. (B) Flat panel with distributed absorption. (C) Monocylinder. (D) Bicylinder. (E) Quadratic-residue diffuser. (*D'Antonio, RPG Diffusor Systems, Inc.*)

and QRD-734 units, built with the primes 19 and 7 and well widths of 2.5 and 3.4 in, are also shown in Fig. 14-20. The diffusion performance of these, while good, is somewhat inferior to the QRD-4311.

Several methods have been devised to measure diffusion and scattering coefficients. The AES-4id-2001 method provides guidelines for evaluating diffusion surfaces and estimating diffusion coefficients. The ISO 17497 standard similarly can be used to obtain diffusion scattering coefficients. Some manufacturers employ these methods, and publish results that quantify the performance of their diffusers. AES-4id-2001 is discussed in more detail in Chap. 23.

Comparison of Gratings with Conventional Approaches

Figure 14-21 compares the diffusing properties of five types of surfaces: (A) flat panel, (B) flat panel with distributed absorption, (C) monocylinder, (D) bicylinder, and (E) quadratic residue diffuser. The left column is for sound at 0° incidence and the right column is for sound at 45° incidence. The "fore-and-aft" scale is diffraction from 90° through 0° to −90°. The horizontal frequency scale ranges 1 to 10 kHz. These three-dimensional plots provide a comprehensive evaluation of diffusion properties.

Regarding these measurements, Peter D'Antonio has made these notes: The first six energy-frequency curves contain artifacts of the measurement process which should be disregarded because they are not in the anechoic condition. For 0° incidence, the specular properties of the flat panel with distributed absorption are quite evident by the pronounced peak at 0°, the specular angle. The good spatial diffusion of the monocylinder is illustrated by the relatively constant energy response from 90° to −90°. The bicylinder shows two closely spaced peaks in the time response. Although the spatial diffusion appears good, there is appreciable equal-spaced comb filtering and broadband high-frequency attenuation. This accounts, in part, for the relatively poor performance of cylindrical diffusers. Quadratic residue diffusers maintain good spatial diffusion even at 45° incidence. The dense notching is uniformly distributed across the frequency spectrum and energy is relatively constant with scattering angle, indicating good performance.

CHAPTER 15

Adjustable Acoustics

I f an audio room is used for only one purpose and one type of music, it can be treated with some precision. However, economics may dictate that a given room must serve more than one purpose. Such a multipurpose room carries with it some acoustical compromises. In some cases, a room's acoustical character must be variable to accommodate different types of music. For example, a studio may record a rock band in the morning and a string quartet in the afternoon. In either case, it is necessary to weigh any compromises arising from multiuse or variability against the ultimate sound quality of the design. For this chapter, it is desirable to cast aside the impression that acoustical treatment is an inflexible exercise and consider means of acoustical adjustability.

Draperies

As radio broadcasting developed in the 1920s, draperies on the wall and carpets on the floor were often used to "deaden" studios. It became more apparent that this radio studio treatment was quite unbalanced, absorbing middle- and high-frequency energy but providing little absorption at lower frequencies. As proprietary acoustical materials became available, hard floors became common and drapes all but disappeared from studio walls.

A decade or two later, acoustical engineers, interested in adjusting the acoustical environment of the studio, turned with renewed interest to draperies. A good example of this return to draperies was illustrated in 1946 in the rebuilding of Studio 3A of the National Broadcasting Company in New York City. This studio was redesigned with the goal of providing optimum conditions for recording music for home playback and for recording transcriptions for broadcast. The reverberation-frequency characteristics for these two applications differ. By using drapes and hinged panels (considered later), the reverberation time was made adjustable over more than a two-to-one range. The heavy drapes were lined and interlined and were hung some distance from the wall to make them more absorbent at lower frequencies (see Figs. 12-15 through 12-18). When the drapes were withdrawn, polycylindrical elements having a plaster surface were exposed.

If due regard is given to the absorption characteristics of draperies, there is no reason, other than cost, why they should not be used. The effect of the fullness of the drape must be considered. The acoustical effect of an adjustable element using drapes can thus be varied from that of the drape itself when closed, to that of the material behind when the drapes are withdrawn into the slot provided, as shown in Fig. 15-1. The wall treatment behind the drape could be anything from hard plaster for minimum sound absorption to resonant structures having maximum absorption in the low-frequency

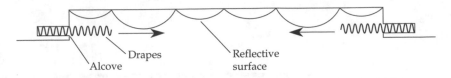

region, more or less complementing the effect of the drape itself. Acoustically, there would be little point to retracting a drape to reveal material having similar acoustical properties.

Adjustable Panels: Absorption

Portable absorbent panels offer a certain amount of flexibility in adjusting listening room or studio acoustics. The simplicity of such an arrangement is illustrated in Fig. 15-2. In this example, a shallow wooden cabinet holds a perforated hardboard facing with acoustically transparent cloth covering, a glass-fiber layer, and an interior air cavity. This type of panel can be easily mounted on a wall or removed as needed. For example, panels may be introduced to decrease reverberation for voice recording or removed to obtain a live effect for instrumental music recording.

Figure 15-3 shows one method for mounting panels on the wall with beveled cleats; the panels can be easily removed by lifting them off the cleats. Hanging such units on the wall adds absorption, and contributes somewhat to sound diffusion. There is some compromising of the effectiveness of the panels as low-frequency resonators because the units hang loosely from a mounting strip. Leakage coupling between the cavity and the room tends to degrade the resonant effect.

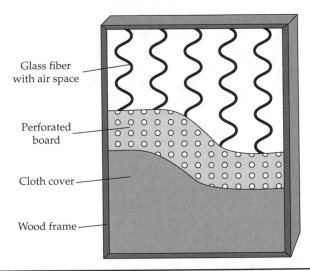

Glass fiber with air space

Perforated board

Cloth cover

Wood frame

FIGURE 15-2   Removable panels can be used to adjust the reverberation characteristics of a room. For maximum variability, unused panels should be removed from the room entirely.

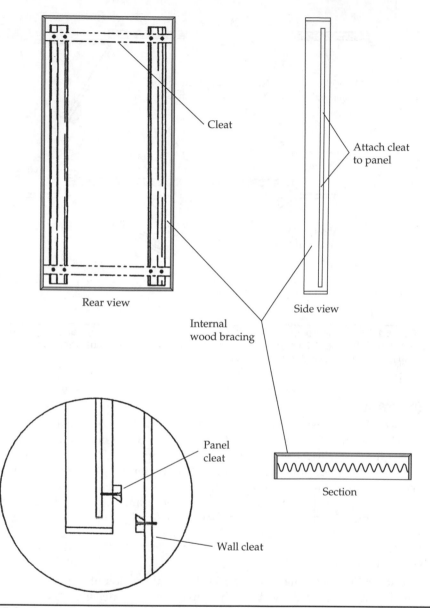

Rear view

Cleat

Internal
wood bracing

Attach cleat
to panel

Side view

Panel
cleat

Wall cleat

Section

FIGURE 15-3   One method for mounting panels on walls using beveled cleats so that the panels
are removable.

Freestanding acoustical flats, sometimes called gobos, are useful studio accessories.
A typical flat consists of a frame of 1×4 in lumber with plywood back filled with a low
density (e.g., 3 lb/ft^3) glass-fiber board faced with a fabric. Arranging a few such flats
strategically can give a certain amount of local control of acoustics at mid to high frequencies, and also provide some isolation at higher frequencies.

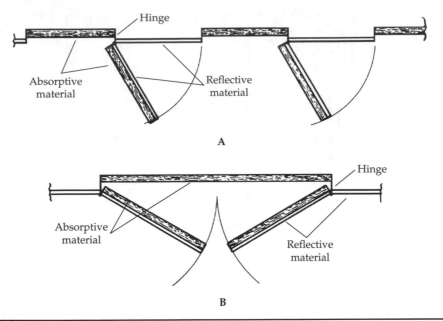

FIGURE 15-4   An inexpensive and effective method of incorporating variability in room acoustics is through the use of hinged panels, reflective on one side and absorbent on the other. (A) Single hinged panels. (B) Double hinged panels.

Hinged Panels

One of the least expensive and most effective methods of adjusting studio acoustics is the hinged panel arrangement of Fig. 15-4A and B. When closed, all surfaces are reflective (plaster, plasterboard, or plywood). When opened, the exposed surfaces are absorptive (glass fiber or carpet). For example, 3 lb/ft^3 density glass-fiber boards 2 to 4 in thick could be used to cover the absorptive surfaces. These panels could be covered with acoustically transparent cloth to improve appearance. Spacing the glass fiber from the wall would improve absorption at low frequencies.

Louvered Panels

Multiple louvered panels can be adjusted by the action of a single lever in the frames, as shown in Fig. 15-5A. Behind the louvers is a low-density glass-fiber board or batt. The width of the panels determines whether they form a series of slits (Fig. 15-5B) or seal tightly together (Fig. 15-5C). Opening the louvers of the panels in Fig. 15-5C slightly would acoustically approach the slit arrangement of Fig. 15-5B, but it might be mechanically difficult to arrange for a precise slit width.

The louvered panel arrangement is very flexible. The glass fiber can be of varying thickness and density and fastened directly to the wall or spaced out different amounts. The louvered panels can be of reflective material (glass, hardboard) or of more absorptive material (soft wood) and they can be solid, perforated, or arranged for slit-resonator operation. In other words, almost any absorption-frequency characteristic can be matched with the louvered structure with the added feature of adjustability.

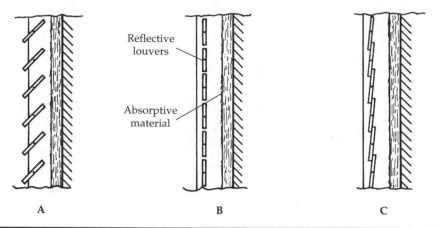

FIGURE 15-5 Louvered panels allow a wide range of acoustic variability. (A) Louvered panels may be opened to reveal absorbent material within. (B) Panels are closed to present a reflective surface. (C) Short louvers can change from a slat resonator (closed) to reveal absorbent material within (open).

Absorption/Diffusion Adjustable Panels

The Abffusor is an example of a commercially available adjustable panel. It combines broadband absorption in the far field with horizontal or vertical diffusion in the near field down to 100 Hz for all angles of incidence. The panel works on the absorption phase grating principle using an array of wells of equal width separated by thin dividers. The depth of the wells is determined by a quadratic residue sequence of numbers to diffuse what sound is not absorbed.

The panels measure approximately 2 × 4 or 2 × 2 ft. They can be mounted in ceiling grid hardware or as independent elements. The absorption characteristics of the Abffusor for two mountings are shown in Fig. 15-6. Mounted directly on a wall, the absorption coefficient is about 0.42 at 100 Hz. With 400 mm of airspace between the panel and the surface, the coefficient is doubled. The latter is approximately the performance with the panel mounted in a suspended ceiling grid. Near perfect absorbance is obtained above 250 Hz. The unit thus provides both mid/high-band sound absorption and sound diffusion. The Abffusor is manufactured by RPG Diffusor Systems, Inc.

Variable Resonant Devices

Resonant structures for use as sound absorbing elements have been used extensively. An example using hinged perforated panels is shown in Fig. 15-7A. Varying the panel position shifts the resonant peak of absorption, as shown in Fig. 15-7B. The approximate dimensions in this example are: panel width 2 ft, panel thickness $^3/_8$ in, hole diameter $^3/_8$ in, hole spacing $1^3/_8$ in on centers.

A most important element of the absorber is a porous cloth having the proper flow resistance covering either the inside or outside surface of the perforated panel. When the panel is in the open position, the mass of the air in the holes and the compliance or "springiness" of the air in the cavity behind act as a resonant system. The cloth offers a resistance to the vibrating air molecules, thereby absorbing energy. When the panel in

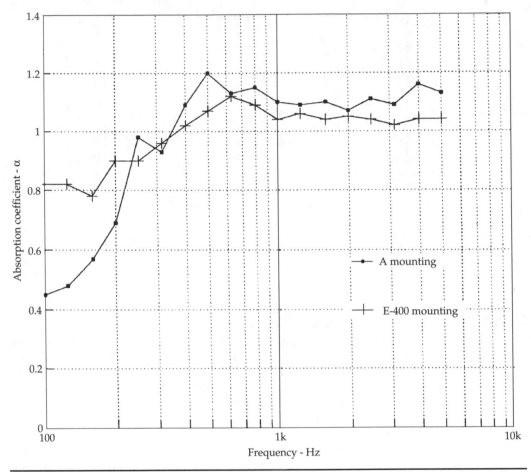

Figure 15-6   The absorption characteristics of an Abffusor panel for two mountings: Directly on wall (A mounting), and with a 400-mm airspace (E-400 mounting).

this example is closed, the cavity virtually disappears and the resonant peak is shifted from about 300 to about 1,700 Hz (Fig. 15-7B). In the open condition, the absorption for frequencies higher than the peak remains remarkably constant out to 5 kHz.

One Hollywood dubbing studio uses another interesting resonant device. Movie dialogue looping requires variable voice recording conditions to simulate the many acoustic situations of motion picture scenes portrayed. In this studio, reverberation time had to be adjustable over a two-to-one range for the 80,000-ft³ stage.

Both side walls of the stage used the variable arrangement of Fig. 15-8, which shows a cross section of a typical element extending from floor to ceiling with all panels hinged on vertical axes. The upper and lower hinged panels of 12 ft length are reflective on one side (2 layers of ³/₈-in plasterboard) and absorbent on the other (4-in glass-fiber board). When open, they present their absorbent sides and reveal slit resonators (1 × 3 slats spaced ³/₈ in to ³/₄ in with glass-fiber board behind), which utilize the space behind the canted panels. In some areas, glass fiber is fastened directly to the wall. Diffusion is less of a problem when only highly absorbent surfaces are exposed but when the reflective

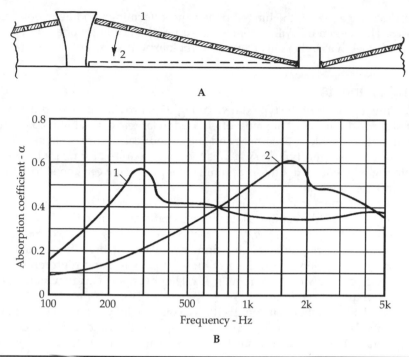

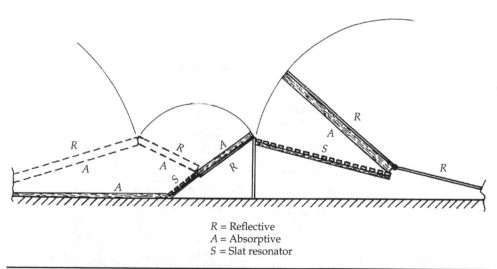

FIGURE 15-7 Hinged, perforated panels can be used to vary absorption. (A) The panels can be placed in two positions. A porous cloth of the proper flow resistance covers one side of the perforated panel. (B) Changes in the absorption characteristic can be realized by shifting the panel from one position to another.

R = Reflective
A = Absorptive
S = Slat resonator

FIGURE 15-8 Variable acoustical elements in a mixing-looping stage. Reflective areas are presented when the doors are closed. Absorbent areas and slat resonators are presented when opened.

surfaces are exposed, the hinged panels meet, forming good geometric diffusing surfaces. This design by William Snow illustrates the flexibility offered in combining different types of absorbers in an effective yet inexpensive arrangement.

Rotating Elements

Rotating elements of the type shown in Fig. 15-9 provide unique adjustability; because of size constraints, they are most often used in larger rooms. In this particular configuration, the flat side is absorbent and the cylindrical diffusing element is reflective. A disadvantage of this type of system is the space required for rotation. The edges of the rotating element fit tightly to minimize coupling between the studio and the space behind the elements.

A music room may be designed with a series of rotating cylinders partially protruding through the ceiling. The cylinder shafts are ganged and rotated with a rack-and-pinion drive in such a way that sectionalized areas of the exposed cylinder gives moderate low-frequency absorption increasing at high frequencies, good low-frequency absorption decreasing at high frequencies, and high reflection absorbing little energy at low or high frequencies. However, such arrangements are expensive and mechanically complex.

The Triffusor, shown in Fig. 15-10, is a commercially available system. Is a rotatable equilateral-triangular prism with absorptive, reflective, and diffusive sides. A nonrotating form of the Triffusor is available with two absorptive sides and one diffusive side, especially adapted for use in corners. The nominal dimensions of the unit are: height 4 ft, faces 2 ft across. In a normal mounting, the edges would be butted and each unit supplied with bearings for rotation. In this way, an array of these units could provide all absorptive, all diffusive, all reflective, or any desired combination of the three surfaces. The Triffusor is manufactured by RPG Diffusor Systems, Inc.

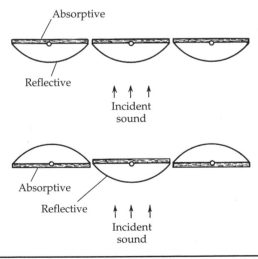

FIGURE 15-9 Rotating elements can vary the reverberation characteristics of a room. They have the disadvantage of requiring considerable space to accommodate the rotating elements.

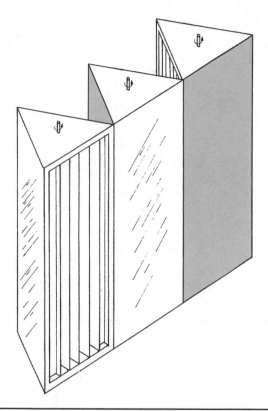

FIGURE 15-10   The Triffusor may be used in groups to provide variable acoustics in a space. Rotation of the individual units can bring diffusing, absorbing, or reflecting surfaces into play.

Portable Units

The Tube Trap is a commercially available, modular, low-frequency absorber. The construction of the trap is shown in Fig. 15-11. It is a cylindrical unit available in 9-, 11-, and 16-in diameters and 2- and 3-ft lengths. Smaller-diameter units may be stacked on top of larger-diameter units, and placed in the corner of a room. A quarter-round adaptation may also be used. The trap is a simple cylinder of 1-in glass fiber, given structural strength by an exoskeleton of wire mesh. A plastic sheet designated as a "limp mass" covers half of the cylindrical surface. For protection and appearance, a fabric cover is added.

As with any absorber, the total absorption is given by: (area)(coefficient) = sabins of absorption. With absorption modules, it is expedient to rate the sabins of absorption contributed by each tube. The absorption characteristics of the 3-ft-long Tube Traps and of the 9-, 11-, and 16-in-diameter models are shown in Fig. 15-12. Appreciable absorption, especially with the 16-in trap, is achieved below 125 Hz.

When tubes are stacked in each corner behind loudspeakers, the limp mass, which covers only half of the area of the cylinder, provides reflection for midrange and higher frequencies. Moreover, low-frequency energy passes through this limp mass and is absorbed. Through reflection of mid/high frequencies, it is possible to control the

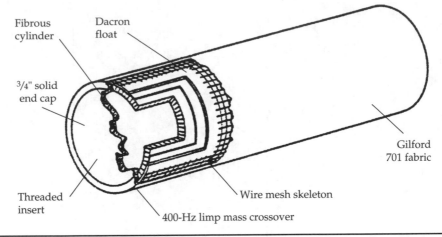

FIGURE 15-11 The construction of the Tube Trap. It is a cylinder of 1-in glass fiber with structural support. A plastic "limp mass" covers half of the cylindrical surface, which reflects and diffuses sound energy above 400 Hz. (*Acoustic Sciences Corporation*)

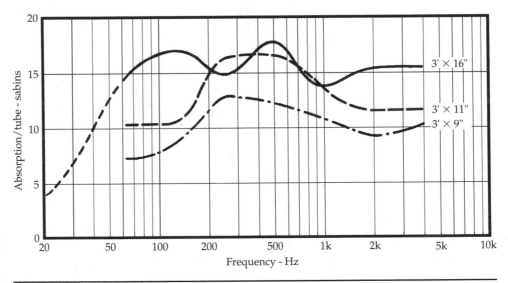

FIGURE 15-12 Absorption characteristics of three sizes of Tube Traps. The 16-in unit provides good absorption down to about 50 Hz.

brightness of the sound at the listening position. Figure 15-13 shows the two positions of the tubes. If the reflector faces the room (Fig. 15-13B), the tube fully absorbs the lower frequency energy while the listener receives brighter sound. Mid/high-frequency sound is diffused by its cylindrical shape. If less bright sound is preferred, the reflective side is placed to face the wall. This may introduce timbral changes resulting from the cavity formed by the intersecting wall surfaces and the cylindrical reflective panel. By placing absorptive panels on the wall surfaces, as indicated in Fig. 15-13A, this timbral change can be controlled.

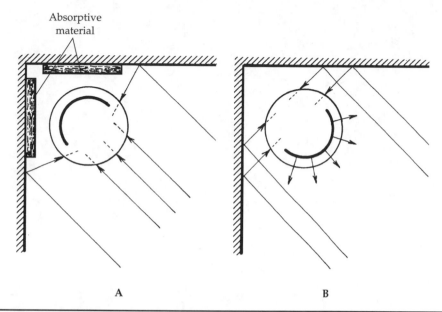

Absorptive material

A   B

FIGURE 15-13 The positioning of the limp mass reflector in a Tube Trap gives some control over the brightness of sound in the room. (A) With the limp mass facing the corner, the absorbent side of the cylinder absorbs sound over a wide range. (B) If the limp mass faces the room, frequencies above 400 Hz are reflected.

Tube Traps can also be placed in the rear two corners of the room if experimentation indicates it is desirable. Two tubes may be stacked in a corner, the lower, larger one absorbing lower frequencies and the upper, smaller one absorbing moderate lows and midrange energy. Half-round units can be used to control sidewall reflections or provide general absorption elsewhere. Whether these types of adjustable modules plus carpet, furnishings, structural absorption (wall, floor, ceiling) combine to provide the proper overall decay rate (liveness, deadness) must be determined either by listening, calculation, or measurement. The Tube Trap is manufactured by Acoustic Sciences Corporation.

A special form of Triffusor with two absorptive sides and one diffusive side is called the Korner Killer. Placed with the absorptive faces into the corner, it directs the diffusive face toward the room. This helps control normal modes and adds diffusion to the room. The diffused reflections are reduced 8 to 10 dB in the diffusion process, minimizing their possible contribution to perceptual confusion of the stereo image. This is in contrast to the higher level limp mass reflections of the Tube Trap in Fig. 15-13B, which as early reflections, may tend to confuse the stereo image. The Korner Killer is manufactured by RPG Diffusor Systems, Inc.

CHAPTER **16**

Control of Interfering Noise

The background noise levels in recording studios, listening rooms, concert halls, and other acoustically sensitive spaces must be minimized if these rooms are to be used in their intended way. Hums, buzzes, rumbles, vibrations, aircraft noise, traffic noise, plumbing noise, dogs barking, or office sounds are most incongruous, if audible. Such sounds might not be noticed outside these spaces when they are a natural part of everyday soundscapes, but during a pause or a quiet musical or speech passage, they are intolerable.

The task of controlling interfering noise is perhaps the most challenging problem in architectural acoustics. Interior room treatment is critically important, but even the most expertly designed room treatment is worthless if noise intrudes into its space. Likewise, in many cases, sound originating in the space and interfering with neighboring spaces can be a serious issue. It is a difficult fact of room design that low ambient noise levels are an important prerequisite for most applications. To be useful, a design must suppress any internally generated noise, such as from air handling equipment, and also provide isolation from any external noise, such as from road traffic. Equally vexing is the fact that quiet air is expensive; there are almost no economical shortcuts in obtaining the low ambient noise levels required for high-quality audio recording, playback, or listening. Still, an understanding of the behavior of noise sources and noise transmission will enable the designer to reach even the most demanding noise criteria, and minimize the effects of unwanted sound.

Approaches to Noise Control

There are five basic approaches to reducing noise in an acoustically sensitive space:

- Locating the room in a quiet place.
- Reducing the noise output of the offending source.
- Interposing an insulating barrier between the noise and the room.
- Reducing the noise energy within the room.
- Both airborne and structureborne noise must be considered.

Locating an acoustically sensitive room away from outside interfering sounds is an elegant solution, but a rare luxury because of the many factors (other than acoustical) involved in site selection. Clearly, sites near airports, railroads, highways, or other noise sources are always problematic. It is useful to remember that doubling the distance from a noisy street or other sound source reduces the level of airborne noise approximately 6 dB. Whenever possible, floor plans should place noise sensitive rooms away from noise sources such as interior machinery rooms, or exterior noise sources such as roads. If the room in question is a listening room or home studio which is part of a residence, due consideration must be given to serving the other needs of the occupants. If the room is a professional recording or broadcast studio, it may be part of a multipurpose complex and the noises originating from business machines, air handling equipment, foot traffic within the same building, or even sounds from other studios, may dominate the situation.

When optimal site planning is not available, the next best solution is to reduce the noise output of the offending source. This is usually the most logical and efficient approach. Sometimes this is possible, but sometimes it is not. For example, a noisy machine may be placed successfully in a sound-insulating box, but it is probably impossible to make traffic engine and tire noise softer. Many techniques are available to reduce the noise output of sources. For example, the noise output of a ventilating fan might be reduced 20 dB by installing a pliant mounting or by separating a metal air duct with a decoupling collar. Installing a carpet in a hall might solve a foot traffic noise problem, or a rubber pad might reduce an equipment vibration problem. In most cases, working on the offending source and thus reducing its noise output is far more productive than corrective measures between the source and the receiving room or within the receiving room in question.

Although it is often difficult and expensive, a common solution is to establish a noise-insulation barrier between the noise source and an acoustically sensitive space. A barrier such as a wall offers a transmission loss to sound transmitted through it, as discussed later. An impinging noise level would be reduced by the transmission loss of the wall. The walls, floor, and ceiling of an acoustically sensitive room must all give at least the required transmission loss to outside noises, reducing them to acceptable levels inside the room. However, a wall with some transmission loss can only reduce the noise level by that amount if no flanking or bypassing of the wall by other paths is present.

Protecting a room from street traffic noise can be difficult. For example, a barrier may be built along a highway to shield nearby residences from traffic noise. Shrubbery and trees can modestly help in shielding from street sounds; for example, a cypress hedge 2-ft thick gives about a 4-dB reduction.

It is sometimes useful to apply treatment to the interior of a noise-source room or to a noise-receiving room, to reduce overall ambient noise levels. The level of noise in a room can be modestly reduced by introducing sound-absorbing material into the room. For example, if a sound-level meter registers an ambient noise level of 45 dB inside a studio, this level might be reduced to 40 dB by covering the walls with quantities of absorbing materials. Going far enough to reduce the noise significantly, however, would probably make the reverberation time too short. The control of reverberation must take priority. The amount of absorbent installed in the control of reverberation will reduce the noise level only slightly, and beyond this we must look to other methods for further noise reduction.

Noise can permeate a studio or other room when it is transmitted through air pathways (airborne) or transmitted through solid structures (structureborne). Noise can also be radiated by diaphragm action of large surfaces or a combination of these three. The sound of a truck driving by is mainly heard as an airborne noise, but some vibration from the tires may be passed through the ground. Likewise, the sound of a plane taking off, a band's loud jam session, and a crying baby are all airborne sounds. Quieter jet engines, reduced volume controls, and pacifiers would all be examples of noise source attenuation. Many structureborne noises are caused by vibration or an impact from a piece of machinery; the energy can be conveyed through a building's structure and reradiated as audible airborne sound. The building structure efficiently converts the vibration into sound, in much the same way that a guitar string by itself produces very little sound, but when attached to a guitar body, the sound can be quite loud.

Airborne Noise

If an airway exists, then sound will easily travel though the air. It is therefore easy for sound to pass through an otherwise highly soundproof barrier. For example, a heavy metal plate with holes occupying 13% of the total area can transmit as much as 97% of the sound impinging on it. Furthermore, increasing the mass of the metal plate will have little or no effect. A good deal of sound can pass through a small crack or aperture in an otherwise solid wall because the air leak has a transmission loss of zero. Similarly, any flanking path will allow sound to travel around a barrier, severely compromising its transmission loss. For example, sound can easily travel from room to room through a common plenum or air-handling ducts.

A crack under a door or a wall penetration created by a loosely fitting electrical outlet box can compromise the insulating properties of an otherwise excellent partition. Air-tightness is absolutely necessary to insulate against airborne noises. For this reason, construction elements such as louvered doors and windows must be avoided, masonry walls should be painted when possible, and it is critically important to seal any holes or leaks in partitions with a flexible, nonhardening sealant. Similarly, rubber gaskets must be used around doors or other openings. Even a careful design can be compromised by sloppy construction that will lead to degraded transmission loss below the design intent.

Transmission Loss

The purpose of a sound-insulating wall or barrier is to attenuate impinging sound, and thus insulate the interior from the outside noise. The wall's ability to attenuate noise passing through it can be specified by its transmission loss (TL). TL is the loss as sound passes through a barrier. In particular, TL can be defined as difference between sound-pressure level (SPL) on the source side of the barrier, and the SPL on the receiver side:

$$TL = SPL_{source side} - SPL_{receiver side} \qquad (16\text{-}1)$$

For example, as shown in Fig. 16-1, if a wall has a transmission loss of 45 dB, an outside noise level of 80 dB would be reduced to 35 dB, that is, (80 dB – 45 dB = 35 dB). A wall with a 60-dB TL would reduce the same noise level to 20 dB. The higher the TL value, the greater the attenuation provided by a material. However, a TL rating is only effective if no flanking or bypassing of the wall by other paths is present.

It is important to note that absorption coefficients (α) are based on a linear scale, and TL values are based on a logarithmic scale. So, comparing them can be misleading. We define τ as the transmission coefficient, the amount of sound that passes through a material where $\tau = 1 - \alpha$. We relate τ to TL as:

$$TL = 10 \log \frac{1}{\tau} \qquad (16\text{-}2)$$

So, for example, a glass fiber material might have a high absorption coefficient of 0.9 at 500 Hz which would yield a τ of 0.1, that is, (1 – 0.9 = 0.1). And TL of the glass fiber would be 10, that is, 10 log (1/0.1), which is quite poor. This explains, among other

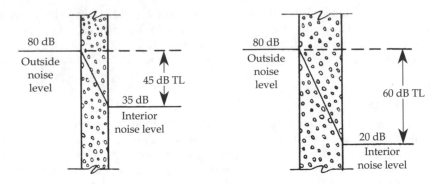

FIGURE 16-1 The difference between the outside noise level and the inside noise level determines the transmission loss (TL) of the wall.

things, why porous absorbers are poor sound barriers, especially at low frequencies. Recalling that porous absorbers allow airflow, this result is not surprising. In fact, as we shall see, solid massive barriers provide the best sound insulation. The total amount of sound power that passes through a partition is proportional to $S\tau$, where S is its area and τ is the transmission coefficient.

Effect of Mass and Frequency

For insulating against outside airborne sounds, the general rule is the heavier the wall the better. The more massive the wall, the more difficult it is for sound waves in air to move it. Figure 16-2 shows how the transmission loss of a rigid, solid wall is related to the density of the wall. The wall weight in Fig. 16-2 is expressed in pounds per square foot

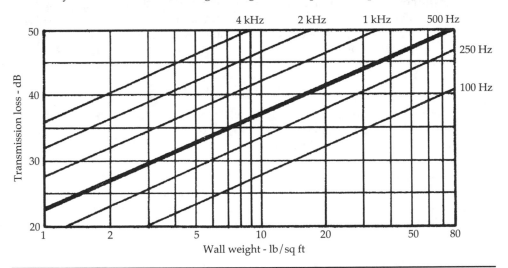

FIGURE 16-2 The mass of the material in a barrier rather than the type of material determines the transmission loss of sound passing through the barrier. The mass law predicts that a doubling of wall mass will increase TL by 6 dB (in practice, 5 dB is usually achieved). The transmission loss also depends on frequency, although values at 500 Hz are commonly used in casual estimates. The wall weight is expressed in pounds per square foot of wall surface.

of surface, sometimes called the surface density. For example, if a 10 × 10 ft concrete-block wall weighs 2,000 lb, the wall weight would be 2,000 lb per 100 ft², or 20 lb/ft². The thickness of the wall is not directly considered.

The transmission loss increases as the mass of the barrier increases. Also, transmission loss increases as the frequency increases. In a single-layer panel (such as concrete or brick wall), these effects can be approximated as follows:

$$TL = 20 \log (fm) - 33 \tag{16-3}$$

where f = frequency of sound, Hz
   m = surface mass of barrier, lb/ft² or kg/m²

Note: In metric units, change 33 to 47.

From these equations, since 20 log (2) equals 6 dB, we see that when mass is doubled, it theoretically results in a 6-dB increase in TL. This is sometimes called the theoretical mass law; it predicts that every doubling of mass will increase TL by about 6 dB. Moreover, we also see that any doubling of frequency (an octave increase) will also result in a 6-dB increase in TL. Because it assumes that the barrier has zero stiffness, this is more accurately known as the limp mass law. But in addition to mass, stiffness and damping also affect transmission loss. For example, real-world barriers have some stiffness; the stiffer the panel, the lower the TL. Moreover, the thicker the panel (which increases mass), the greater the stiffness (which reduces TL). So, the 6-dB prediction is not completely accurate. In practice, a doubling of mass usually yields about a 5-dB increase in TL.

Theoretically, a barrier with a 4-in thickness may have a TL of 40 dB at 500 Hz. If the thickness (and mass) is doubled to 8 in, the new TL is about 45 dB. However, a law of diminishing returns is at work. To achieve another 5-dB increase in TL, the barrier's thickness would have to be increased to 16 in, and another 5 dB would demand 32 in, and so on. Mass is extremely useful for sound insulation, but is not always the best approach. Also, as we will see, other factors such as the coincidence effect will affect TL.

The transmission losses indicated in Fig. 16-2 are based on the mass of the material rather than the type of material. We see that materials with greater mass are more effective at sound insulation. It is mainly the mass of the material that matters, not the material itself. For example, the transmission loss through a layer of lead of certain thickness can be matched by a plywood layer about 95 times thicker.

Coincidence Effect

At lower frequencies, the TL of a barrier is mainly determined by mass; TL increases by about 5-dB/octave. But at some frequency region above this, the stiffness of the barrier will yield a resonance; the barrier will flex and bend at the same wavelength of the incidence sound, and as a result, it will more easily transmit sound. At and around the resonance frequency, the TL will drop by 10 to 15 dB; this is called the coincidence-frequency dip. Above this frequency region, the TL will again rise with frequency according to the mass law, or even exceed the 5-dB/octave slope.

For a given material, the coincidence frequency is inversely proportional to the panel's thickness. Therefore, the coincidence frequency can be raised by decreasing thickness; this can be advantageous because it may move the coincidence-frequency dip to a frequency that is high enough to be above our frequency range of interest, for example, above speech frequencies. However, when thickness is decreased, overall TL is decreased as well.

Different materials exhibit very different coincidence frequencies: An 8-in-thick concrete wall at about 100 Hz; a $\frac{1}{2}$-in-thick plywood panel at about 1.7 kHz; $\frac{1}{8}$-in glass at about 5 kHz; $\frac{1}{8}$-in lead at about 17 kHz. Materials with greater damping have a smaller coincidence-frequency dip. In some applications, a damping layer can be added to a material with little inherent damping. For example, laminated glass (but not thermal glass) has more damping than an ordinary glass pane. A discontinuous structure such as bricks set in mortar conducts sound less efficiently (providing higher TL) than a more homogeneous material like steel.

From Fig. 16-2 we can also see that the higher the frequency, the greater the transmission loss, or in other words, the better the wall is as a barrier to outside noise. In the figure, the line for 500 Hz is made heavier than the lines for other frequencies because it is common to use this frequency for comparisons of walls of different materials. However, below 500 Hz the wall is less effective as a sound barrier, and for frequencies greater than 500 Hz it is more effective.

Separation of Mass

Ideally, transmission loss could be greatly improved by positioning two barriers between the noise source and the receiving room. This is because masses that are separated by an unbridged air cavity are extremely effective isolators. For example, the combined TL of two identical walls would theoretically be twice the individual TLs. This would be much more effective than simply doubling the surface weight of one wall. That is because two panels separated by an unbridged air cavity provide better TL than one panel. For example, an 8-in concrete wall may have a TL of 50 dB, and a 16-in concrete wall may have a TL of 55 dB. Two separated 8-in concrete walls will outperform one 16-in wall. However, although ideal, completely unbridged cavities are unattainable; thus, the combined TL will be much less than 80 dB (40 + 40).

Only in the case of two separate structures, each on its own foundation, is an unbridged condition approached. Walls are usually at least connected at the footer and headers. The depth of an air cavity between panels affects the system's stiffness; the larger the cavity, the lower the coincidence-effect resonant frequency. Generally, the cavity depth between masses should be as great as allowable; very narrow cavities can yield poor performance. The coincidence-frequency dip of two separated panels can be reduced by using two panels of different surface weight. For example, performance of a double-pane window can be improved by using panes of different thickness.

Composite Partitions

The total sound power passing through a composite partition, for example, a wall with a door and a window, is the sum of the sound powers transmitted by each element at some specified frequency. Each element's sound power is calculated by multiplying its area S_i by its transmission coefficient τ_i. The total transmitted sound power is thus $\Sigma S_i \tau_i$. Alternatively, STC may be used instead of τ, as described later. Ideally, to avoid a weak link in a composite partition, each element should convey the same transmitted sound power. The TL of different elements is not necessarily equal, but their $S\tau$ products ideally should be equal. For example, if a wall has a large window comprising 20% of the total area, and a door comprising 3.5% of the area, the respective TL values for the wall, window, and door should be: 33.6, 27.8, and 20.2 dB. This would yield a composite TL of 30 dB at some specified frequency.

A weak link in a composite partition is an issue of real concern. For example, suppose that a brick wall provides a TL of 50 dB. When a window with a TL of 20 dB, occupying 12.5% of the total area, is introduced, the composite TL drops to 29 dB. The damage has been done, and a larger-sized window is not much worse; for example, when the window occupies half of the partition, the composite TL is 23 dB. The most serious weak link is an air leak. For example, a hole measuring 1 in^2 will transit as much sound as an entire 100-ft^2 gypsum-board partition. As another example, suppose that an opening 10-in wide allows noise to enter a room to yield a 60-dB noise level. If the opening is reduced tenfold to 1 in, ignoring diffraction, the noise level would still be 50 dB. If the opening is reduced to one-hundredth of its original size to 0.1 in, the noise level would still be 40 dB. In other words, even a small air leak is highly detrimental.

Porous Materials

Porous materials such as glass fiber (rock wool, mineral fiber) are excellent sound absorbers and good heat insulators. However, as noted, they are of limited value in insulating against sound when used as standalone absorbers or when placed on wall surfaces. Using glass fiber to reduce sound transmission will help, but only moderately. The transmission loss for porous materials is directly proportional to the thickness traversed by the sound. This loss is about 1 dB (100 Hz) to 4 dB (3 kHz) per inch of thickness for a dense, porous material (rock wool, density 5 lb/ft^3) and less for lighter materials. This direct dependence of transmission loss on thickness for porous materials is in contrast to the relatively high transmission loss for solid, rigid walls. As described later, porous absorbers can improve sound insulation when placed inside wall cavities.

Building insulation installed within a wall increases its transmission loss a modest amount, primarily by reducing cavity resonance that would tend to couple the two wall faces at the resonance frequency of the cavity. A certain increase in the transmission loss of the wall can also be attributed to attenuation of sound in passing through the glass fiber material, but this loss is small because of the low density of the material. Considering all mechanisms, the transmission loss of a staggered-stud wall with a layer of gypsum board on each side can be increased about 7 dB by adding 3.5 in of glass fiber insulation. A double wall might show as much as a 12 dB increase by adding 3.5 in and 15 dB with 9 in of insulation.

Porous absorbers are useful for reducing reflected or ambient sound in a noise-source room. Additional absorption is effective at reducing sound levels when a room is relatively reflective; in some cases, a 10-dB reduction may be achieved. Clearly, when a worker is near a loud noise source, absorption on the walls will not reduce direct sound reaching the worker. When possible, noise sources should be placed closer to the center of the room, and away from walls. Absorption is also somewhat helpful in a noise-receiving room, to lower ambient sound levels. The amount of sound transmitted through a barrier also depends on the surface area of the barrier. In particular, the noise reduction between two adjacent rooms is given by:

$$\text{NR} = \text{TL} + 10 \log \left(\frac{A_{\text{receiving}}}{S} \right) \tag{16-4}$$

where NR = noise reduction, dB
    TL = transmission loss of barrier
  $A_{\text{receiving}}$ = absorption in receiving room, sabins
    S = surface area of common barrier, ft^2

Sound Transmission Class

The solid line of Fig. 16-3 is a replotting of data from the mass law graph of Fig. 16-2 for a wall weight of 10 lb/ft². If the mass law were perfectly followed, we would expect the transmission loss of a practical wall of this density to vary with frequency, as shown by the solid line. However, actual measurements of transmission loss of this wall might be more like the broken line of Fig. 16-3. These deviations reflect resonance (such as the coincidence effect) and other factors of the wall panel, which are not included in the simple mass law concept.

Because of such commonly occurring irregularities, it is of practical value to use a single number that gives a reasonably accurate indication of the sound transmission loss characteristics of a wall. This is done with a procedure specified by the American Society for Testing and Materials to determine the sound transmission class (STC) of a sound insulating wall. The ASTM E-413 standard states that STC is designed to correlate with subjective impressions of sound isolation from normal sources in homes and offices. Although STC values are very helpful, they are not designed to describe more "industrial" applications such as recording studios, in particular when loud music is the noise source. This is because STC is intended mainly for use in the speech-frequency region.

To determine the STC of a wall, the measured graph of the wall is compared to a reference graph (STC contour) using a special procedure. This standard contour covers the frequency range from 125 Hz to 4 kHz. The contour comprises three line segments with different slopes: 3 dB per $\frac{1}{3}$-octave from 125 to 400 Hz; 1 dB per $\frac{1}{3}$-octave from 400 Hz to 1.25 kHz; flat from 1.25 to 4 kHz. The first segment has a rise of 15 dB, and the second segment has a rise of 5 dB. After comparing the measured graph with the standard contour, the STC of the panel is the TL value of the standard contour at 500 Hz. The higher the STC, the higher the TL.

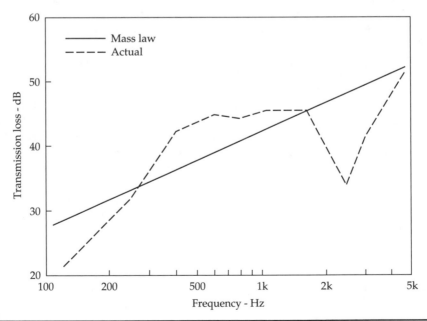

FIGURE 16-3 Actual measurements of transmission loss in walls often deviate considerably from the mass law (Fig. 16-2) because of resonances and other effects.

The results of such classification have been applied to walls of various types for ready comparison. An STC rating of 50 (dB is omitted) for a wall would mean that it is better in insulating against sound than a wall of STC 40. In practice, given a quiet receiving room, a wall with an STC of 30 would allow speech to be heard on the other side. An STC of 50 would block most loud speech, but music could be heard. An STC of 70 would block all speech, but some music, particularly bass frequencies, may be heard.

It is not proper to consider STC ratings as averages but the procedure does escape the pitfalls of averaging dB transmission losses at various frequencies. As noted, the STC value only covers the speech frequency range, so it has limited applications when music is the sound source. In particular, a high STC rating will not ensure good low-frequency attenuation. In fact, the STC rating is not intended for external partitions where, for example, traffic noise may be an issue. Also, the STC does not account for specific dips in transmission loss. In some cases, the ceiling sound transmission class (CSTC) or the outdoor-indoor transmission class (OITC) is cited. These are similar to STC but are determined by somewhat different procedures. In both cases, the higher the value, the greater the transmission loss.

Comparison of Wall Structures

Figure 16-4 shows the measured performance of a 4-in concrete-block wall as a sound barrier. A concrete-block wall provides good sound isolation with a mild coincidence-frequency dip. It is interesting to note that plastering both sides increases the transmission loss of the wall from STC 40 to 48. Figure 16-5 shows a

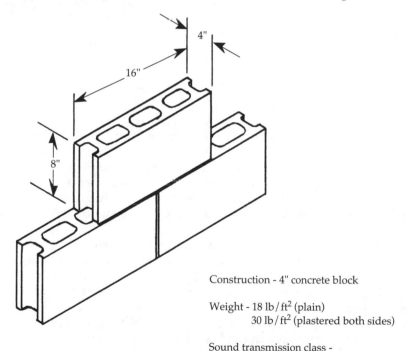

Construction - 4" concrete block

Weight - 18 lb/ft² (plain)
          30 lb/ft² (plastered both sides)

Sound transmission class -
          STC 40 dB plain
          STC 48 dB plastered both sides

Figure 16-4   Concrete-block (4-in) wall construction. (*Solite Corporation*)

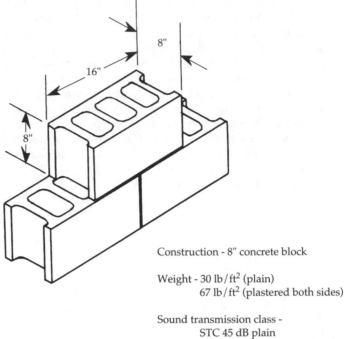

Construction - 8" concrete block

Weight - 30 lb/ft² (plain)
           67 lb/ft² (plastered both sides)

Sound transmission class -
           STC 45 dB plain
           STC 56 dB plastered both sides

FIGURE 16-5   Concrete-block (8-in) wall construction. (*Solite Corporation, and LECA*)

considerable improvement in doubling the thickness of the concrete-block wall. In this example, the STC 45 is improved 11 dB by plastering both sides. By adding gypsum-board partitions to both sides of a concrete-block wall, STC can be increased to about 70. Figure 16-6 shows the very common 2 × 4 frame construction with ⅝-in gypsum-board covering. This kind of partition can also offer good isolation, albeit with a coincidence-frequency dip; the two sides of the partition provide separation of mass. The STC of 34 without glass fiber in the wall cavity is improved only 2 dB by filling the cavity with glass fiber material, a meager improvement that would probably not justify the added cost. Of course, glass fiber may be warranted for thermal insulation or other reasons.

Figure 16-7 shows an acoustically efficient and inexpensive staggered-stud wall construction. The studs are alternately connected to the panels, but tied to the same headers and footers. Here the inherently low coupling between the two independent wall diaphragms is further improved by filling the space with glass fiber. Ideally, the top and bottom of the partition should be decoupled from the ceiling and floor. Attaining the full STC 52 rating would require careful construction to ensure that the two wall surfaces are truly independent and not acoustically "shorted out" by back-to-back electrical conduits, outlet boxes, or other devices placed in the same cavity.

The last wall structure to be described is the double wall construction of Fig. 16-8. The two walls are entirely separate, each having its own 2 × 4 plate. Without glass fiber this wall is only 1 dB better than the staggered-stud wall of Fig. 16-8, but by filling the

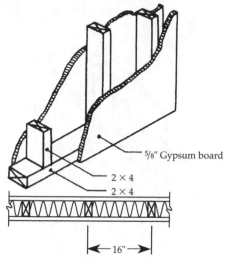

5/8" Gypsum board

2 × 4

2 × 4

16"

Construction - Standard-stud partition

Weight - 7.3 lb/ft^2

Sound transmission class -
STC 34 dB without glass fiber
STC 36 dB with 3^1/2" glass fiber

FIGURE 16-6 Standard-stud partition. (*Owens-Corning Corporation*)

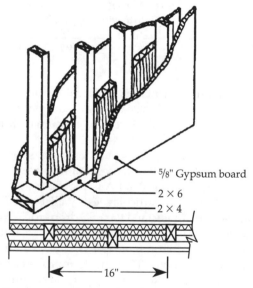

5/8" Gypsum board

2 × 6

2 × 4

16"

Construction - Staggered-stud partition

Weight - 7.2 lb/ft^2

Sound transmission class -
STC 42 dB without glass fiber
STC 46 to 52 dB with glass fiber

FIGURE 16-7 Staggered-stud partition. (*Owens-Corning Corporation*)

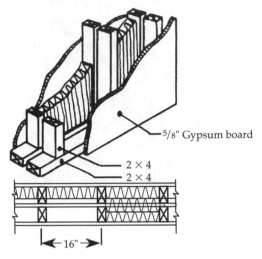

5/8" Gypsum board

2 × 4
2 × 4

←— 16" —→

Construction - Double wall

Weight - 7.1 lb/ft^2

Sound transmission class -
    STC 43 dB without glass fiber
    STC 55 dB with 3^1/$_2$" glass fiber
    STC 58 dB with 9" glass fiber

FIGURE 16-8   Double-wall partition. (*Owens-Corning Corporation*)

inner space with glass fiber insulation, STC ratings up to 58 dB are possible. In some designs, QuietRock is specified instead of gypsum board. QuietRock is an internally damped wallboard; it is more expensive than gypsum board, but is superior in terms of STC rating.

To demonstrate the principle of separation of masses, consider the four gypsum-board partitions in Fig. 16-9. Each wall uses double-plate construction with wood studs, insulation, and no cross bracing. Wall A is probably the most common construction, yielding an STC of 56 in this particular test. Wall B adds an internal gypsum-board layer, but this increases coupling, and the STC drops to 53. Wall C adds another internal layer, and STC drops further to 48 dB. Mass is increased, but the separating air cavity is degraded by the additional layers. In wall D, the additional layers are moved to the outside surfaces of the partition; STC is much improved at 63. In all cases, for best performance gypsum-board joints should be offset and carefully taped. This example demonstrates that how materials are arranged can be more important than the quantity of materials used. In this example, we see that the technique of separating masses is effective at isolation.

It was stated earlier that porous sound absorbing materials are of limited value in insulating against sound. This is true when normal transmission loss is considered, but in structures such as those in Figs. 16-7 and 16-8, porous materials helpfully absorb sound energy in the cavity, and the added damping decreases the coincidence-frequency dip. Cavity absorption can improve the transmission loss in some wall

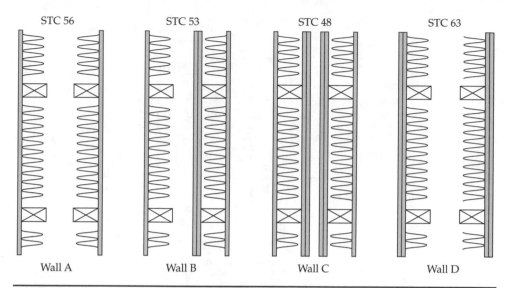

FIGURE 16-9 Four designs for gypsum-board partitions with glass fiber fill. Wall A has STC of 56; wall B has STC of 53; wall C has STC of 48; wall D has STC of 63. (*Berger and Rose*)

structures by as much as 15 dB, principally by reducing resonances in the space between the walls, while in others the effect is negligible. The low-density mineral fiber batts commonly used in building construction are as effective as high-density boards, and they are much cheaper. Mineral fiber batts within a wall may also meet certain fire-blocking requirements in building codes. Any absorption material must be placed loosely in the cavity; tightly packed material can decrease TL.

The staggered-stud wall and the double wall, on the basis of mass alone, would yield a transmission loss of only about 35 dB (Fig. 16-2). The isolation of the inner and the outer walls from each other and the use of insulation within can increase the wall isolation effectiveness by 10 to 15 dB.

Sound-Insulating Windows

If a window is placed in the wall between a control room and studio, or in a wall facing loud outdoor ambient sound levels, the window's sound transmission loss should be comparable to that of the wall itself. A window with insufficient TL would be a weak link that would seriously compromise the acoustical isolation of the wall. A well-built staggered-stud or double-stud wall might have an STC of 50 dB and provide sufficient isolation, as would a concrete-block wall. To approach this performance with a window requires very careful design and installation. A typical single pane $1/8$-in glass may have an STC of 25. Prefabricated double-pane, sound-insulating windows are available commercially and may, for example, provide an STC rating of 50 or more. On the other hand, a double-pane window designed for thermal insulation may have an STC value less than a single-pane window.

In most critical applications, a sound-isolating double window is indicated; a triple window adds little to the TL. The mounting must minimize coupling from one wall to the other. One source of coupling is the window frame, another is the stiffness of the air

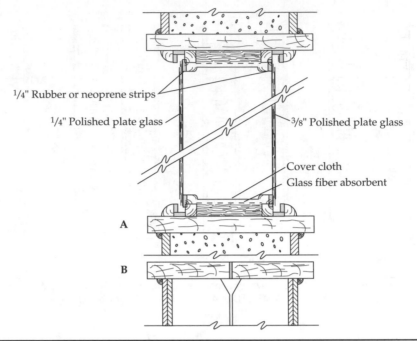

¹/₄" Rubber or neoprene strips

¹/₄" Polished plate glass

³/₈" Polished plate glass

Cover cloth
Glass fiber absorbent

A

B

Figure 16-10   Wall construction with double-pane window.

between the glass panels. The plan of Fig. 16-10 shows a practical solution to the double-window problem for concrete-block walls. Figure 16-10B is an adaptation to the staggered-stud construction. In the latter there are, in effect, two entirely separate frames—one fixed to the inner and the other to the outer staggered-stud walls. A felt strip may be inserted between them to ensure against accidental contact. As in any construction, air leakage must be prevented.

Heavy plate glass should be used, and the heavier the better. However, even heavy glass will exhibit a coincidence-frequency dip. As noted, there is a slight advantage in having two panes of different thickness. If desired, one glass can be inclined to the other to control light or external sound reflections, but this angle will have negligible effect on the transmission loss of the window itself. The glass should be isolated from the frame by rubber or other pliable strips. The spacing between the two glass panels has an effect. The greater the spacing, the greater the transmission loss. However, there is little gain in going beyond 8 in, nor serious loss in dropping down to 4 or 5 in. However, a small spacing in a double-pane window, perhaps less than 1 in, can yield lower STC than a single-pane window.

The absorbent material between the pane edges in the design of Fig. 16-10 discourages resonances in the airspace. This adds significantly to the overall insulation efficiency of the double window, and it should extend completely around the periphery of the window cavity. If the double window of Fig. 16-10 is carefully constructed, sound insulation should approach that of an STC 50 wall but will probably not quite reach it, particularly if the window area is large. For the staggered-stud wall in which a double window is to be placed, the use of a 2 × 8 plate instead of the 2 × 6 plate will simplify mounting of the inner and outer window frames.

Sound-Insulating Doors

Doors can present a difficult design challenge. A fixed, sealed window can provide good isolation, but a door, by nature, is not fixed and is more difficult to seal. In fact, whenever possible, doors should not be placed in partitions where high transmission loss is desired. The transmission loss of a door is determined by its mass and stiffness, and by the air-tightness of its seals. An ordinary household panel door (hollow core) hung in the usual way might offer less than 20 dB of sound insulation. Increasing the door weight (solid core) and taking reasonable precautions on seals might gain another 5 or 10 dB, but a door that matches a 50-dB wall requires great care in design, construction, and maintenance. Steel doors or patented acoustical doors yielding specified values of transmission loss are available commercially but they are quite expensive. To avoid the expense of high transmission-loss doors, sound locks are commonly used. These small vestibules with two doors of medium transmission loss are very effective. Sound locks also offer the practical advantage of partial isolation when one door (but not both) is momentarily opened.

Doors with good insulating properties can be constructed if the requirements of mass, stiffness, and airtightness are met. Figure 16-11 suggests one inexpensive approach to the mass requirement, filling a hollow door with sand. Heavy plywood (¾ in) is used for the door panels. Mounting hardware and framing must be upscaled as needed to support the added weight.

Achieving a good seal around a so-called soundproof door can be very difficult. Great force is necessary to seal a heavy door. Wear and tear on pliant sealing strips can destroy their effectiveness, especially at the floor where footwear is a problem. The detail of Fig. 16-11 shows one approach to the sound leakage problem in which an absorbent edge built around the periphery of the door serves as a trap for sound traversing the gap between door and jamb. This absorbent trap could also be embedded in the door jamb. Such a soft trap could also be used in conjunction with one of the several types of seals.

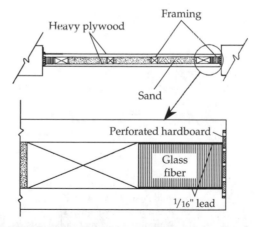

FIGURE 16-11 A reasonably effective and inexpensive acoustic door. Dry sand between the plywood faces adds to the mass and thus the transmission loss. Sound traveling between the door and jamb tends to be absorbed by the absorbent door edge.

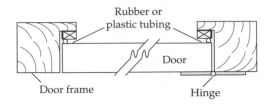

Figure 16-12   A door can be sealed by compressible rubber or plastic tubing held in place by a fabric wrapper.

Some door designs use hinged cams that slightly lift the door as it opens, then lowers the door into its frame when it is closed. In this way, the weight of the door itself presses it tightly against its seals, improving isolation. In some door designs, a seal automatically drops down to a threshold when the door is closed, and lifts when the door is opened; this helps prevent wear and tear on the seal.

Figure 16-12 shows a simple door seal that is reasonably satisfactory. This seal comprises a soft rubber or plastic tubing an inch or less in outside diameter with a wall thickness of about $^3/_{32}$ in. The wooden nailing strips hold the tubing to the door frame by means of a pliant wrapper. A raised sill is required at the floor if the tubing method is to be used all around the door (or another type of seal such as weatherstripping could be used at the bottom of the door). An advantage of tubing seal is that the degree of compression of the tubing upon which the sealing properties depend is available for inspection.

A complete door plan is shown in Fig. 16-13. It is based upon a 2-in-thick solid slab door and utilizes a magnetic seal such as used on refrigerator doors. The magnetic material is barium ferrite in a PVC (polyvinyl chloride) rod. In pulling toward the mild steel strip, a good seal is achieved. The aluminum strip C decreases sound leakage around the periphery of the door.

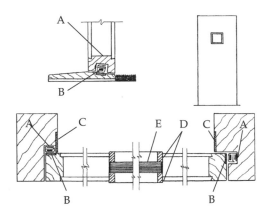

A - PVC magnetic seal
B - Mild steel $3/4" \times 1/8"$ flat
C - Aluminum cover strip $1 1/2" \times 1/8"$
D - Aluminum glazing beads $3/4" \times 3/8"$
E - Polished plate glass - $5/4"$

Figure 16-13   A door design utilizing magnetic seals of the type used on refrigerator doors.

It is possible to obtain a very slight acoustical improvement by padding both sides of a door. A plastic fabric over 1-in foam rubber sheet can be quilted with upholstery tacks. However, wear and tear may be a drawback.

Structureborne Noise

Sound travels very efficiency (with little attenuation) through dense materials, including such building materials as concrete and steel. Structureborne noise such as vibration from outside traffic or HVAC (heating, ventilating, and air-conditioning) units, or even the impact of footsteps in a distant part of a building can easily reach an acoustically sensitive room. Structureborne vibrations can cause walls and floors to vibrate, reradiating energy as airborne noise. Unwanted sounds can invade a room by mechanical transmission through solid structural members such as wood, steel, concrete, or masonry. Air handling noises can be transmitted to a room by the sheet metal of the ducts (as well as by the air in the ducts), or both. Water pipes and plumbing fixtures, unfortunately, have excellent sound-carrying capabilities. Structureborne noise is thus most efficiently controlled at the source of the noise. Massive, rigid partitions such as concrete walls are most useful for attenuating airborne noise, but offer little resistance to structureborne noise. On the other hand, lightweight materials offer little protection against airborne noise, but can be used to decouple elements of structures, and are thus effective against structureborne noise.

It is difficult to make a solid structure vibrate through airborne noise falling upon it because of the inefficient transfer of energy from tenuous air to a dense solid. On the other hand, a motor bolted to a floor, a slammed door, or a machine on a table with legs on the bare floor can cause the structure to vibrate significantly. These vibrations can travel great distances through solid structure with little loss. With wood, concrete, or brick beams, longitudinal vibrations are attenuated only about 2 dB in 100 ft. Sound travels easily in solids; for example, sound travels in steel about 20 times farther than in air for the same loss. Although joints and cross-bracing members increase the transmission loss, TL is still very low in common structural configurations. An efficient way to minimize structureborne transmission is through decoupling. For example, vibration from a machine placed directly on a concrete floor will pass easily through the floor, but if the machine is mounted on springs on an isolation pad, which in turn is also decoupled from the floor, the vibration pass-through can be greatly reduced.

Structureborne noise is often generated by an impact on a structural surface, or by a vibration that enters the structure. Even a brief impact can impart tremendous energy into a structure. For example, footsteps on a wooden floor can transmit sound loudly to the room below. The impact insulation class (IIC) is a single value rating that can be used to quantify impact noise in a floor/ceiling. Measurements are taken at sixteen $1/_3$-octave bands from 100 to 3,150 Hz and are overlaid on a standard contour. The higher the IIC value, the better the isolation, and the lower the received noise level. Because the standard contour does not fully account for poor performance at low frequencies, the IIC can sometimes overrate lightweight floors.

Perhaps the most efficient way to minimize impact noise in a floor/ceiling is to install a soft floor covering. For example, a carpet and underlay could be placed on a concrete floor. This greatly reduces impact. The IIC value might improve by 50 points; however, the STC value would remain about the same. Carpet is somewhat less effective on wooden floors, but still provides significant improvement. As described below, a floating floor can also be used to improve IIC as well as STC.

Noise Transmitted by Diaphragm Action

Although very little airborne sound energy is transmitted directly to a rigid structure, airborne sound can cause a wall to vibrate as a diaphragm and the wall, in turn, can transmit the sound through the interconnected solid structure. Such structureborne sound might then cause another wall at some distance to vibrate, reradiating noise into the space we are interested in protecting. Thus, two walls interconnected by a solid structure can serve as a coupling agent between exterior airborne noise and the interior of the listening room or studio itself.

Floating Floors

As noted, the best way to reduce the transmission of structureborne noise is by decoupling one structure from another; any discontinuity will help interrupt the transmission path. A decoupled "room-within-a-room" construction technique is often used to minimize the problem of structureborne noise. The design begins with floating floors, upon which are built floating walls and floating ceilings, all decoupled from the potentially vibrating structural room construction. A floating floor consists of an additional floor raised and decoupled from the structural floor. The floating floor may be built of wood or concrete. In either case, it is isolated from the structural floor by using springs, isolation blocks, or other elements. It is important to verify that the structural floor can withstand the added weight of the floating floor. A well-constructed floating floor, particularly a floating concrete slab, can provide a high IIC value as well as a high STC value.

A floating floor acts as a mechanical low-pass filter system, attenuating low-frequency noise and vibration above its cutoff frequency. Clearly, the cutoff frequency should be lower than the frequency of any undesired noise or vibration. The cutoff frequency of the system is:

$$f_0 = \frac{3.13}{d^{1/2}}$$ (16-5)

where f_0 = cutoff frequency, Hz
  d = static deflection of the isolator, in or cm

Note: In metric units, change 3.13 to 5.

The static deflection is the decrease in the height of the isolating pads or springs when the floating floor is placed on them. The cutoff frequency should be at least half (preferably one-quarter) of the lowest frequency to be attenuated.

Construction of a concrete floating floor begins by placing compressed glass fiber board (perhaps 1-in thick) around the perimeter of the room to isolate the floating floor from the structure, as shown in Fig. 16-14. A strip of wood (perhaps $\frac{1}{2}$-in thick) is set on top of the perimeter board. Compressed glass fiber cubes, molded neoprene cubes, or other isolation means are distributed across the structural floor so that compression of the mounts and the loading are matched. Sheets of plywood (perhaps $\frac{1}{2}$-in thick) are laid on the mounts and the edges are fastened together with metal straps and screws. The plywood is covered by a plastic-sheet vapor barrier, overlapped the edges by at least a foot, and running up the perimeter boards. When preparing a form for a poured floating floor, it is essential to ensure that concrete from the pour cannot leak down to

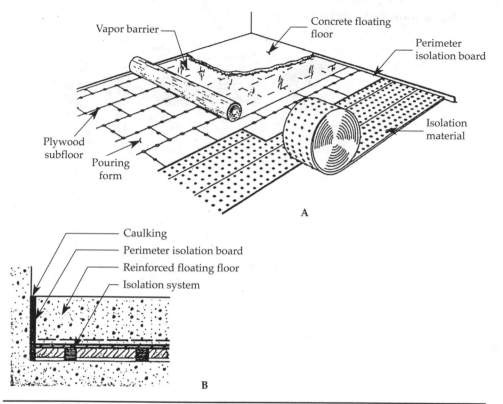

FIGURE 16-14   Example of a concrete floating floor construction. (A) Overall floor assembly. (B) Detail of floor perimeter. (*Kinetics Noise Control*)

touch the structural floor; this would "short circuit" the intended discontinuity and seriously compromise the performance of the floating floor.

Welded screen reinforcing mesh is blocked to be positioned in the center of the floating slab. The floating slab is then poured. Care is taken to ensure that the floating slab does not touch the structural elements of the room. After the slab is cured, the plastic sheet can be cut, and the wood strip removed. The resulting perimeter gap is sealed with a nonhardening sealant. In some designs, the isolation pads have metal housings and an internal screw. The floating slab is poured directly onto the vapor barrier resting on the structural slab, and after it is cured, the floating slab is raised using floor jacks from above.

A floating floor can also be constructed of plywood, and placed over a concrete or plywood structural floor, as shown in Fig. 16-15. Construction of the floating floor begins by placing compressed glass fiber blocks or boards, or other isolation means on the structural floor. The perimeter of the floating floor is isolated from the surrounding walls. Wood sleepers are placed over the pads and glass fiber is placed between the sleepers. One or more layers of plywood panels are secured to the sleepers. This type of floor can be constructed in residential housing, and provide good performance. Carpeting will improve the IIC value. However, the construction is not particularly good at insulating low-frequency impact noise.

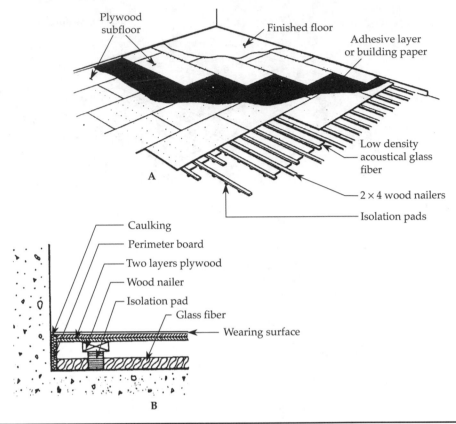

FIGURE 16-15 Example of a wooden floating floor construction. (A) Overall floor assembly. (B) Detail of floor perimeter. (*Kinetics Noise Control*)

Floating Walls and Ceiling

Floating walls and ceilings are isolated from the structural floor, walls, and ceiling to isolate the inner room from structureborne noise. Floating walls can be constructed with metal studs placed on the floating floor. The top of each wall section is secured by resilient sway braces connected to the structural wall. The walls can be faced with two layers of standard gypsum drywall. Glass fiber insulation can be used. It is important to ensure than any penetrations in the wall such as pipes or ducts are isolated from the floating wall. Floating (suspended) ceilings are constructed of metal frames that are supported by wires on isolation hangers hung from the structural ceiling. Drywall is used to cover the frame. An acoustical sealant is used around the periphery of the ceiling.

Noise and Room Resonances

Room resonances can affect the problem of outside noise in a studio. Any prominent modes persisting in spite of acoustic treatment make a room very susceptible to interfering noises having appreciable energy at these frequencies. In such a case, a feeble interfering sound could be augmented by the resonance effect to a disturbing level. Increased isolation or increased absorption around that modal frequency is required.

Noise Standards and Specifications

Hearing damage is serious occupational hazard. Factory workers, truck drivers, and many others are subject to noise levels that are potentially harmful; over time, with repeated exposure, hearing loss may occur. The federal Occupational Safety and Health Administration (OSHA) in the Department of Labor maintains noise exposure limits in the workplace. Noise exposure is measured in daily noise doses for an 8-hour day. Table 16-1 lists the permissible daily noise exposure, measured with the slow response of a standard sound-level meter. The maximum allowed dose is 100% of the daily limit. A dose is calculated as the time a worker is exposed to different noise levels, relative to the maximum exposure time permitted at that level. For example, a worker may be exposed to a maximum of a 90-dBA noise for 8 hours, a 100-dBA noise for 2 hours, or a 115-dBA noise for 15 minutes. When the daily exposure is due to two or more noise levels, the total noise dose is given by:

$$D = \frac{C1}{T1} + \frac{C2}{T2} + \frac{C3}{T3} + \cdots \cdot$$

(16-6)

where C = duration of exposure, hours
T = noise exposure limit, hours

For example, when a worker is exposed to a noise level of 100 dBA for 1.5 hours, and a 95-dBA level for 0.5 hour, the noise dose is: $D = 1.5/2 + 0.5/4 = 0.90$. Thus, the worker has been exposed to 90% of the maximum permissible noise.

The time-weighted and A-weighted sound level in dBA, sometimes referred to as TWA, may also be computed as TWA = $105 - 16.6 \log (T)$. A subsequent hearing conservation measure calls for a TWA given as $100 - 16.6 (T)$.

Exterior environmental noise sources such as traffic, aircraft, and industrial noise are often difficult to quantify because they usually vary over time. The equivalent steady sound level (L_{eq}) can be used to specify noise levels; it is measured in dB or dBA. It is the sound-pressure level; if it was constant, that would equal the sound energy in a varying noise. Measurement periods of 1 and 24 hours are sometimes used. The day-night equivalent sound level (L_{dn}) measures noise levels over a 24-hour period. A value of 10 dB is

Time Duration per Day T (hours)	Maximum Permissible Exposure (dBA)
8	90
6	92
4	95
3	97
2	100
1.5	102
1	105
0.5	110
0.25 or less	115

TABLE 16-1 OSHA Permissible Noise Exposure Limits

added to levels measured at nighttime between 10 p.m. and 7 a.m. because of their greater annoyance at night.

The Federal Aviation Administration may use L_{dn} to specify aircraft noise levels. In some cases, a community noise equivalent level (CNEL) is employed; it differs from L_{dn} by adding weighting factors for evening hours between 7 and 10 p.m. The L_{eq} for evening hours is increased by 5 dBm, and the L_{eq} for nighttime hours is increased by 10 dB. A variety of other metrics are used to quantify aircraft noise, including perceived noisiness (PN), judged or calculated perceived noise level (PNL), tone-corrected perceived noise level (TPNL), effective perceived noise level (EPNL), and others.

When a room's purpose is primarily for spoken word, it may be useful to specify the acceptable noise level in terms of the speech interference level (SIL). SIL is calculated as the average of the SPL levels in octaves centered at 500; 1,000; 2,000; and 4,000 Hz. For calculation purposes, the female SIL is set at 4 dB below that of the male SIL.

Other noise exposure regulations have been devised by the Environmental Protection Agency, Department of Housing and Urban Development, Workman's Compensation, and other agencies and nongovernment groups. These regulations are subject to frequent change.

CHAPTER **17**

Noise Control in Ventilating Systems

I n any acoustically sensitive room, interfering sounds can come from adjacent rooms, from equipment in other areas, or can intrude from the outside. Some rooms have their own noise problems, such as noise generated by equipment cooling fans, hard-disk drives, and so on. But there is one source of noise that is common to almost all acoustically sensitive rooms, and that is the noise coming from the HVAC (heating, ventilating, and air-conditioning) system, including its motors and fans, ducts, diffusers, and grilles. The minimization of these noises is the subject of this chapter.

The control of heating and air-conditioning noise and low-frequency vibrations can be expensive. A noise specification in an air-handling contract for a new structure can escalate the price. Alterations of an existing air-handling system to correct high noise levels can be even more expensive. It is important for studio designers to have an understanding of potential noise problems in air-handling systems so that adequate control and supervision can be exercised during planning stages and installation. This applies equally to the most ambitious professional studio and to a home listening room. Quiet air can be obtained by considering five factors: careful specification and installation of noise sources such as motors, fans, and grilles; use of sound absorbers in machine rooms and in air ducts to reduce noise; minimizing air turbulence noise in ducts and at duct openings; minimizing acoustical crosstalk between rooms; reducing fan speed and air velocity.

Selection of Noise Criterion

One of the first decisions concerning background noise is the selection of a noise-level goal. The essential question of "How quiet should it be?" is complicated by the fact that the noise level across the entire audio spectrum must be considered. Moreover, because our hearing response is not flat with regard to amplitude or frequency, we must assume that the noise floor criteria will not be flat either. To help quantify the specification of noise criteria, and to more easily communicate it, several noise criteria standards have been devised.

One approach to the question of noise specification is embodied in the family of balanced noise criteria (NCB) curves, as shown in Fig. 17-1. The selection of one of these NCB contours, ranging from NCB-10 to NCB-65, and patterned after human equal-loudness contours, establishes the maximum allowable constant noise-pressure level in each octave band. Putting the noise goal in this form makes it easily measurable. The downward slope of these contours reflects both the lower sensitivity of the human ear at low frequencies and the fact that most noises with distributed energy

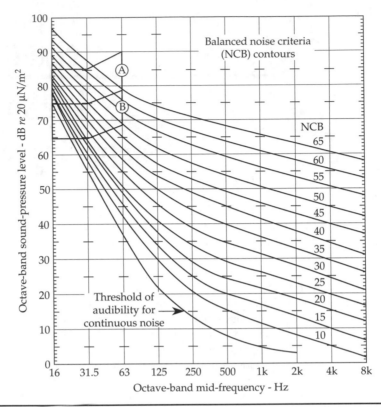

FIGURE 17-1   Balanced noise criteria contours for occupied rooms. (*Beranek*)

drop off with frequency. The NCB-0 curve represents the threshold of audibility for continuous sound. The NCB curves are described in the ANSI S12.2 standard.

To determine whether the noise in a given room meets the contour goal selected, sound-pressure level readings are made in each octave at 16-Hz to 8-kHz center frequencies, using a sound-level meter equipped with octave filters, in an unoccupied building. Results are plotted on the standard graph of Fig. 17-1, and a "curve fitting" procedure is used to compare readings to the family of standard curves. The nearest NCB contour that is above the plotted readings is shifted downward, so that the shifted NCB contour is tangent to the plotted readings. The amount of the shift is noted in decibels. The rating of the measured noise level is the shifted NCB contour minus the shift. For example, a NCB-25 contour shifted down by 2 dB yields a measured rating of NCB-23. If the NCB-25 contour had been specified as the highest permissible sound-pressure level in an air-conditioning contract, the HVAC installation in this example would be acceptable.

NCB values are further defined by comparing the perceived balance between low- and high-frequency sounds. The value is then rated as N (neutral), H (hissy), or R (rumbly), depending on the predominant spectral response. Neutral represents a balanced noise spectrum. Hissy defines a noise that contains excessive energy in the 1- to 8-kHz region. Rumbly defines noise that contains excessive energy in the 16- to 500-Hz region. For example, a NCB-25(R) value would indicate a quiet room, but with some rumble below 500 Hz that exceeds the NCB contour by at least 3 dB. The region below 63 Hz (A and B of Fig. 17-1) is further specified according to RV (vibration). Octave-band sound-pressure levels of the

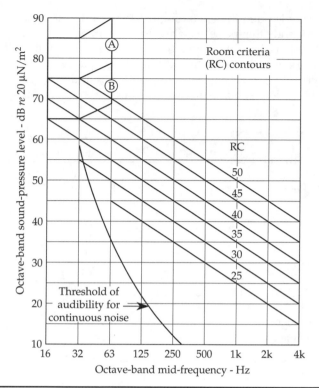

FIGURE 17-2 Room criteria contours for occupied rooms.

magnitudes indicated may induce rattles or vibrations in lightweight partitions and ceiling constructions. As with any single-value audio specification, NCB values can sometimes be misleading because they do not account for specific frequency-response differences. Two rooms might have the same NCB, but have very different acoustic spectra.

In some cases, room criteria (RC) values are specified; this method is similar to the NCB method. RC contours are measured from 16 Hz to 4 kHz, and range from RC-25 to RC-50, as shown in Fig. 17-2. The contours are a series of straight-line –5 dB/octave slopes. The contours are not associated with equal-loudness contours, but instead represent perceived neutral background noise. Each line is given an RC value that corresponds to the SPL level at 1 kHz. For example, the RC-25 line passes through 25 dB SPL at 1 kHz. To make an RC determination, octave measurements are taken, an average is computed, and values are compared to the standard RC contours. Similar to NCB, RC values are further defined as N (neutral), H (hissy), or R (rumbly), depending on the predominant spectral response. The region from 16 to 63 Hz is scrutinized for RV (vibration). If vibrations are present, V is added to the R designation; for example, RC-30(RV). RC curves are also standardized in ANSI S12.2.

What criteria level should be selected as the allowable limit for background noise in an acoustically sensitive space such as a recording studio? This depends on the general studio quality level to be maintained, the use of the studio, and other factors. The advent of digital recording changed our view of which RC (or NCB) contour to select as a goal. Signal-to-noise ratios of more than 100 dB must be accommodated. This means a lower acoustical noise floor. A lower noise floor means stricter construction practices and HVAC contract noise specifications that substantially increase the cost.

There is little point in demanding RC-15 from the air-handling system when intrusion of traffic and other noise is higher than this. In general, RC-20 should be the highest contour that should be considered for a recording studio or listening room, and RC-15 is suggested as a practical and attainable design goal for the average studio. RC-10 would be excellent but it would require additional effort and expense to reduce ambient noise to this level. Table 17-1 lists a number of types of rooms and suggestions for RC values.

Type of Room	Recommended RC Level (RC Curve)	Equivalent Sound Level, (dBA)
Apartments	25–35 (N)*	35–45
Assembly halls	25–30 (N)	35–40
Churches	30–35 (N)	40–45
Concert and recital halls	15–20 (N)	25–30
Courtrooms	30–40 (N)	40–50
Factories	40–65 (N)	50–75
Legitimate theaters	20–25 (N)	30–65
Libraries	35–40 (N)	40–50
Motion picture theaters	30–35 (N)	40–45
Private residences	25–35 (N)	35–45
Recording studios	15–20 (N)	25–30
Restaurants	40–45 (N)	50–55
Sport coliseums	45–55 (N)	55–65
TV broadcast studios	15–25 (N)	25–35
Hospitals/clinics—private rooms	25–30 (N)	35–40
Hospitals/clinics—operating rooms	25–30 (N)	35–40
Hospitals/clinics—wards	30–35 (N)	40–45
Hospitals/clinics—laboratories	35–40 (N)	45–50
Hospitals/clinics—corridors	30–35 (N)	40–45
Hospitals/clinics—public areas	35–40 (N)	45–50
Hotels/motels—individual rooms or suites	30–35 (N)	35–45
Hotels/motels—meeting or banquet rooms	25–35 (N)	35–45
Hotels/motels—service and support areas	40–45 (N)	45–50
Hotels/motels—halls, corridors, lobbies	35–40 (N)	50–55
Offices—conference rooms	25–30 (N)	35–40
Offices—private	30–35 (N)	40–45
Offices—open-plan areas	35–40 (N)	45–50
Offices—business machines/computers	40–45 (N)	50–55
Schools—lecture and classrooms	25–30 (N)	35–40
Schools—open-plan classrooms	30–40 (N)	45–50

*Neutral (N). The levels in the octave bands centered at 500 Hz and below must not exceed the octave-band levels of the reference spectrum by more than 5 dB at any point in the range; the levels in the octave bands centered at 1 kHz and above must not exceed the octave-band level of the reference spectrum by more than 3 dB at any point in the range.

TABLE 17-1   Recommended RC Levels for Different Types of Rooms

The objective in any room design is to achieve a noise level appropriate for the functions of the space—not the lowest possible noise level. Overdesign is as unforgivable as underdesign.

Fan Noise

In noise control, the most efficient solution is to minimize the noise at the source. Machinery and fan noise are thus important issues to address. Fans are a chief contributor to HVAC noise (but they are not the only culprit). Fan noise will travel through a duct, gradually decreasing in intensity as energy is absorbed. Some fan noise will leave the duct openings, while other noise will radiate along the length of the duct itself. The sound power output of a fan is largely fixed by the air volume and pressure required in the installation, but there are variations between types of fans. Figure 17-3 gives the specific sound power output of two types of fans: the airfoil centrifugal and the pressure blower. Specific sound power level means that the measurements have been reduced to the standard conditions of 1 ft³/minute and a pressure of 1 inch of water. On this basis, noise of various types of fans can be compared equitably. The centrifugal fan is one of the quietest types of fans available. Large fans are generally quieter than small fans. This is also true of the pressure blower fan, as shown in Fig. 17-3.

Aside from the term "specific," sound power in general is foreign to the audio field. All the acoustical power radiated by a piece of machinery must flow out through a hemisphere. The manufacturer is the source of noise data on any particular fan. Sound power is usually evaluated by sound-pressure level readings over the hemisphere.

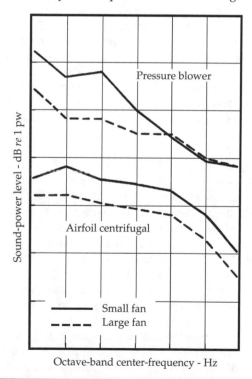

FIGURE 17-3 Noise sound-power output of airfoil centrifugal and pressure blower fans commonly used in HVAC systems.

Sound power is proportional to sound pressure squared. Noise ratings of fans in terms of sound power can be converted back to sound-pressure levels applicable to a given room by means of the following formula:

$$L_{\text{pressure}} = L_{\text{power}} - 5 \log V - 3 \log f - 10 \log r + 25 \text{ dB} \qquad (17\text{-}1)$$

where L_{pressure} = sound-pressure level
L_{power} = sound-power level
V = room volume, ft³
f = octave-band center frequency, Hz
r = distance source to reference point, ft

There is usually a tone generated by the fan such that: Frequency = (rev/sec)(number of blades). This tone adds to the level of the octave band in which it falls. To account for the contribution of the fan tone, 3 dB for centrifugal fans and 8 dB for pressure fans should be added to the one-octave band level.

Machinery Noise and Vibration

A building may have HVAC infrastructure equipment such as pumps, compressors, chillers, and evaporators. In a new room design, the first step in the mitigation of HVAC noise and vibration is to carefully consider the location of the HVAC machinery. HVAC problems can be severe if the HVAC equipment is placed near an acoustically sensitive room. For example, if the equipment room is adjacent to or on the roof directly above a studio, the common wall or roof panel will vibrate like a giant diaphragm, efficiently radiating noise into the studio. Therefore, although cost may be increased because of longer duct lengths and thermal losses, it is important to locate the equipment as far removed from the acoustically sensitive areas as possible. HVAC equipment should be located on the opposite side of the building, or in a room or structure that is detached from any sensitive rooms. If an air-handling unit must be placed on a roof directly over a sensitive room, it should be positioned over a load-bearing wall, and not in the middle of a roofing span, to minimize diaphragmatic action. Similarly, to minimize vibration, equipment should preferably be placed on the ground, as opposed to a higher floor.

The next step is to consider some form of source isolation against structureborne vibration. In most cases, air-handling equipment is placed in a machinery room. If the equipment is to rest on a concrete slab shared with the building, the machinery-room floor slab should be isolated from the main-floor slab as a floating-floor slab. Compressed and treated glass fiber strips should be used to separate slabs during pouring. Vibration sources such as motors must be decoupled from the building structure by placing the motor on rubber or fiberglass pads, coiled springs, or similar devices. In many cases, the motor is mounted on an isolation block, which in turn is isolated with pads or springs from a floating floor. This isolation block, sometimes called an inertia base, is usually made of concrete. It increases the total mass of the system, which reduces vibrations and in particular, lowers the natural frequency (described below) of the system. This in turn improves the isolation efficiency of the system. Care must be taken to ensure that any isolation method provides complete decoupling; if a solid piece mechanically "short circuits" the decoupling, then isolation can be severely compromised. An example of a mounting method for machinery is shown in Fig. 17-4.

Additionally, air-handling units should be placed within noise-control enclosures. Pipes and ducts should be mounted from isolation hangers, and penetrations where

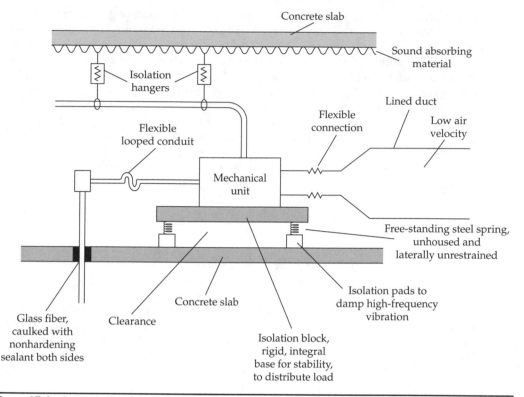

FIGURE 17-4 An example of techniques used to mitigate noise from a machine. Emphasis is placed on isolation mounting and decoupling.

they leave the machinery room should be flexibly sealed. Interior walls and ceiling of the machinery room should be covered with sound-absorbing materials such as glass fiber boards to reduce the ambient noise level in the machinery room, and thus reduce the required transmission loss of the enclosure. The noise reduction achieved by adding absorption can be calculated from:

$$\text{NR} = 10 \log \frac{A}{A_0} \, \text{dB} \qquad (17\text{-}2)$$

where A_0 = absorption before treatment
        A = absorption after treatment

For example, if an enclosure initially has 100 sabins of absorption at 500 Hz, and after treatment has 1,000 sabins of absorption, then a 10-dB noise level reduction is achieved.

In practice, regardless of the mounting and installation method, some energy is imparted from the vibration source to the building. This can be viewed in terms of transmissibility (*T*), which can be expressed as the force transmitted to the building structure divided by the force applied by the sound source. If no isolation is used, the transmissibility may be 1.0. Ideally, no energy from the motor would be transmitted to the building structure, and transmissibility would be zero. In any case, the lower the transmissibility, the better. Looked at in another way, transmissibility can be measured by its efficiency; for example, if the transmissibility is 0.35, its efficiency is 65%.

A vibration source such as a motor, mounted on isolation pads or similar means, can be considered as a mass on a spring. As such, it will have a natural frequency of vibration f_n, where it will freely vibrate. The value of f_n is a function of the stiffness of the spring and mass. In addition, f_n is a function of the static deflection of the system. In particular, f_n equals a constant divided by the square root of the static deflection. The static deflection is the decrease in the height of the isolation pad or spring when the equipment is placed on it. The value of deflection is usually provided by the manufacturer. When the deflection is measured in inches, the constant is 3.13. A vibration source will also have its own forcing frequency f. For example, a motor rotating at 1200 rpm (revolutions per minute) will exhibit a forcing frequency of 20 Hz (cycles per second). When equipment vibrates at several frequencies, the lowest frequency is used in designing the isolation system.

Transmissibility can also be calculated by using f and f_n:

$$T = \left| \frac{1}{(f/f_n)^2 - 1} \right| \tag{17-3}$$

where T = transmissibility

f = driving frequency of source, Hz

f_n = natural frequency of system, Hz

When designing an isolation system, attention must be paid to minimizing T. This is done by increasing the value of f/f_n. This, in turn, is usually accomplished in the design by reducing the natural frequency f_n of the system, for example, by specifying a high value for static deflection. Figure 17-5 shows this design goal as a graph of T versus f/f_n. A value of T that is greater than 1.0 would mean that the isolation system is quite unsuccessful; it is amplifying the vibration (and is in resonance when f/f_n = 1.0). To

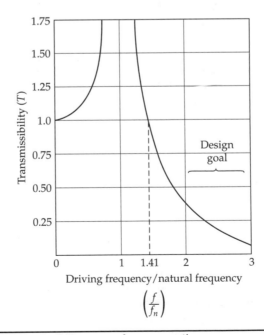

FIGURE 17-5   Graph of transmissibility versus frequency ratio.

avoid that, we see that f/f_n must be greater than 1.41. Many designers aim for an f/f_n value of about 2 or 3, providing a successful efficiency of about 80%.

Air Velocity

In air distribution systems, the velocity of airflow is a very important factor in keeping HVAC noise at a satisfactorily low level. The higher the airflow velocity, the higher the turbulence noise. Other factors can also introduce turbulence noise. Noise generated by airflow varies approximately as the 6th power of the velocity. As air velocity is doubled, the sound level will increase about 16 dB at the room outlet. Some references state that airflow noise varies as the 8th power of the velocity and give 20 dB as the figure associated with doubling or halving the air velocity. In any case, low air velocity is a prerequisite for low noise.

A basic design parameter of an HVAC system is the quantity of air the system is to deliver. There is a direct relationship between the quantity of air, air velocity, and size of duct. The velocity of the air depends on the cross-sectional area of the duct. For example, if a system delivers 500 ft³/minute and a duct has 1 ft² of cross-sectional area, the velocity is 500 ft/minute. If the area is 2 ft², the velocity is reduced to 250 ft/minute; if 0.5 ft², velocity is increased to 1,000 ft/minute. An air velocity maximum of 500 ft/minute is suggested for broadcast studios, recording studios, and other critical spaces. Specifying a large duct size (allowing low air velocity) in initial design will eliminate problems later.

High-pressure, high-velocity, small-duct systems are generally less expensive than low-velocity systems. Budget systems commonly yield noise problems in studios because of high air velocity and its resulting high noise. Compromise can be made by flaring out the ducts just upstream from the outlet grille, and using grilles with large louvered openings, or no grilles at all, in an effort to minimize air turbulence at the grille. The increasing cross-sectional area of the flare results in air velocity at the grille being considerably lower than in the duct feeding it.

Even if fan and machinery noise are sufficiently attenuated, by the time the air reaches the acoustically sensitive room, air turbulence associated with nearby 90° bends,

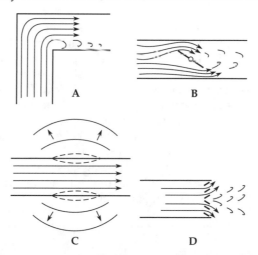

Figure 17-6 Air turbulence caused by discontinuities in the flow path can be a serious source of noise. (A) 90° miter bend. (B) Damper used to control quantity of air. (C) Sound radiated from duct walls set into vibration by turbulence or noise inside the duct. (D) Grilles and diffusers.

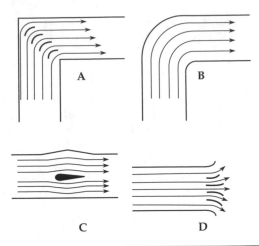

FIGURE 17-7   Air turbulence noise can be materially reduced by using several techniques. (A) Use of deflectors. (B) Radius bends. (C) Airfoils. (D) Carefully shaped grilles and diffusers.

dampers, grilles, and diffusers can be serious noise producers, as suggested by Fig. 17-6. Duct transitions and bends should be as rounded or smooth as possible to reduce turbulence. Similarly, vanes should be used to direct air smoothly through branch take-offs.

Natural Attenuation

When designing an air distribution system, certain helpful natural attenuation effects should be taken into account. Otherwise, if they are neglected, expensive overdesign may result. When a plane wave sound passes from a small space, such as a duct, into a larger space, such as a room, some noise is reflected back toward the source. The effect is greatest for low-frequency sound. Research has also indicated that the effect is significant only when a straight section of ductwork, three to five diameters long, precedes the duct termination. Any terminal device, such as a diffuser or grille, tends to nullify this attenuation effect. A 10-in duct flowing air into a room without a grille can provide a 15-dB reflection loss in the 63-Hz octave. This is about the same attenuation that a 50- to 75-ft run of lined ductwork would give. Figure 17-7 shows several methods of attenuation.

A similar loss occurs at every branch or take-off. The attenuation depends on the number of branches and their relative area. Two branches of equal area (each one-half the area of the feeding duct) would yield a 3-dB attenuation in each branch. There is also an attenuation of noise in bare, rectangular sheet ducts due to wall flexure amounting to 0.1 to 0.2 dB/ft at low frequencies. Round elbows introduce attenuation, especially at higher frequencies. Right-angle elbows also introduce attenuation, as shown in Fig. 17-8. All of these losses are built into the air-handling system and serve to attenuate fan noise and other noise traveling along the duct.

Duct Lining

In noise control, it is often important to reduce noise along the path it travels. The application of sound-absorbing materials to the inside surfaces of ducts is a standard method of reducing noise levels. The attenuation or insertion loss is measured in decibels per length of duct; attenuation is generally poor at low frequencies. Such lining comes in

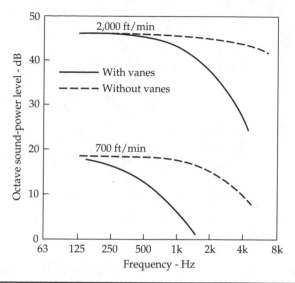

FIGURE 17-8   Noise produced by 12 × 12 in square cross section 90° miter elbow with and without deflection vanes and at 2,000 and 700 ft/minute air velocities. Calculations follow ASHRAE procedures.

the form of rigid boards and blankets and in $\frac{1}{2}$ to 2 in thicknesses. Typical lining is made of glass fiber coated with a resilient material; in some cases, use of this material is not allowed. Such acoustical lining also serves as thermal insulation when it is required. Generally, the large the duct size, the lower the insertion loss; this is because there is relatively less material for a given airflow. The approximate insertion loss offered by 1-in duct lining in typical rectangular ducts depends on the duct size, as shown in Fig. 17-9. The approximate insertion loss of round ducts is given

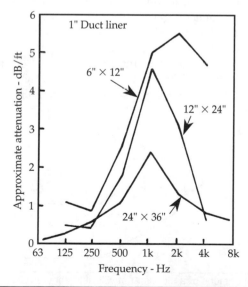

FIGURE 17-9   Attenuation offered by 1-in duct lining on all four sides of rectangular duct. Dimensions shown are the free area inside the duct with no airflows. Calculations follow ASHRAE procedures.

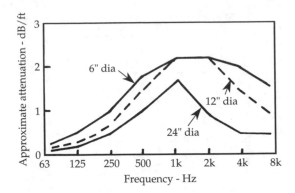

Figure 17-10   Attenuation offered by spiral wound round ducts with perforated spiral wound steel liner. Dimensions shown are the free area inside the duct, with no airflow. Calculations follow ASHRAE procedures.

in Fig. 17-10. Insertion loss is much lower in round ducts than in lined rectangular ducts with comparable cross-sectional areas.

Care must be taken when ductwork passes through a noisy room on its way to a quiet room. Noise from the noisy room will enter the duct (either through openings or through the duct wall itself) and pass through it to the quiet room, easily violating the transmission loss of the common wall. Noise travels equally upstream or downstream to airflow in the duct. The duct in the noisy room must be insulated to minimize this acoustical crosstalk and attenuation must be provided in the duct between rooms.

Plenum Silencers

A sound-absorbing plenum is an economical device for achieving significant attenuation. Figure 17-11 shows a modest-sized plenum chamber, which if lined with 2 in thickness of 3 lb/ft³ density glass fiber, will yield a maximum attenuation of about 21 dB. The attenuation characteristics of this plenum are shown in Fig. 17-12 for two thicknesses of lining. With a lining of 4 in of fiberboard of the same density, quite uniform absorption is obtained across the audible band. With glass fiberboard 2 in thick, attenuation falls off below 500 Hz. It is apparent that the attenuation performance of a plenum of given size is determined primarily by the lining.

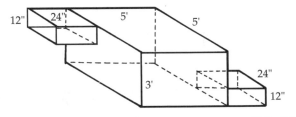

Figure 17-11   Dimensions of a plenum that will yield about of 21-dB attenuation throughout much of the audible spectrum. (*ASHRAE Handbook*)

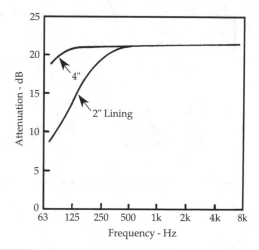

FIGURE 17-12   Calculated attenuation characteristics of the plenum of Fig. 17-11 lined with 2-in and 4-in glass fiber of 3 lb/ft^3 density. Calculations follow ASHRAE procedures.

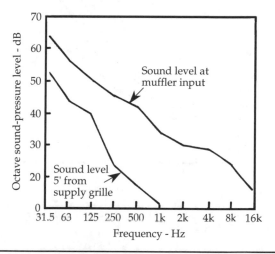

FIGURE 17-13   Noise sound pressure levels measured at two locations: plenum muffler input, and 5 ft from the supply grille. This plenum muffler is approximately the size of that of Fig. 17-11, but only half the height and contains baffles.

Figure 17-13 gives measurements for a practical, lined-plenum muffler approximately the same horizontal dimensions as that of Fig. 17-11, but only half the height and with baffles inside. Attenuation of 20 dB or more above the 250-Hz octave is realized in this case.

Plenum performance can be improved by increasing the ratio of the cross-sectional area of the plenum to the cross-sectional area of the entrance and exit ducts, and by increasing the amount or thickness of absorbent lining. A plenum located at the fan discharge can be an effective and economical way to decrease noise entering the duct system. In some cases, unused attic spaces can be used as plenums for noise attenuation.

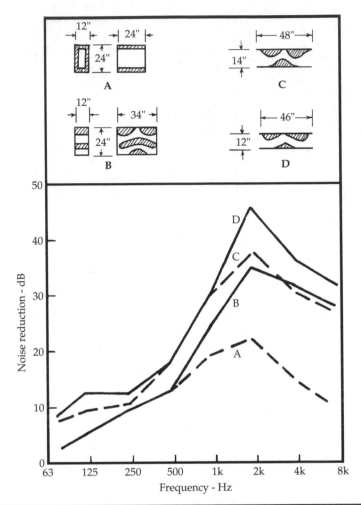

FIGURE 17-14   Attenuation characteristics of three packaged silencers compared to that of the lined duct, curve A. (*Doelling*)

Packaged Attenuators

Numerous proprietary noise attenuators are available. Cross sections of several types are shown in Fig. 17-14 with their performance plotted below. For comparison, the attenuation of a simple lined duct is given in curve A. Some attenuators have no line-of-sight through them; that is, the sound must be reflected from the absorbing material to traverse the unit and hence will have somewhat greater attenuation. In these packaged attenuators, the absorbing material is usually protected by perforated metal sheets. The attenuation of such units is very high at midband speech frequencies but not as good at low frequencies.

Reactive Silencers

Some silencers use the principle of an expansion chamber, as shown in Fig. 17-15. These attenuators function by reflecting sound energy back toward the source, thereby cancel-ing some of the sound energy (similar to an automobile muffler). Because there is both

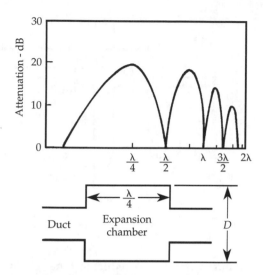

FIGURE 17-15 Attenuation characteristics of a reactive silencer using an expansion chamber. Sound is attenuated by virtue of the energy reflected back toward the source, canceling some of the oncoming sound. (*Sanders*)

an entrance and exit discontinuity, sound is reflected from two points. The destructive interference (attenuation nulls) alternates with constructive interference (attenuation peaks) across the frequency band, the attenuation peaks becoming lower as frequency is increased. These peaks are not harmonically related; therefore, they do not produce high attenuation for a noise fundamental and all of its harmonics, but rather attenuate bands of the spectrum. By tuning, however, the major peak can eliminate the fundamental, while most of the harmonics, of much lower amplitude, would receive some attenuation. Two reactive silencers can be placed in series and tuned so that one cover the nulls of the other. Thus, continuous attenuation can be realized throughout a wide frequency range. No acoustical material, such as an absorber, is required in this type of silencer.

Tuned Silencers

The resonator silencer, illustrated in Fig. 17-16, is a tuned stub that provides high attenuation at a narrow band of frequencies. Even a small unit of this type can produce 40- to

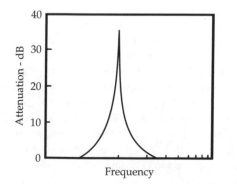
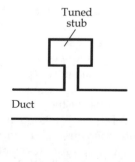

FIGURE 17-16 Attenuation characteristics of the tuned-stub silencer. (*Sanders*)

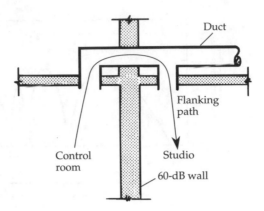

FIGURE 17-17   A high-transmission-loss wall can be bypassed by sound traveling over a short duct path from grille to grille.

60-dB attenuation and is effective at reducing tonal HVAC noise, such as fan tones. However, this type of silencer offers little constriction to airflow, which can be a drawback to other types of silencers.

Duct Location

Careful design is needed when an air-handling system is common between two acoustically sensitive rooms. Care must be taken to minimize duct transmission between them. For example, it would be a design error to specify an STC-60 dB wall between studio and control room, and then serve both rooms with the same supply and exhaust ducts closely spaced, as in Fig. 17-17. The duct would create a short path speaking tube from one room to the other, quite effectively nullifying the transmission loss of the 60-dB wall. With an untrained air-conditioning contractor doing the work, not sensitive to acoustical problems, such errors can easily happen. To obtain as much as 60-dB attenuation in the duct system to match the construction of the wall requires the application of many of the principles discussed earlier. Figure 17-18 suggests two approaches to the problem; to separate grilles as far as possible if they are fed by the same duct, or better yet, to serve the two rooms with separate supply and exhaust ducts.

ASHRAE

The ASHRAE (American Society of Heating, Refrigerating, and Air-Conditioning Engineers) organization is prolific resource of HVAC data for the acoustical designer. In this chapter it is impractical to go deeply into specific design technologies, but the ASHRAE handbooks can be useful guides. They introduce fundamental principles, including the source/path/receiver concept; basic definitions and terminology; and acoustical design goals. Other topics include noise control for outdoor equipment installations, system noise control for indoor air-conditioning systems, general design considerations for good noise control, mechanical-equipment-room noise isolation, vibration isolation and control, and troubleshooting for noise and vibration problems. ASHRAE also provides standard methods to estimate octave noise levels in rooms generated by air-handling systems.

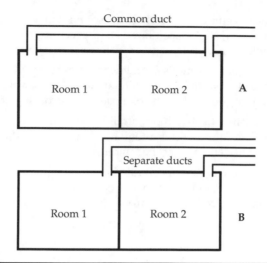

FIGURE 17-18   Two possible solutions to the problem of Fig. 17-17. (A) Separate grilles as far as possible. (B) Feed each room with a separate duct. Both approaches increase duct path length and thus duct attenuation.

Active Noise Control

Noise can be cancelled by radiating a replica of the noise in inverse phase and combining them. The cancelling noise is equal in frequency and level, but out of phase with the original unwanted noise, resulting in destructive interference and noise attenuation. A sensing microphone inputs the original noise to a processor, which analyzes the noise and generates the inverse signal, which is applied to a loudspeaker. Active noise control works fairly well in the immediate vicinity of an industrial area of heavy noise, but it is more generally used in very controlled spaces such as ventilation ducts (or headphones). The technique is most effective for continuous, low-frequency noise. The prospect of cost-effective active noise control in sound-sensitive areas such as home listening rooms, recording studios, or control rooms remains remote.

Some Practical Suggestions

- The most effective way to control airflow noise is to increase duct size so as to decrease airflow velocities. The economy of smaller ducts, however, may be sufficient to pay for silencers to bring the higher noise of higher-velocity flow to acceptable levels.

- Right-angle bends and dampers create noise due to air turbulence. Locating such fittings 5 to 10 diameters upstream from the outlet allows the turbulence to abate.

- Noise and turbulence inside a duct cause the duct walls to vibrate and radiate noise into surrounding areas. Rectangular ducts are worse offenders than round ones. Such noise increases with air velocity and duct size, but can be controlled with external treatment of isolating material.

- Acoustical ceilings are not good sound barriers; hence in an acoustically sensitive area, the space above a drop ceiling should not be used for high-velocity terminal units.

- The ear can detect sounds far below the prevailing NCB or RC contour noise. The goal should be to reduce noise to an acceptable level. For example, in a studio, noise cannot be heard on the playback of a recording at normal listening level.

- Plenums are effective and straightforward devices adaptable to quieting programs, and they offer attenuation throughout the audible spectrum. They are especially effective at the fan output.

- Some noise energy is concentrated in high frequencies, some in low frequencies. There must be an overall balance in the application of silencers so that the resulting noise follows the proper NCB or RC contour. Otherwise overdesign can result.

CHAPTER **18**

Acoustics of
Listening Rooms

This chapter considers that portion of a home designated for the playback of recorded music. This may be a room used for music playback, an audio/video home theater, or a professional audio work room. Some homes have a room that is dedicated to the enjoyment of music or films. Other homes have a multipurpose space, typically a living room serving also as a listening room. The playback equipment can range from a simple stereo system to an elaborate home cinema. In any case, the room acoustics will play an important role in the sound quality of the audio playback signal.

For both the professional acoustician and the critical audiophile, the listening room is as much of a challenge as the design of a professional recording studio. All the major acoustical problems are involved in the design of a listening room, as well as other small audio rooms. The discussion of the acoustics of listening rooms in this chapter are therefore considered as an introduction to the acoustics of other types of small audio spaces in following chapters. In particular, we will consider the essential problem of how loudspeaker playback is influenced by room acoustics.

Playback Criteria

The acoustics of a space is a vital part of both the recording and reproducing process. In every acoustical event, there is a sound source and a receiving device with an acoustical link between the two. The source/receiver may be a musical instrument/microphone or a loudspeaker/listener. Audio recordings contain the imprint of the acoustics of the recording environment. For example, if the sound source is a symphony orchestra and the recording is made in a concert hall, the reverberation of the hall is very much a part of the orchestral sound. If the reverberation time of the hall is 2 seconds, a 2-second tail is evident on every impulsive sound and sudden cessation of music, and it affects the fullness of all the music. In playing this recording in a home listening room, what room characteristics will best complement this type of music?

Another recording may be of popular music. This music was probably recorded in a relatively dead studio with overdubbing. Rhythm sections playing in this dead studio and well separated acoustically are laid down on separate tracks. During subsequent sessions the other instruments and vocals are recorded on other tracks. Finally, all are combined at appropriate levels in a stereo or multichannel mix down. In the mix down, many effects, including artificial reverberation, are added. What listening room characteristics are best for playback of this recording?

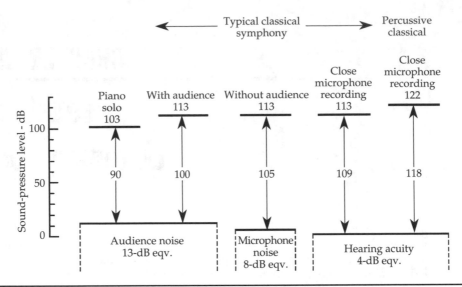

FIGURE 18-1   A dynamic range of up to 118 dB may be necessary for subjectively noise-free reproduction of music. (*Fielder*)

If the taste of the audiophile is highly specialized, the listening-room characteristics can be designed for relative optimum results for one type of music. If the listener's taste is more universal, the acoustical treatment of the listening room may require compromises to make it suitable for different types of music.

The dynamic range of reproduced music in a listening room is determined by the softest sounds played, and the loudest. The loudest level depends primarily at the maximum amplifier power, and the power-handling ability and efficiency of the loudspeakers. The softest levels of the dynamic range are limited by environmental or electrical noise. For example, ambient household noise may determine the lower dynamic-range limit. The usable range between these two extremes is usually far less than the range of an orchestra in a concert hall. If peak instantaneous sound levels are considered when determining dynamic range requirements, much greater ranges are required.

Fiedler used musical performances of high peak levels in a quiet environment to show that a dynamic range of up to 118 dB is necessary for subjectively noise-free reproduction of music. The results are summarized in Fig. 18-1. He considered the peak instantaneous sound level of various sources, as shown at the top of the figure, and the just-audible threshold for white noise added to the program source when the listener is in a normal listening situation, as shown at the bottom of the figure. This suggests that for realistic playback, a listening room requires extensive sound isolation to ensure a low ambient noise level, and a high-power playback system capable of generating high sound-pressure levels with low distortion.

Peculiarities of Small-Room Acoustics

The 10-octave spread of the audible spectrum is so great that the acoustical analysis of any room is problematic. This is even truer for small rooms, which perform quite differently from large rooms. The reason is apparent when room size is considered in terms of the wavelength of sound. The 20-Hz to 20-kHz audio band covers sound wavelengths from 56.5 to 0.0565 ft (0.68 in). Below about 300 Hz (wavelength 3.8 ft), the average listening room must be considered as a resonant cavity. It is not the room structure that resonates, it is the air confined within the room. As frequency increases above 300 Hz, the wavelengths become smaller and smaller with the result that sound may be considered as rays.

Reflections of sound from the enclosing surfaces dominate both the low- and high-frequency regions. At lower frequencies, reflections result in standing waves and the room becomes a chamber resonating at many different frequencies. These standing waves dominate low-frequency response in a small room. Reflections of sound from the room surfaces also dominate at the midband and higher audible frequencies, without the cavity resonances, but with specular reflections as the major feature.

Room Size and Proportion

Problems are inevitable if sound is recorded or reproduced in spaces that are too small. For example, Gilford states that studio volumes less than approximately 1,500 ft^3 are so prone to sonic coloration that they are impractical. Rooms smaller than this produce sparse modal frequencies with probable wide spacings between modes, which are the source of audible distortions. All other things being equal, it is generally true that the larger the room, the better the sound quality.

As we observed in Chap. 13, it is possible to select room proportions that yield the most favorable distribution of room modes. With new construction it is strongly advised to use these proportions as a guide, confirming any dimensions by calculation and study of the spacing of axial-mode frequencies.

In most cases, in the home listening room, the room shape and size are already fixed. The existing room dimensions should then be used for axial mode calculations. A study of these modal frequencies will then reveal the presence of coincidences (two or more modes at the same frequency) or isolated modes spaced 25 Hz or more from neighbors. Such faults pinpoint frequencies at which colorations may occur, and steps may be taken to remedy the problem.

There is no such thing as perfect room proportions. It is easy to place undue emphasis on a mechanical factor such as this. In reality, one should be informed on the subject of room resonances and be aware of the consequences. In striving to attain high sound quality, reducing sound colorations by attention to room modes is one of many factors to consider.

Reverberation Time

Reverberation time is one of several determinants of acoustical quality of small rooms. The amount of overall absorbance in a listening room establishes the general listening conditions. If the room is excessively absorptive or excessively reverberant, music quality will deteriorate and most listeners will grow fatigued of the unnatural sound field. There is no optimal value for reverberation time in a listening room. A simple conversation test is often all that is needed to ascertain that reverberation time is suitable. If a talker at the location of the center-channel speaker can comfortably be heard by a listener at the listening position, and speech quality sounds natural, then the reverberation time is approximately correct—the room is neither too live nor too dead.

The Sabine equation for reverberation time makes it possible to estimate the amount of absorbing material required for a reasonable reverberant condition. It is expedient to assume a reasonable reverberation time, say about 0.3 second, for the purposes of these calculations. From this, the total number of sabins of absorption can be estimated, which would result in suitable listening conditions. In many home listening rooms, the structure and the furnishings often supply most of the absorbance required. However, in some cases, absorbing wall panels or carpets may be needed. Careful listening tests must determine the degree of room ambience most suitable for a favorite type of music.

Low-Frequency Considerations

Consider the bare room of Fig. 18-2 as the starting point in analyzing low-frequency response. The room is 21.5 ft long and 16.5 ft wide, with a 10-ft ceiling. These dimensions fix the axial mode resonances and their multiples (harmonics). Following discussions in Chap. 13, axial mode effects will be emphasized and tangential and oblique modes will be neglected. Axial mode frequencies to 300 Hz are calculated for the length, width, and height dimensions and tabulated in Table 18-1. These axial modal frequencies are arranged in ascending order of frequency, irrespective of the dimensional source (length, width, or height). Spacings between adjacent modes are entered in the right-hand column. No coincidences are noted; only a single pair is as close as 1.5 Hz. The dimensional ratios 1:1.65:2.15 are well within the Bolt area.

These calculated axial modal frequencies are now applied to the listening room. This is done graphically in Fig. 18-3, following a style used by Toole. The positions of the nulls locate null-lines, which are drawn through the listening room. Lines representing the length-mode nulls are drawn through both the elevation view and the plan view because these nulls actually form a null plane that extends from floor to ceiling. The position of the listener's seat can be moved to avoid these particular nulls at 26, 53, and 79 Hz, but observe that there are eight more nulls below 300 Hz.

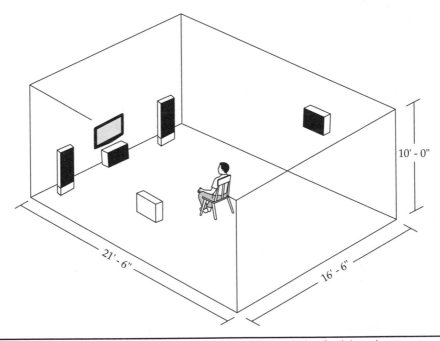

10' - 0"

21' - 6"

16' - 6"

FIGURE 18-2 Dimensions of a listening room assumed for analysis of axial modes.

| Room Dimensions = 21.5 × 16.5 × 10.0 ft | | | | |
| Axial Mode Resonances (Hz) | | | | |
Length $L = 21.5$ ft $f_1 = 565/L$ (Hz)	Width $W = 16.5$ ft $f_1 = 565/W$ (Hz)	Height $H = 10.0$ ft $f_1 = 565/H$ (Hz)	Arranged in Ascending Order	Axial Mode Spacing (Hz)
f_1 26.3	34.2	56.5	26.3	7.9
f_2 52.6	68.4	113.0	34.2	18.4
f_3 78.9	102.6	169.5	52.6	3.9
f_4 105.2	136.8	226.0	56.5	11.9
f_5 131.5	171.0	282.5	68.4	10.5
f_6 157.8	205.2	339.0	78.9	23.7
f_7 184.1	239.4		102.6	2.6
f_8 210.4	273.6		105.2	7.8
f_9 236.7	307.8		113.0	18.5
f_{10} 263.0			131.5	5.3
f_{11} 289.3			136.8	21.0
f_{12} 315.6			157.8	11.7
			169.5	1.5
			171.0	13.1
			184.1	21.1
			205.2	5.2
			210.4	15.6
			226.0	10.7
			236.7	2.7
			239.4	23.6
			263.0	10.6
			273.6	8.9
			282.5	6.8
			289.3	18.5
			307.8	

Mean axial mode spacing = 11.7 Hz.
Standard deviation = 6.9 Hz.

TABLE 18-1 Axial Modes

The three lowest axial-mode nulls associated with the height of the room (56, 113, and 170 Hz), are sketched on the elevation view of Fig. 18-3. These nulls are horizontal planes at various heights. The head of the listener in the elevation view lies between two nulls and at the peak of the 79-Hz resonance.

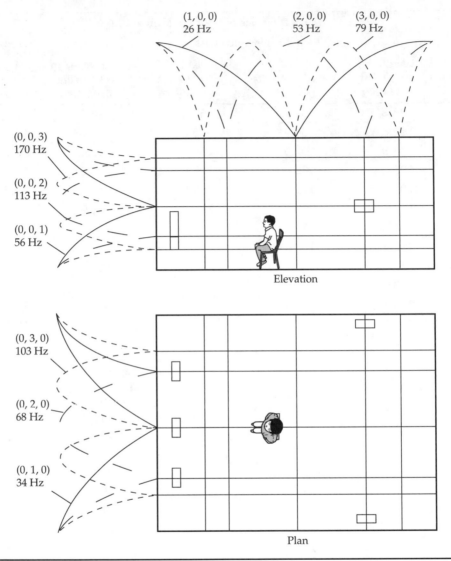

FIGURE 18-3   Plan and elevation views of the listening room of Fig. 18-1 showing the low-frequency modal pressure distributions of the first three axial modes of Table 18-1.

The three axial modes of lowest frequency are sketched on the plan view. The nulls in this case are vertical planes extending from floor to ceiling. The listener, situated on dead center of the room, intercepts the nulls of every odd axial mode. Therefore, this would be a poor choice for the primary listening position.

The resonance nulls have been sketched because their location is definite, but between any two nulls of a given axial mode, a peak exists. Although nulls are capable of removing a sizeable chunk of spectrum, the low-frequency acoustics of the room is dominated by the wide, relatively flat peaks.

The complexity of the modal structure of listening room acoustics is apparent. Only the first three axial modes of length, width, and height have been shown. All of the axial modal frequencies listed in Table 18-1 have an important part to play in the low-frequency acoustics of the space. These axial modes exist only when they are excited by the low-frequency sound of the signal being reproduced in the room. The spectrum of the music is continually shifting; therefore, the excitation of the modes is also continually shifting. For example, the length axial mode at 105.2 Hz (Table 18-1) is excited only as spectral energy in the music hits 105.2 Hz. And all this represents only the sound energy below about 300 Hz.

Modal Anomalies

An obvious source of low-frequency anomalies results from the momentary deviations from flatness of the room response, which are caused by concentrations of modes or large spacings between modes. For example, transient bursts of music energy result in unequal, forced excitation of the modes. As the transient excitation is removed, each mode decays at its natural (and often different) frequency. Beats can occur between adjacent decaying modes. Energy at new and different frequencies is injected, which create audible distortion of the signal.

Control of Modal Resonances

The low-frequency sound field at the listener's ears is made up of the complex vectorial sum of all axial, tangential, and oblique modes at that particular spot in the room. The loudspeakers energize the modal resonances prevailing at their locations. The modes that have nulls at a loudspeaker location cannot be energized, but those having partial or full maxima at this location will be energized proportionally. The interaction of low-frequency resonances in the listening room at the loudspeaker and listening positions is complex and transient, but they can be understood if broken down into the contributions of individual modes.

The position of both the sound sources and the listener should avoid nulls in the modal (axial in particular) distribution. Loudspeaker positions should be considered tentative, moving them slightly if necessary to improve sound quality. The same is true for listening position. Shifts in position, including height, will affect frequency response and sound quality at the listening position.

Bass Traps for the Listening Room

In many cases, it is not practical to acoustically treat each mode separately. A general low-frequency treatment can more efficiently control "room boom" and other resonance anomalies. In fact, when choosing a listening room, consider that a room of wood-frame and drywall construction could very well inherently have sufficient structural low-frequency absorption to provide all the general modal control necessary. In contrast, a room constructed of concrete blocks may lack low-frequency absorption and would require additional bass-trap treatment.

In addition to the adjustment of the general ambience of the room, low-frequency absorption in corners of the room near the loudspeakers can have an important effect on stereo or surround imaging. Figure 18-4 suggests four ways that such absorption can be obtained. Absorbers are placed in corners because modal analysis shows that bass energy is concentrated there. The first unit, shown in Fig. 18-4A, is a Helmholtz

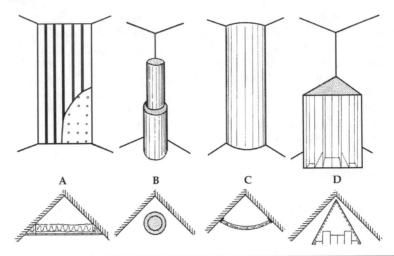

Figure 18-4   Four possible ways of providing low-frequency absorption for the corners of a listening room. (A) Built-in corner resonator. (B) Stacked cylindrical resonators. (C) Polycylindrical resonator. (D) Resonator with diffuser.

resonator trap, built into the corner of the room, floor to ceiling. This could employ either a perforated face or spaced slats. Some design frequency must be assumed, perhaps 100 Hz, and an average depth of the triangular shape must be estimated. By varying trap depth, a more broadband absorption can be obtained. A diaphragmatic absorber could also be used.

The cylindrical absorber of Fig. 18-4B provides the necessary absorption by stacking a smaller diameter unit on top of a larger diameter unit, resulting in a total height of perhaps 6 ft. If necessary, an even larger diameter base could be used, for very low-frequency absorption. These traps are fibrous cylinders with wire mesh skeleton, together acting as resonant cavities. Half of the periphery is a reflective, limp-mass sheet. Low-frequency energy (below 400 Hz) readily penetrates this sheet, while high-frequency energy is reflected from it. Used in a corner, the cylinder would be rotated so that this reflective side faces the room. The cylinder thus contributes to diffusion of the room as well as to deep-bass absorption.

The corner treatment of Fig. 18-4C uses tracks of $1 \times \frac{1}{2}$ in J-metal installed in the corner to hold the edges of a panel. The sheet is bent and then snapped in place. The airspace behind the acoustic panel provides low-frequency absorption. A curved membrane reflector strip within the panel provides wide-angle reflection above 500 Hz.

Another possible corner treatment is the acoustic module shown in Fig. 18-4D. The nominal dimensions of the module are height 4 ft, faces 2 ft. The module has two absorptive sides and one quadratic residue diffuser side. With the absorptive sides facing the corner for modal control, the diffusive side faces the room. The diffuser face diffuses the sound energy falling on it, and also reduces the amplitude of the energy returned to the room.

With any of the devices of Fig. 18-4 in the corners of the listening room near the loudspeakers, the chances are good that sufficient modal control is introduced to minimize stereo image problems resulting from room resonances. Corners of the room without loudspeakers that could be treated similarly in case more modal control is required.

Mid-High Frequency Considerations

The propagation of sound of shorter wavelengths, above about 300 Hz, can be considered as rays that undergo specular reflection. Figure 18-5 shows a listening room and listener for the consideration of the mid- to high-frequency reflections of sound from loudspeakers. Sound from the front right loudspeaker is studied in detail as characteristic of the symmetrical room.

The first sound to reach the listener's ears is the direct sound D, traveling the shortest distance. Reflection F from the floor arrives next. Reflections from the ceiling C, the near side wall $S1$, and the far side wall $S2$ arrive later. Another early reflection E from the front wall is from the diffusion of sound from the edges of the loudspeaker cabinet. The direct ray D carries important information concerning the signal being radiated. If

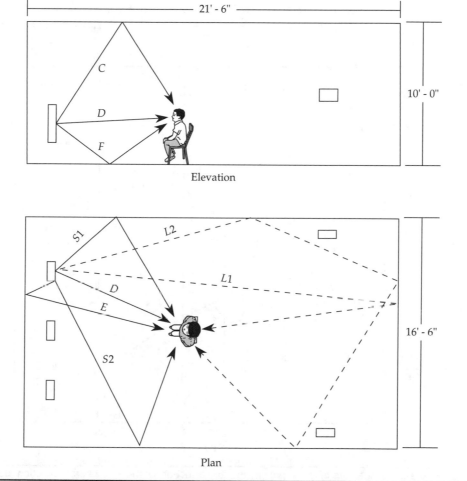

FIGURE 18-5 Plan and elevation views for a listening room showing the direct sound (D) and early reflections from the floor (F), the ceiling (C), the side walls ($S1$ and $S2$), and diffusion edges of the loudspeaker cabinet (E). The later reflections $L1$ and $L2$ are the beginning of the reverberant component.

it is obscured by strong early reflections, for example, the perception of imaging and localization may be diminished.

These beams constitute the direct sound and early reflections, as contrasted to reflections from the rear surfaces of the room and general reverberation, arriving much later. For example, late reflections $L1$ and $L2$ are representative of reflections that form the reverberant component.

Olive and Toole investigated the effect of simulated lateral reflections in an anechoic environment with speech as the test signal. Their work is partially summarized in Fig. 18-6 (a repeat of Fig. 6-11 for convenience). The variables are reflection level and reflection delay. When the reflection level is 0 dB, the reflection is the same level as the direct signal. When the reflection level is –10 dB, the reflection level is 10 dB below the direct. Curve A is the threshold of audibility of the echo. Curve B is the image-shift threshold. Curve C is the threshold at which the reflection is heard as a discrete echo.

For any delay, the reflection is not heard for reflection levels below curve A. For the first 20 msec, this threshold is essentially constant. At greater delays, progressively lower

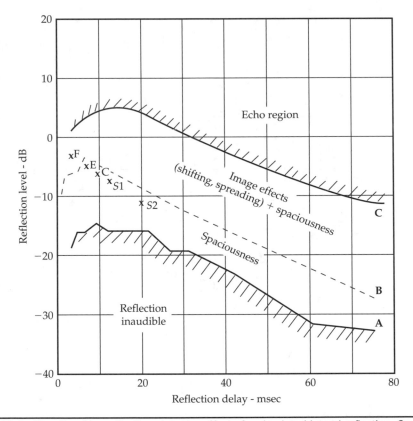

FIGURE 18-6 Results of investigations into the effect of a simulated lateral reflection. Curve A is the absolute threshold of detection of the reflection. Curve B is the image-shift threshold. Curve C is the threshold at which the reflection is heard as a discrete echo. The early room reflections calculated in Table 18-2 are plotted. (*Composite of results: Curves A and B from Olive and Toole, Curve C from Meyer and Lochner*)

Path Sound Paths	Path Length (ft)	Reflection Direct (ft)	Reflection Direct Level* (dB)	Delay† (msec)
D (Direct)	8.0	0.0	—	—
F (Floor)	10.5	2.5	−2.4	2.2
E (Diffraction)	10.5	2.5	−2.4	2.2
C (Ceiling)	16.0	8.0	−6.0	7.1
S1 (Side wall)	14.0	6.0	−4.9	5.3
S2 (Side wall)	21.0	13.0	−8.4	11.5
L1 (Rear wall)	30.6	22.6	−11.7	20.0
L2 (Rear wall)	44.3	36.3	−14.9	32.1

$$^*\text{Reflection level} = 20 \log \frac{\text{Direct path}}{\text{Reflected path}}$$

$$^†\text{Reflection delay} = \frac{(\text{Reflected path}) - (\text{Direct path})}{1,130}$$

TABLE 18-2   Reflections: Their Amplitudes and Delays

reflection levels are required for a just-audible reflection. For a home listening room or other small room, delays in the 0- to 20-msec range are significant. In this range, the reflection audibility threshold varies little with delay.

It is instructive to compare the levels and delays of the early reflections of Fig. 18-5 with those of Fig. 18-6. Table 18-2 lists estimates of the level and delay of each of the early reflections (F, E, C, S1, S2) identified in Fig. 18-5 (assuming perfect reflection and inverse square propagation). These reflections, plotted in Fig. 18-6, all fall within the audible region between the reflection threshold and the echo-production threshold. The direct signal is immediately followed by competitive early reflections of various levels and delays, producing comb-filter distortion.

The level of the competing reflections should be reduced relative to the direct signal, except for the lateral reflections. Figure 18-6 is a study of a direct signal and a single lateral reflection. Allowing a single lateral reflection of adjustable level would permit control of the spaciousness and image effects of the room. Therefore, when designing the acoustical conditions of this listening room, the early reflections should be eliminated, except those lateral reflections from the left- and right-side walls; these subsequently will be adjusted for optimum sound quality.

Identification and Treatment of Reflection Points

One method for reducing the levels of the early reflections is to treat the front portion of the room with sound absorbing material. However, this may also absorb the lateral reflections and may make the room too dead. Instead, the principle recommended here is to add a minimum of absorbing material to treat only the specific surfaces responsible for unwanted reflections.

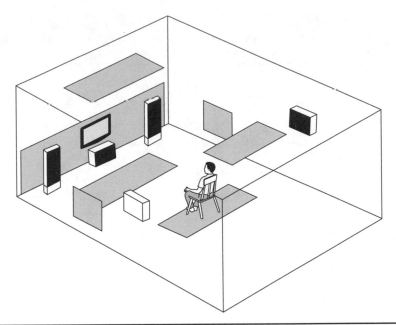

FIGURE 18-7   The room of Fig. 18-2 with minimum sound absorbing treatment to reduce the level of the early reflections from room surfaces. The lateral reflectivity of the side-wall absorbers may be adjusted to control spaciousness and image effects in the listening room. Additional absorbing material may be needed to adjust the average reverberant character of the room for best listening.

Locating these reflection points is easy with the help of an assistant equipped with a mirror. For example, with the listener seated at the sweet spot, the assistant moves a mirror on the floor until the observer can see the tweeter of the front right loudspeaker reflected in it. This is the point where the floor reflection strikes the floor. This point is marked and the procedure repeated for the tweeter of the front left loudspeaker, and the second floor reflection spot is marked. A small throw rug covering these two marks should reduce the floor reflections to inaudibility. The same procedure is carried out for locating the reflection points for the left- and right-side walls (however, see discussion later) and the ceiling reflections. Each of these points should also be covered with enough absorbing material (such as light cloth, heavy cloth, velour, acoustical tile, glass fiber panel) to ensure sufficient high-frequency absorption at the reflection points.

The point of reflection for the sound energy diffracted from the edges of the loudspeaker cabinet is more difficult to locate. Installing an absorber on the wall behind the loudspeakers should mitigate reflections due to cabinet diffraction. Potential reflections from all the tweeters in all the loudspeakers in the playback system are similarly analyzed.

When all the covered reflection points are in place, as shown in Fig. 18-7, imaging will probably be much clearer and more precise now that the early reflections have been reduced.

Lateral Reflections and Control of Spaciousness

The lateral reflections from the side walls have been essentially eliminated by the absorbing material placed on the wall. A critical listening test should be performed with the side-wall absorbers temporarily removed, but with the floor, ceiling, and diffraction absorbers still in place. The recommendations of Fig. 18-7 can now be tested. Does the strong lateral reflection give the desired amount of spaciousness, or does it cause unwanted image spreading? The adjustment of the magnitude of the lateral reflections can be explored by using sound absorbers of varying absorbance (light cloth, heavy cloth, velour, acoustical tile, glass fiber panel) on the side-wall reflection points. For example, the lateral reflections can be reduced somewhat by hanging a light cloth instead of velour.

Techniques such as these provide the ability to adjust the lateral reflections to achieve the desired spaciousness and stereo and surround imaging effect to suit the individual listener or to optimize conditions for different types of music. The discussion of the acoustical design techniques for small rooms is continued in the following chapters.

Loudspeaker Placement

Speaker placement depends greatly on the room geometry. For example, a rectangular room will require different placement from a room with more complex floor plan. Very generally, stereo speakers should be placed in front of the "sweet spot" listening position so there is a ±30° angle between them. A surround-sound system may follow the guidelines provided in the ITU-R BS.775-2 standard. In a 5.1-channel system, it recommends center-channel speaker at 0°; front left and right speakers at ±30°; surround speakers at ±110° to ±120°, all relative to the listening position. This is shown in Fig. 18-8. Some listeners prefer the surrounds slightly forward or backward. These positions may vary, for example, with the screen size, or whether the surround speakers are bidirectional speakers. In 7.1-channel systems, four surround speakers may be placed at ±60° and ±150°. Exact placement depends on room acoustics, speaker type, and listener preference.

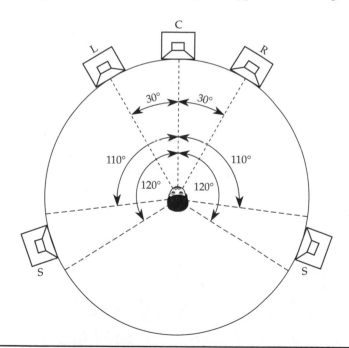

FIGURE 18-8   The ITU-R BS.775-2 standard. For 5.1-channel playback, it recommends center-channel speaker at 0°; front left and right speakers at ±30°; surround speakers at ±110° to ±120°, all relative to the listening position.

The front speakers (particularly the tweeters) should be placed at approximately the same height. The front speakers (particularly the tweeters) should also be placed at approximately ear level when seated and aimed at the listening position. Surround speakers are often placed 2 or 3 ft above ear height when seated.

Because of low-frequency standing waves, subwoofer placement very much depends on room acoustics and listening position in the room. A subwoofer placed in a corner will excite the maximum number of room modes, and provide the most bass. Placement along one wall, or away from a wall, will excite relatively fewer modes. Conversely, a listening position, depending on location in the room, will be receptive to greater or fewer bass modes. In some cases, corners will already be occupied by bass-trap absorbers.

One subwoofer set-up technique is to place the subwoofer in the listening chair, at seated ear level if possible, and play music and movies. Then you move around the room (on hands and knees, placing ear level where the subwoofer will be) and listen for the location that provides the best bass response for a variety of music and movies. Then, place the subwoofer at that location. In some cases, two or more subwoofers in different locations may be needed to provide satisfactory bass response, especially over a larger listening area. Be sure to identify and eliminate any buzzes and rattles caused by the subwoofer or satellite speakers.

CHAPTER 19

Acoustics of Small Recording Studios

arge recording studios are the undisputed leaders in the professional music recording industry. But, because of their availability and economy, small studios play an increasingly important role. Many musicians use small studios to develop their skills and to make both demo and commercial recordings. Music recorded in small studios can be suitable for release on disc or can be downloaded and sold. Many small studios operated by not-for-profit organizations produce material for educational, promotional, and religious purposes. Small studios are required for the production of campus and community radio, television, and cable programs. All of these have limited budgets and limited technical resources. The operator of these small studios is often caught between a desire for top quality and the lack of means. This chapter is primarily intended for small studio owners, although the principles expounded are more widely applicable, even to large-studio design.

What is a good recording studio? There is only one ultimate criterion—the acceptability of the sound recorded in it for its intended audience. In a commercial sense, a successful recording studio is one fully booked and making money. If the public likes the music, the studio passes the test. There are many factors influencing the acceptability of a studio beside sound studio quality, but sound quality is vital, at least for success on a substantial, long-term basis.

This chapter focuses on the acoustics of the "studio" room in a recording studio, that is, the room where music is performed and recorded. The acoustical response, as picked up by microphones, is paramount. Chapter 20 discusses the acoustics of control rooms, that is, the room in recording studios where sound is played back and mixed. In this room, the acoustics of loudspeaker playback is most critical.

Ambient Noise Requirements

To be useful, any studio must have low ambient noise levels. In fact, studios rank among the quietest work spaces. This can be quite difficult to achieve. In many cases, a room is "quiet," except for a few occasional noises. Clearly, this is unacceptable. A studio must be quiet at all times.

Many noise and vibration problems can be avoided entirely by choosing a site in a quiet location. Is the site close to a train track or busy intersection? Is it under the approach path (or worse yet, the take-off pattern) of an airport? The maximum external noise spectrum must be reduced to the background noise criteria goal within the space by the transmission loss of the walls, windows, and doors. Floating floors may be required.

If the space being considered is within a larger building, other possible noise and vibration sources within the same building must be identified and evaluated. Is there a machine shop on the floor above? A noisy elevator? A dance studio? A reinforced concrete building, while providing high transmission-loss partitions, also efficiently conducts noises throughout via structural paths. The structure of the walls determines the degree to which external noise is attenuated. This is equally true of the ceiling and floor structures in regard to attenuation of noise from above and below.

A professional studio, even a small one, might require support activities such as duplication, accounting, sales, reception, shipping, and receiving each with its own noise generation potential. Review of such noise sources concentrates attention on transmission loss characteristics of internal walls, floors, and ceilings.

The HVAC (heating, ventilating, and air-conditioning) system must be designed to provide the required noise criteria goals. Noise and vibration from motors and fans, ducts, diffusers, and grilles must all be minimized for good results. Noise control and HVAC design is discussed in Chaps. 16 and 17.

Acoustical Characteristics of a Studio

Sound picked up by a microphone in a studio consists of both direct and indirect sound. The direct sound is the same as would exist in a free field or in an anechoic chamber. The indirect sound, which immediately follows the direct, is the sound that results from all the various non-free-field effects characteristic of an enclosed space. The latter is unique to a particular room and comprises the acoustical response. Everything that is not direct sound is indirect, reflected sound.

Direct and Indirect Sound

Figure 19-1 shows how sound level varies with distance from a source, for different room absorbencies. The source could be someone talking, a musical instrument, or a loudspeaker. Assume a sound-pressure level of 80 dB measured 1 ft from the source. If all surfaces of the room are 100% reflective, as shown by contour A, the sound pressure level in this reverberation chamber would be 80 dB everywhere in the room because no sound energy is absorbed. There is essentially no direct sound; it is all indirect. Contour B represents the fall off in sound-pressure level with distance from the sound source with all surfaces 100% absorptive. In this case all the sound is direct; there is no indirect component. The best anechoic chambers approach this condition. This is a true free field, and for this condition the sound-pressure level decreases 6 dB for each doubling of distance.

Role of Room Treatment

Between the indirect "all-reverberation" case of contour A and the direct "no-reverberation" case of contour B lie a multitude of other possible "some-reverberation" cases that depend on room treatment. In the area between these two extremes lies the real world of studio acoustics. The room represented by contour C is much deader than that of D. In practice, the direct sound is observable a short distance away from the source, but farther away, the indirect sound dominates. A transient sound

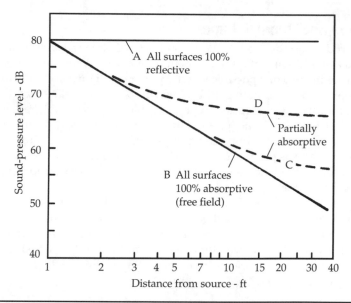

FIGURE 19-1   The sound-pressure level in an enclosed space varies with distance from the source of sound according to the absorbency of the space.

picked up by a microphone would, for the first few milliseconds, be dominated by the direct component, after which the indirect sound arrives at the microphone as reflections from room surfaces. These are spread out in time because of the different path lengths traveled.

Another component of indirect sound results from room resonances, which in turn are the result of reflected sound. The direct sound from the source excites these resonances, bringing into play all the effects described in Chap. 13. When the source excitation ceases, each mode dies away at its own natural frequency and at its own rate. Sounds of very short duration might not last long enough to fully excite room resonances.

Distinguishing between reflections and resonances is an acknowledgment that neither a reflection concept nor a resonance concept will fully describe room acoustic behavior throughout the entire audible spectrum. Resonances dominate the low-frequency region in which wavelengths of the sound are comparable to room dimensions. The ray concept works for higher frequencies and their shorter wavelengths. The 300- to 500-Hz region is a transition zone. But remembering the limitations of the analysis, the components of sound in a small studio can be characterized.

Indirect sound also depends on the materials of construction, such as doors, windows, walls, and floors. These too are set into vibration by sound from the source, and they too decay at their own particular rate when excitation is removed. If Helmholtz resonators are used in room treatment, sound not absorbed is reradiated.

The sound of the studio, embracing all of these components of indirect sound plus the direct sound, has its counterpart in musical instruments. In fact, it is helpful to consider a studio as a musical instrument. It has its own characteristic sound, and skill is required to extract its full potential.

Room Modes and Room Volume

The short dimensions of a small room almost guarantee acoustical anomalies resulting from excessive spacing of room resonant frequencies. A cubical room distributes modal frequencies in the worst possible way, by piling up all three fundamentals, and each trio of multiples with maximum gaps between modes. Having any two dimensions in multiple relationship results in this type of problem. For example, a height of 8 ft and a width of 16 ft mean that the second harmonic of 16 ft coincides with the fundamental of 8 ft. This emphasizes the importance of proportioning the room for best distribution of axial modes. This is particularly important in small rooms; as we will see, large rooms have the inherent advantage of more low-frequency modes, and hence smoother low-frequency response.

If there are fewer low-frequency axial modes than desired in the room under consideration, and this is always the case in a small room, modes must be distributed as uniformly as possible. This can be achieved by choosing a favorable room dimension ratio. For example, Sepmeyer suggests a dimension ratio of 1.00:1.28:1.54. Table 19-1 shows the selected dimensions, based on ceiling heights of 8, 12, and 16 ft, resulting in room volumes of 1,000; 3,400; and 8,000 ft^3.

Mode Analysis for Different Room Sizes

Axial mode frequencies were calculated after the manner of Table 13-6 and plotted in Fig. 19-2. A visual inspection of Fig. 19-2 shows the increase in the number of axial modes as volume is increased. This results in the benefit of closer low-frequency mode spacing, and hence smoother low-frequency response.

The number of axial modes below 300 Hz varies from 18 for the small studio to 33 for the large studio, as shown in Table 19-2. The low-frequency response of the two smaller studios at 45.9 and 30.6 Hz is inferior to the large studio at 22.9 Hz. The average spacing of modes, based on the frequency range from the lowest axial mode to 300 Hz, is also listed in Table 19-2. The average spacing varies from 14.1 Hz for the small studio to 8.4 Hz for the large studio. Room volume is thus an important factor in recorded sound quality; small studios have an inherent limitation in this regard.

Modes other than axial are present. The major diagonal dimension of a room also represents the lowest frequency supported by room resonances due to oblique modes. Thus, the frequency corresponding to the room diagonal listed in Table 19-2 is another measure of the low-frequency capability of a room in addition to its lowest axial frequency. This approach gives the lowest frequency for the large room as 15.8 Hz, compared to 22.9 Hz for the lowest axial mode.

	Ratio	Small Studio	Medium Studio	Large Studio
Height	1.00	8.00 ft	12.00 ft	16.00 ft
Width	1.28	10.24 ft	15.36 ft	20.48 ft
Length	1.54	12.32 ft	18.48 ft	24.64 ft
Volume	NA	1,000 ft^3	3,400 ft^3	8,000 ft^3

TABLE 19-1   Studio Dimensions

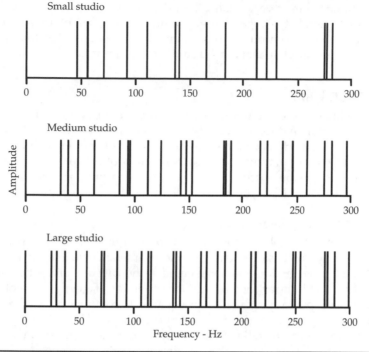

FIGURE 19-2 Comparison of the axial-mode resonances of a small (1,000 ft³), a medium (3,400 ft³), and a large (8,000 ft³) studio, all having the proportions 1.00:1.28:1.54.

The reverberation times listed in Table 19-2 are assumed, nominal values for the respective studio sizes. Given these reverberation times, the mode bandwidth is estimated from the expression $2.2/RT_{60}$. Mode bandwidth varies from 3 Hz for the large studio to 7 Hz for the small studio. The advantage of closer spacing of axial modes in the large studio tends to be offset by its narrower mode bandwidth. So, we see conflicting factors at work as we realize the advantage of the mode skirts overlapping each other. In general, however, the greater number of axial modes for the large studio, coupled with the extension of room response in the low frequencies, produces a response superior to that of the small studio.

	Small Studio	Medium Studio	Large Studio
Number of axial modes below 300 Hz	18	26	33
Lowest axial mode (Hz)	45.9	30.6	22.9
Average mode spacing (Hz)	14.1	10.4	8.4
Frequency corresponding to room diagonal (Hz)	31.6	21.0	15.8
Assumed reverberation time of studio (sec)	0.3	0.5	0.7
Mode bandwidth ($2.2/RT_{60}$)	7.3	4.4	3.1

TABLE 19-2 Studio Resonances

A studio having a very small volume has fundamental response problems in regard to room resonances; larger studio volume yields smoother response. Frequency-response anomalies encountered in studios having volumes less than 1,500 ft³ are sometimes severe enough to make small studio rooms impractical.

Reverberation Time

Reverberation is the composite effect of all three types of indirect sound. Measuring reverberation time does not directly reveal the nature of these individual components of reverberation. Herein is the weakness of reverberation time as an indicator of a room's acoustical quality. The important action of one or more of the reverberation components may be obscured by the compositing process. This is why reverberation time is one indicator of acoustical conditions, but not the only one.

Reverberation in Small Rooms

Some acousticians feel it is inaccurate to apply the concept of reverberation time to relatively small rooms. It is true that a genuine reverberant field may not exist in small spaces. The Sabine reverberation equation is based on the statistical properties of a random sound field. If such an isotropic, homogeneous distribution of energy does not prevail in a small room, is it proper to apply the Sabine equation to compute the reverberation time of the room? The answer is a purist "no," but a practical "yes." Reverberation time is a measure of decay rate. A reverberation time of 0.5 second means that a decay of 60 dB takes place in 0.5 second. Another way to express this is 60 dB/0.5 second = 120 dB/second decay rate. Whether the sound field is diffuse or not, sound decays at some particular rate, even at the low frequencies at which the sound field is least diffuse. The sound energy stored at the modal frequencies decays at some measurable rate, even though only a few modes are contained in the band being measured. It would seem to be practical to utilize the Sabine equation in small-room design to estimate absorption needs at different frequencies. At the same time, it is important to remember the limitations of the process.

As noted, technically, the term "reverberation time" should not be associated with relatively small spaces in which the sound field is not random. However, a designer must calculate the amount of absorbent needed to establish the acoustical character of a room. While reverberation time is useful for this purpose, the values of reverberation time in a small studio do not have the same significance as in a large space.

Optimal Reverberation Time

If the reverberation time is too long, speech syllables and music phrases are masked and a deterioration of speech intelligibility and music quality results. If reverberation time is too short, music and speech lose character and suffer in quality, with music suffering more. These effects are not definite and precise; there is no specific optimum reverberation time, because many other factors are involved. For example, is it a male or female voice, slow or fast talker, English or German language (they differ in the average number of syllables per minute), vocal or instrumental, solo flute or string ensemble, hard rock or a waltz? In spite of such variables, there is a body of experience from which we can extract helpful information. Figure 19-3 is an approximation rather than a true optimum, but following it will result in reasonable, usable conditions for many types of recording. In particular, the shaded area of Fig. 19-3 represents a compromise in studios used for both speech and music recording.

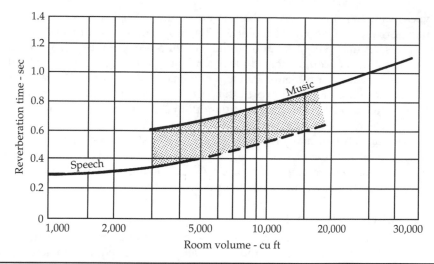

FIGURE 19-3 Suggested reverberation times for recording studios. The shaded area is a compromise region for studios in which both music and speech are recorded.

Diffusion

Diffusion contributes a feeling of spaciousness through the spatial multiplicity of room reflections and the control of resonances. Splaying walls and the use of geometrical protuberances have a modest diffusing effect. Distributing the absorbing material is a useful means of not only achieving some diffusion, but increasing the absorbing efficiency as well.

Modular diffraction grating diffusing elements provide efficient diffusion and are easily installed in small studios. For example, 2 × 4 ft modular units offer diffusion and broadband absorption (0.82 coefficient at 100 Hz, for example), all within a 2-in thickness. In practice, it would be difficult to provide too much diffusion in a studio room.

Noise

A small studio is subject to the same rules of acoustic isolation as any other room. Walls, floors, and ceilings must be constructed for high transmission loss and decoupling from external noise and vibration sources. This helps ensure the low ambient noise levels required for good recording quality. Likewise, it isolates neighboring spaces from the potentially loud music levels present in the studio. The partition between the studio and the control room is particularly critical in a recording studio. Among other functions, the partition must provide good isolation so that engineers in the control room will only hear sound from the monitor loudspeakers, that is, what is actually being recorded, and will not be mislead by other live sound leaking through the partition.

Studio Design Example

The following is an example of the methodology that can be used to design the principal acoustical aspects of a small studio. Consider a room measuring 25 × 16 × 10 ft. Let us assume that the primary interest is conventional music produced by small vocal and instrumental ensembles. For a volume of 4,000 ft³, we may select a rever-beration time of about 0.6 second. The proportions are not optimum for modal response. A more suitable ratio would call for a room length of 23.3 ft instead of the existing 25 ft. We could shorten the room, but in this case, a length of 25 ft creates no serious axial mode problems. In many practical designs, compromises such as this are inevitable, and not always worth the cost to remedy. Indeed, it is the designer's job, from an acoustical and budgetary standpoint, to understand when a compromise is acceptable, and when it is not.

We can consider many different materials and structures capable of giving the low frequency absorption needed in a small studio. Because it is primarily a music studio, we may consider paneling and because we are seeking good diffusion and brilliance for a recording studio, we might consider polycylindrical diffusers.

Absorption Design Goal

We begin the design process with some general thinking about absorption. A reverbera-tion time of 0.6 second is desired. What absorption is required to achieve this? For a studio of 4,000 ft³ volume, we compute this to be 327 sabins. Moreover, we will need 327 sabins at each of six frequency points. Polycylindrical modules can provide a large part of the low-frequency absorption and provide some needed diffusion as well. How many polys are needed to do this? In the 100- to 300-Hz region, the absorption coeffi-cient is 0.3 to 0.4; we choose an average of 0.35 as we will use polys of different sizes. To estimate the area of polys required, we use the relationship:

$$A = S\alpha \qquad (19\text{-}1)$$

where A = absorption units, sabins
S = surface area, ft²
α = absorption coefficient

Thus, we observe that the required surface area $S = 327/0.35 = 934$ ft². For broad-band absorption, polys should be of different chord lengths; let's pick three of the four shown in Fig. 12-26: polys A, B, and D.

Not all the low-frequency absorption will come from polys. The ceiling alone has 400 ft², so it would appear that we have enough surface area to accommodate 934 ft² of polys. For high-frequency absorption let us consider common acoustical tile. Ideally, we should have some of each acoustical material applied to every surface of the studio. Practically, we can only approach this. For one thing, the only practical acoustical material for the floor is carpet and having the floor covered with carpet would give more high-frequency absorption than required.

A floor area of 400 ft² and a carpet absorption coefficient of about 0.7 gives 280 sabins in the high frequencies for the carpet alone, and the polys have coefficients of 0.2 in the high frequencies also. Even if we used only 500 ft² of polys, they would contribute another 100 sabins. Carpet plus polys would give 280 + 100 = 380 sabins and we only want 327. So, carpet cannot be used.

Proposed Room Treatment

The proposed room treatment is shown in Fig. 19-4. On the north and south walls, the cross section of the ceiling polys is visible. A large poly A is in the center, flanked with a poly D on each side. Between the D polys and the outside B polys is space for lighting fixtures running

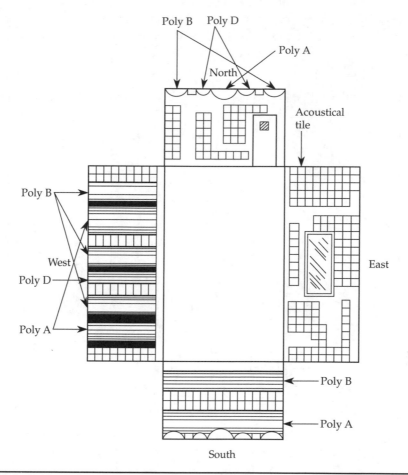

FIGURE 19-4 Fold-out plan of a music studio measuring 25 × 16 × 10 ft. Polycylindrical absorbers and acoustical tiles are used for absorption. The axes of the polys on the walls and ceiling are mutually perpendicular.

the length of the room. This poly treatment on the ceiling opposes the hard, reflective floor covered with vinyl tile and should prevent flutter echo in the vertical dimension.

On the west wall two A polys, three B polys, and one D poly are arranged vertically, which places them perpendicular to the ceiling polys as they should be. Four vertical tiers of $^3/_4$-in acoustical tile are worked into this west wall; two of them are placed at the corners where absorbers are especially effective (see discussion below). This breaks up the west wall so that we would expect no flutter echo between it and the east wall.

The south wall has a single A and a single B poly arranged horizontally so that the axis of each set of polys is perpendicular to the other two sets. Two horizontal tiers of acoustical tile are placed between 4 and 6 ft from the floor (which is head height for a standing person). Good high-frequency absorbers are placed at this height on all walls, except the one to which a performer would turn his back, the west wall.

With the polys as selected, and using the appropriate absorption coefficients, the absorption at the six standard frequencies was calculated, as shown in Table 19-3. The absorption attributed to the polys is plotted in Fig. 19-5.

Size 25 × 16 × 10 ft
Floor vinyl tile
Walls plaster
Volume (25) (16) (10) = 4,000 ft³

	Material	S ft²	125 Hz α	125 Hz Sα	250 Hz α	250 Hz Sα	500 Hz α	500 Hz Sα	1 kHz α	1 kHz Sα	2 kHz α	2 kHz Sα	4 kHz α	4 kHz Sα
EMPTY	Poly A	232	0.41	95.1	0.40	92.8	0.33	76.6	0.25	58.0	0.20	46.4	0.22	51.0
	Poly B	271	0.37	100.3	0.35	94.9	0.32	86.7	0.26	75.9	0.22	59.5	0.22	59.6
	Poly D	114	0.25	28.5	0.30	34.5	0.33	37.6	0.22	25.1	0.20	22.8	0.21	23.9
	Total			223.9		222.2		200.9		159.0		128.8		134.5
FILLED	Poly A	232	0.45	104.4	0.57	132.2	0.38	88.2	0.25	58.0	0.20	46.4	0.22	51.0
	Poly B	271	0.43	116.5	0.55	149.1	0.41	111.1	0.28	75.9	0.22	59.6	0.22	59.6
	Poly D	114	0.30	34.2	0.42	47.9	0.35	39.9	0.23	26.2	0.19	21.7	0.20	22.6
	Total			255.1		329.2		239.2		160.1		127.7		133.4
	Acoustical title, ¾"	274	0.09	24.7	0.26	76.7	0.78	213.7	0.84	230.2	0.73	200.0	0.64	175.4
	Total sabine EMPTY			248.6		296.9		414.6		389.2		328.8		309.9
	Total sabine FILLED			279.8		405.9		452.9		390.3		327.7		308.8
	Reverberation time EMPTY			0.79		0.68		0.47		0.50		0.60		0.53
	Reverberation time FILLED			0.70		0.48		0.43		0.50		0.80		0.63

TABLE 19-3 Room Conditions and Calculations for a Small Studio

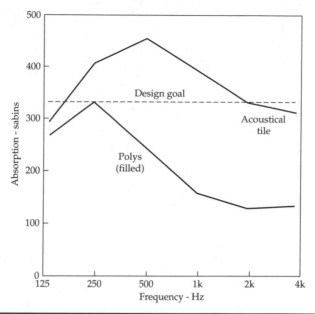

FIGURE 19-5   Relative contributions of the polycylindrical absorbers and acoustical tile to the overall absorption in the music studio of Fig. 19-4.

The $^3/_4$-in acoustical tile offers an absorption coefficient of 0.73 at high frequencies and, following the same procedure as with the polys, we find that we need $200/0.73 = 274$ ft^2 of acoustical tile. Additional tile is then added to the room, distributing it around the studio in patches, until 274 ft^2 is obtained. The upper part of Fig. 19-5 is plotted from the calculations of Table 19-3.

Figure 19-5 shows a slight deficiency of absorption at low frequencies and considerable extra absorption at middle frequencies. Stuffing the polys with glass fiber will increase absorption at low frequencies.

We must always be aware of the lack of precision in calculations such as just presented. The curves appear to be precise, but care must be taken to avoid frequency-response anomalies due to axial modes. And, for example, addition diffusion may be required. In short, successful studio and room design requires more than calculation; it also requires experience and a good ear.

A final and partly unrelated note: when designing a small studio, it is easy to budget for major construction, such as wall and floor/ceiling constructions, but overlook details such as sound lock treatment, doors and their sealing, wiring precautions, illuminating fixtures, observation windows, and other features common to studios. All of these can create serious problems if not handled properly.

Previously, we have considered reverberation and how to compute and measure it (Chap. 11), dissipative and tuned absorbers (Chap. 12), the complications of room resonances (Chap. 13), the need for diffusion and how to obtain it (Chaps. 9 and 14), noise control (Chap. 16), and air-handling equipment (Chap. 17). All of these are also integral parts of any studio design.

CHAPTER 20

Acoustics of Control Rooms

The acoustical design of control rooms is highly specialized and unique in the field of architectural acoustics. Contemporary control-room designs can provide outstanding acoustical performance. A control room has the very specific purpose of providing high-quality loudspeaker playback to a single listening (mixing) position. This is conceptually similar to a high-end home listening room, but with one important difference. In a listening room, the goal is primarily to provide a pleasurable listening experience; while accurate playback is desired, it is secondary to enjoyment. A control room is a work environment where playback accuracy is critical. The engineer bases all recording and mixing decisions on the sound delivered by the monitor loudspeakers, via the room acoustics. If that sound is not accurate, recording decisions will be flawed. For example, if a control room delivers a boosted bass response at the mixing position, then the engineer will be misled by that, and mistakenly compensate for it, and all the recordings from the room will have a bass deficiency.

This chapter focuses on the acoustics of control rooms, that is, the room in recording studios where sound is played back and mixed. In this room, the acoustics of loudspeaker/room interface is critical. Assisting this is the fact that the location of the listener with respect to the monitor loudspeakers is known. The previous chapter discussed the acoustics of the "studio" room in a recording studio, that is, the room where music is performed and recorded via microphones.

Initial Time-Delay Gap

The initial time-delay gap (ITDG) is an acoustical metric used to examine early reflections in a sound field. ITDG is defined as the time between the arrival of the direct sound at a given seat, and the arrival of the first reflection. In smaller, more intimate rooms, with reflective surfaces closer to the listener, ITDG is small. In larger rooms, ITDG is greater. In an extensive analysis of concert halls, Beranek noted that those halls rated the highest by qualified listeners had certain technical similarities, including similar ITDG values. Highly rated halls had a well-defined initial time-delay gap of about 20 msec. In inferior halls, the ITDG was flawed by uncontrolled reflections. The term was originally coined by Beranek to describe this characteristic of concert halls, but the principle can be applied to different kinds of spaces, including recording studios and control rooms.

The factors generating the initial time-delay gap of an untreated control room are illustrated in Fig. 20-1A. The direct sound travels from the loudspeaker to the mixing position. Soon after, reflections from surfaces near the loudspeaker in the front of the room arrive at the mixing position. These early reflections create comb-filter effects in the frequency response at the mixing position. Later, the sound reflected from the floor, ceiling, or other objects arrives at the mixing position. This time gap between the arrival of the direct and reflected components is determined by the geometry of a particular room.

Don Davis and Chips Davis (not related to Don Davis) analyzed the effects of the initial time-delay gap of recording studios and control rooms with a measuring technique known as time-delay spectrometry. Time-delay spectrometry revealed the comb-filter effects associated with early reflections from surfaces near the monitor loudspeakers (and from the console surface). They observed that one way to minimize comb-filter effects was to eliminate or reduce the early reflections in the control room. This clarification of the problem led to the solution of placing absorbing material on the surfaces surrounding the monitor loudspeakers in the front part of the control room, as shown in Fig. 20-1B.

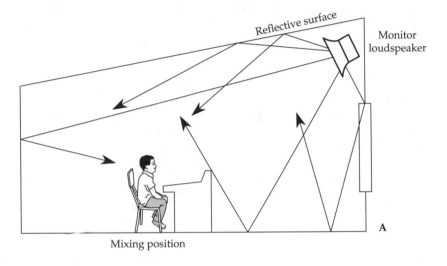

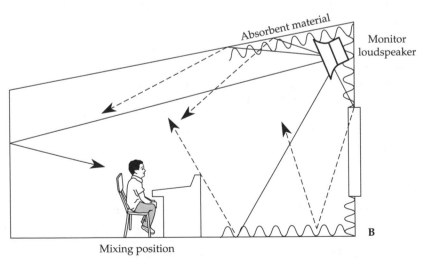

FIGURE 20-1   The initial time-delay gap describes the time between the arrival of direct sound at the mixing position, and the arrival of early reflections. (A) In an untreated control room, the early reflections from the monitor loudspeakers in the front of the room can cause comb-filter effects at the mixing position. (B) In a treated room, with absorbent placed near the monitor loudspeakers, these early reflections are attenuated.

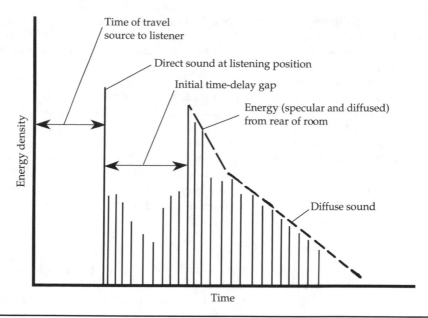

Figure 20-2 In a treated control room, the initial time-delay gap is well defined. Early reflections from the front of the room, although present, have been attenuated by absorbent placed in the front of the room.

In simplified form, Fig. 20-2 shows the energy-time relationships essential for a properly designed and treated control room. At time = 0, the signal leaves the monitor loudspeaker. After an elapsed transit time, the direct sound reaches the mixing position. There follows some insignificant low-level clutter (to be neglected if 20 dB down), after which the first return from the rear of the room arrives. Subsequent delayed reflections constitute the end of the initial time-delay gap and the first signs of an exponential reverberant decay.

Live End-Dead End

Using time-delay spectrometry analysis, Davis and Davis mounted absorbing material on the surfaces of the entire front part of a control room. The results were encouraging; sound clarity in the control room was improved, as was stereo imaging. Also, the ambience of the room was perceptually more spacious. Giving the control room a precise initial time-delay gap gave listeners the impression of a much larger room.

The deadening of the front end of a control room near the observation window is a straightforward procedure—just cover the surfaces with absorbent. There is a resulting improvement of sound quality. Once the front end is made absorbent, attention naturally shifts to the rear end of the control room. If the rear end is deadened, the room will almost certainly be too absorbent overall. Therefore, to obtain a suitable reverberation time, the rear end of the control room is made live. This control-room configuration is known as live end-dead end (LEDE). Late reflections from the rear of the room do not present the same comb-filtering problems as front-end early reflections. Moreover, if the delay time of the rear-end reflections is within the Haas fusion zone, the human auditory system integrates these delayed reflections from the rear wall into the early-arrival sound. The ambience from the rear of the control room is perceived as coming naturally from the sound sources in the front of the room. However, specular reflections from the rear wall surfaces should be avoided; the rear wall surfaces should provide diffused reflections.

The principle of the LEDE control room was originally designed for stereo playback; it naturally lends itself to the assumption that the monitor loudspeakers will be in the front of the room. In rooms configured for multichannel playback, with surround speakers, the LEDE design is more complex. However, the principle of avoiding early reflections is still important. Following the principle, surfaces near the loudspeakers should be deadened to prevent comb-filter effects. Other surfaces should be reflective while providing good diffusion. Specular reflections should be avoided. A further discussion on control-room design is given later; for simplicity, stereo playback is assumed. However, the principles can be extended and modified for surround-sound playback.

Specular Reflections versus Diffusion

In a specular reflection, incident energy is reflected from a surface such that the angle of reflection equals the angle of incidence, much like light reflects from a mirror. The surface is flat and smooth relative to the wavelength of sound so there is no diffusion. In a diffuse reflection, the incident energy is reflected such that it is uniformly scattered in all directions. This is caused when the reflecting surface has irregularities with a size on the order of the wavelength of sound being diffused. As with all diffusion, this is wavelength dependent. For example, a surface with irregularities of about 1 ft will diffuse sound that has a wavelength of about 1 ft (1 kHz). However, for sound with much longer wavelengths, the same 1-ft irregular surface will appear smooth, and sound will be specularly reflected by the entire wall. Also, sound with much shorter wavelengths will specularly reflect from the individual irregularities, at whatever angle they are positioned at.

In a specular reflection, all of the acoustical energy from a given point on the reflector surface arrives in a single instant of time. An example of specular reflection is shown in Fig. 20-3A. If the same sound energy is incident on a reflection phase-grating diffuser, as shown in Fig. 20-3B, the scattered energy is spread out in time. Each element of the

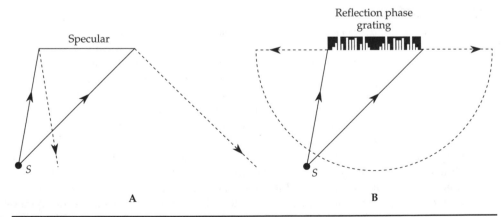

Figure 20-3   A comparison of reflection of sound energy. (A) A flat surface yields specular reflection. (B) A reflection-phase grating yields energy dispersed in a hemidisc.

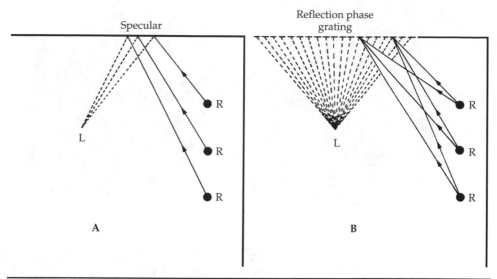

FIGURE 20-4   Side-wall reflections R in a control room return energy to the mixing position, L. (A) If the rear wall is specular, only a single point on the rear wall returns energy to the operator. (B) With a reflection phase-grating diffuser, every element returns energy to the mixing position. (*D'Antonio*)

diffuser returns energy, the respective reflected wavelets arriving at different times. This temporal distribution of reflected (diffused) energy results in a rich, dense, non-uniform mixture of comb filters that the human auditory system perceives as a pleasant ambience. This is in contrast to the sparse specular reflections that combine to form unpleasant wideband frequency-response deviations. For this reason, diffusing elements are placed in the live rear of a live end-dead end control room.

With the reflection phase-grating diffuser, the reflected wavelets are not only spread in time, they are also spread in space. The one-dimensional diffuser spreads its reflected energy in a horizontal hemidisc. By orienting other one-dimensional units, vertical hemidiscs of diffusion are easily obtained. This is in contrast to the specular panel, which distributes reflected energy in only a portion of half-space determined by the location of the source and the size of the panel.

Another feature of the reflection phase grating, illustrated in Fig. 20-4, makes it especially desirable for the live rear end of a control room. Consider the example of three side-wall reflections R that impinge on the rear wall, returning energy to the mixing (listening) position, L. If the rear wall is specular, there is only one point on the surface returning energy from each source to the listener. In contrast, each element of the surface of the grating diffuser sends energy toward the listener. Energy from all sound sources (direct or reflected) falling on the diffuser are scattered to all observation positions. Instead of a small "sweet spot" at the console, a much wider area of good sound quality results.

Low-Frequency Resonances in the Control Room

As we have seen, room modes largely determine the low-frequency response of room, particularly small rooms such as most control rooms. This can be examined with a time, energy, and frequency analysis. Figure 20-5 is a three-dimensional plot showing the

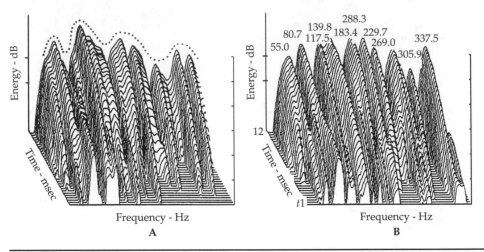

FIGURE 20-5 A three-dimensional energy-time-frequency plot of the sound field of a control room showing the room modes and their decay. (A) The untreated room condition shows uneven response. (B) Modal interference can be reduced by installing a low-frequency diffuser at the rear of the control room. (*D'Antonio*)

relationship between time, energy, and frequency at the mixing position in a control room. The vertical scale is 6 dB between marks. This is not sound-pressure level, but rather true energy. The frequency scale ranges from 9.64 to 351.22 Hz, a region expected to be dominated by modes. Time ranges from the rear, 2.771 msec per step, or about 0.1 second for the entire traverse. The dramatic ridges in Fig. 20-5A are the modal (standing wave) responses of the control room, which together make up the low-frequency acoustical response of the room.

In addition to the modal response, this analysis contains information concerning a second phenomenon of control room design. This is the interference between the direct low-frequency wave from a monitor loudspeaker and its reflection from the rear wall. If the mixing position is 10 ft from the rear wall, the reflection lags the direct sound by the time it takes the wave to travel 20 ft. The delay is $t = (20 \text{ ft})/(1{,}130 \text{ ft/sec}) = 0.0177$ second. The frequency of the first comb-filter notch is $1/2t$ or $1/(2)(0.0177) = 28.25$ Hz (see Chap. 10). Subsequent notches, spaced $1/t$ Hz, occur at 85, 141, 198, 254 Hz, and so on.

The depth of the notches depends on the relative amplitudes of the direct and reflected components. One way to control the depth of the notches is to absorb the direct sound so that the reflection is low. However, this removes useful sound energy from the room. A better way is to build large diffusing units capable of diffusing at these low frequencies. The result of this approach is shown in Fig. 20-5B. A 10-ft-wide, 3-ft-deep, floor-to-ceiling diffuser is placed behind the midband diffusers. This provides a smoother frequency response and smoother decay. These modes decay about 15 dB in the 0.1-second time sweep. This means a decay rate of roughly 150 dB/sec, which corresponds to a reverberation time of 0.4 second, but the variation of decay rate between the various modes is great.

Initial Time-Delay Gaps in Practice

An energy-time display allows a sophisticated evaluation of the distribution in time of acoustical energy in a room and the presence or absence of an initial time-delay gap (ITDG). Figure 20-6 shows three energy-time displays of widely differing spaces. The display of Fig. 20-6A shows the response for a control room in the Master Sound Astoria recording studio in New York. The well-defined initial time-time gap, the exponential decay (a straight line on a log-frequency scale), and the well-diffused reflections are the hallmarks of a well-designed room.

Figure 20-6B shows the energy-time display for the Concertgebouw in Haarlem, the Netherlands. A well-defined initial time-time gap of about 20 msec qualifies it as one of Beranek's quality concert halls.

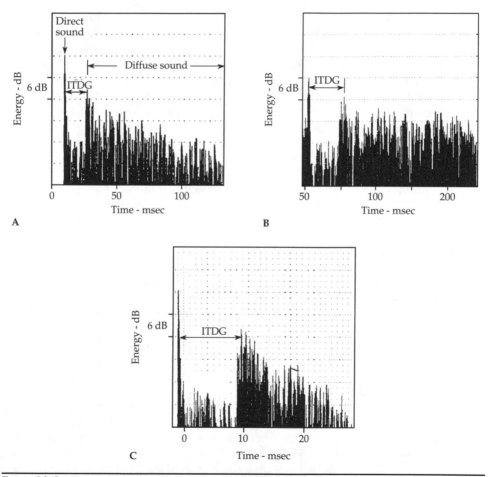

Figure 20-6 Illustrations of initial time-delay gaps from widely different types of spaces, each of high acoustical quality. (A) The control room of Master Sound Astoria, New York. (B) The Concertgebouw concert hall, Haarlem, the Netherlands. (C) A small listening room, Audio electronics Laboratory, Albertson, New York. (*D'Antonio*)

There are many high-quality concert halls in the world, as well as many high-quality control rooms. However, there have been relatively few small listening or recording rooms that could be classified as quality spaces for the simple reason that their small volumes are dominated by normal modes and their associated problems. Figure 20-6C shows the energy-time display for a high-quality small listening room belonging to Audio Electronics Laboratory of Albertson, New York. A beneficial, initial time-delay gap of about 9 msec is made possible by a live end-dead end design in which the narrator and microphone are placed in the dead end of the room facing the live end.

Loudspeaker Placement and Reflection Paths

The management of reflections is a major concern in control room design. Davis recommended making the entire front end dead by applying absorbent to the surfaces. Berger and D'Antonio devised a method that does not depend on absorption but rather on the shaping of surfaces to nullify the negative effect of reflections. In this approach, the front end of the control room may be live, but with a very specific shape to provide specific reflection paths. This configuration places the listener in a reflection-free zone, and the approach is sometimes known as RFZ. The placement of a loudspeaker close to solid boundaries can greatly affect its output. If it is placed close to an isolated solid surface, its power output into half space is doubled, providing an increase of 3 dB. If the speaker is placed close to the intersection of two such surfaces, the power is confined to quarter space, and there is an increase of 6 dB. If placed close to the intersection of three such solid surfaces, the power radiates into a one-eighth space, and there is an increase of 9 dB.

Placing monitor loudspeakers at some specific distance from the trihedral surfaces of a room has been common in control rooms and home listening rooms. If the distances from the loudspeaker to the surfaces are appreciable in terms of wavelength of sound, problems are introduced. The overall power boost effect may be minimized, but frequency response might be affected due to the constructive and destructive combination of direct and reflected waves.

A point source in a trihedral corner has a flat response at an observation point if that observation point is in a reflection-free zone. There are no reflections to contribute to interference effects. D'Antonio extends this observation and recommends placing control-room monitor loudspeakers in trihedral corners formed from splayed surfaces. By splaying the room boundaries, a reflection-free zone can be created around the listener. By splaying the walls and ceiling, it is even possible to extend the reflection-free zone across the entire console, several feet above, and enough space behind, to include the producer behind the mixing position.

One reflection that is uniquely problematic in control rooms is the reflection from the mixing console. The console is placed between the monitor loudspeakers and the mixing engineer, and its angle often allows a reflection to strike the mixing position. This strong reflection arriving from the same direction as the direct sound, combined with the direct sound, can result in a comb-filtered signal at the mixing position. This can result in timbral deviations of the signal. To some extent, the comb-filter response at the mixing position is alleviated by the many other reflections naturally occurring in the room, arriving at different times and from different directions. Even so, the console reflection is a stubborn problem in many control rooms.

The Reflection-Free-Zone Control Room

In their 1980 paper, Don Davis and Chips Davis specified that there should be ". . . an effectively anechoic path between the monitor loudspeakers and the mixer's ears." What they called "anechoic" is today called a reflection-free zone. The most obvious way to achieve an anechoic condition is through absorption, hence the "dead end" designation.

To design a control room with a reflection-free zone (RFZ), image sources must be considered. The contribution of a reflection from a surface can be considered as coming from a virtual source on the other side of the reflecting plane on a line perpendicular to that plane through the observation point and at a distance from the plane equal to that to the observation point. With splayed surfaces in three dimensions, all the virtual sources must be visualized; this is necessary to establish the boundaries of the reflection-free zone.

A floor plan of a reflection-free zone control room is shown in Fig. 20-7. The monitor loudspeakers are flush-mounted as close as possible to the trihedral corner formed with the ceiling intersection. Next, both the front side walls and the front ceiling surfaces are splayed to send reflections away from the volume enclosing the listener. It is

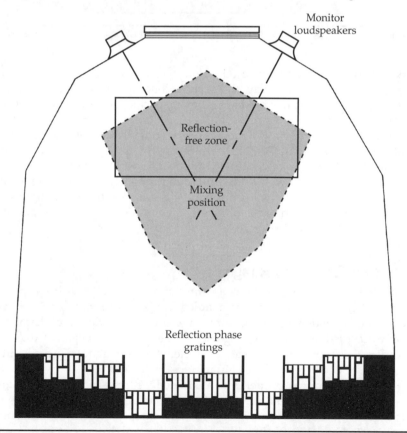

FIGURE 20-7 Plan view of a control room with a live, rather than a dead, front end. Front-end reflections at the mixing position are avoided by shaping the surfaces to create a reflection-free zone. The rear of the room is dominated by a mid- and high-band fractal diffuse and a low-frequency diffuser. (*D'Antonio*)

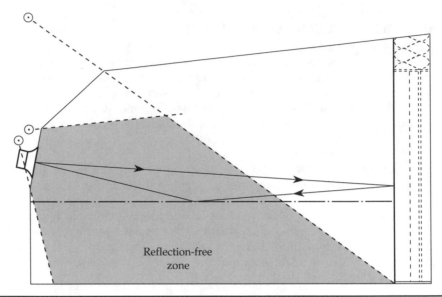

FIGURE 20-8   Elevation view of the reflection-free zone of the control room of Fig. 20-7. (*D'Antonio*)

possible to create an adequate reflection-free zone at the mixing position by proper splaying of walls. In this way, an anechoic condition is achieved without recourse to absorbents. If absorbent is needed to control specific reflections, it can be applied to the splayed surfaces.

The rear end of the control room is provided with a complement of reflection phase-grating diffusers. In Fig. 20-7, the self-similarity principle is employed in the form of fractals. A high-frequency quadratic residue diffuser is mounted at the bottom of each low-frequency diffuser well. Wideband sound energy falling on the rear wall is diffused and directed back to the mixing position. This sound is diffused both in space and time by the hemidisc pattern of the diffusers. An elevation view of the reflection-free zone control room is shown in Fig. 20-8.

Control-Room Frequency Range

Some of the salient constructional features of control rooms address specific acoustical issues. The range of frequencies handled in a control room is very great, and every frequency-dependent component must perform its function over that range. The commonly accepted high-fidelity range of 20 Hz to 20 kHz is a span of 10 octaves, or three decades. This represents a range of wavelengths from about 57 ft to $5/_8$ in. The control room must be built with this fact in mind.

The lowest modal frequency is associated with the longest dimension of a room, which may be taken as the diagonal. Below that frequency there is no modal resonance support for sound. The room's response falls sharply below this frequency, which can be estimated by $1,130/2L$, in which L is the diagonal distance in feet. For a rectangular room following Sepmeyer's proportions having dimensions of $15.26 \times 18.48 \times 12$ ft, the diagonal is 28.86 ft. This places the low-frequency cutoff frequency for this particular room at $1,130/[(2)(28.86)] = 21$ Hz.

From 21 to about 100 Hz (for a room of this size), normal modes dominate and wave acoustics must be applied. From 100 to about 400 Hz in this room is a transition region in which diffraction and diffusion prevail. Above 400 Hz, true specular reflection and ray acoustics provide correct modeling. These frequency zones determine construction of the control room, essentially a massive shell to contain and distribute low-frequency modal energy and an inner shell for reflection control.

Outer Shell and Inner Shell of the Control Room

The size, shape, and proportions of the massive outer shell of the control room determine the number of modal frequencies and their specific distribution, as discussed in Chap. 13. There are two schools of thought: one prefers splaying of walls of the outer shell to break up modal patterns, while the other prefers the rectangular shape. Only a modest deviation from a rectangular shape toward a trapezoidal shape is feasible. Such a shape does not eliminate modal patterns, it just distorts them into an unpredictable form. Others feel that symmetry for both low-frequency and high-frequency sound better fits the demands of stereo and multichannel playback. To contain the low-frequency sound energy associated with control-room activities, thick walls, possibly 12-in-thick concrete, are required.

The purpose of the control-room inner shell is, among other things, to provide the proper reflection pattern at the mixing position. For example, in a reflection-free zone configuration, construction of the inner shell can be relatively light. For the inner shell, shape is more important than mass. However, care must be taken to avoid any buzzing or rattling from lightweight partitions.

CHAPTER 21

Acoustics of Audio/Video Rooms

In a room dedicated to one or a few related functions, the acoustical treatment can be tailored to a few applications, achieving excellent results. However, in many cases, rooms must be used for a diversity of functions. It is not practical to design separate rooms for each function; instead, one room must be designed to work acoustically well, with some compromises, for many functions. This is often the case for rooms used for audio/video postproduction. For example, in these rooms, these activities may be considered: Postrecording, digital sampling, MIDI, editing, sound effects (Foley), dialog replacement, voice over, signal processing, composing, video production, and equipment evaluation.

Design Factors

Because audio/video rooms must support such widely varying functions, it is expedient to consider three essential factors in audio/video room design:

- Any function requiring an open microphone demands good general acoustics in terms of room frequency response, reverberation, diffusion, and so on. Room treatment must be designed to provide this.

- Most audio work will be performed by recording engineers, preferably monitoring their work with high-quality loudspeakers. Thus, the room must allow good audio playback in stereo or multichannel.

- In most applications, quiet conditions are required. Noise from outside the room must be considered; acoustical isolation suitable for the work is required. In addition, noise from inside the room must be considered; noise from production equipment (such as cooling fans and hard-disk drives) must be minimized.

Many of the points applicable to the audio/video work place are covered in earlier chapters. To avoid redundancy, references are made to appropriate earlier coverage. A review of this referenced material is very much a part of this chapter.

Acoustical Treatment

A bare work space obviously needs some sort of treatment. An untreated work space is merely a container of air particles responsible for the propagation of sound. Practically all attenuation of sound (signals as well as noise) takes place at the boundaries of this airspace. The absorption of sound by the air itself is negligible in small rooms. Carpet on the floor boundary, lay-in panels on the ceiling boundary, and absorbent on the wall surfaces will reduce sound energy in the room upon each reflection. The example to follow illustrates how each of these needs can be addressed.

Audio/Video Work Place Example

As stated in Chap. 19, an audio room smaller than 1,500 ft³ is almost certain to produce timbral anomalies. In this example, a medium-sized work place (selected from Table 19-1) will be considered having these dimensions: length of 18 ft, 6 in; width of 15 ft, 4 in; and height of 12 ft. This yields a floor area of 284 ft² and a volume of 3,400 ft³. The proportions are as close to "optimum" as is feasible (see Table 13-5 and Fig. 13-6).

Appraisal of Room Resonances

Even though the room proportions are favorable, it is important to verify axial mode spacing. Only axial modes are considered because the tangential modes are inherently reduced 3 dB and the oblique modes are reduced 6 dB with respect to the more powerful axial modes. The length, width, and height axial mode resonance frequencies to 300 Hz are listed in Table 21-1. These constitute the low-frequency acoustics of this room. The spacings of these modes determine the smoothness of the low-frequency room response. Spacings of 25.8 and 30.6 are the only ones exceeding Gilford's suggested limit of 20 Hz (Chap. 13). Experience suggests that these two spacings are possible sources of frequency-response deviations. However, this is by no means a major flaw of the room. Favorable room proportions can only minimize the potential for modal problems, not eliminate them.

Control of Room Resonances

Most room resonances can be controlled with bass traps. In this example, some bass trap absorption in the 50- to 300-Hz region will be necessary. If the audio/video work space is of frame and drywall construction, an impressive amount of low-frequency absorption will be inherently present in the frame structure. The floor and wall diaphragms vibrate, and absorb low-frequency sound in the process. This is demonstrated in subsequent calculations. However, additional low-frequency absorption might be needed.

Absorption Calculation

For an estimate of the absorption (in sabins) needed in the room, we use the Sabine equation used to calculate reverberation time:

$$RT_{60} = \frac{0.049V}{A} \qquad (21\text{-}1)$$

where RT_{60} = reverberation time, sec
V = volume of room, ft³
A = total absorption of room, sabins

The volume of the room is 3,400 ft³. Assuming a desired reverberation time of 0.3 second, the amount of absorption can be calculated: $A = 0.049V/RT_{60} = (0.049)(3,400)/0.3 = 555$ sabins (absorption units). This is an approximate absorption value that will result in reasonable acoustical performance, and it can be varied at a later time to meet specific needs. The materials used to provide this absorption are described below.

Room Dimensions = 18.5 × 15.3 × 12.0 ft				
Axial Mode Resonances (Hz)				
Length **L = 18.5 ft** $f_1 = 565/L$ (Hz)	**Width** **W = 15.3 ft** $f_1 = 565/W$ (Hz)	**Height** **H = 12.0 ft** $f_1 = 565/H$ (Hz)	**Arranged in Ascending Order**	**Axial Mode Spacing (Hz)**
f_1 30.6	36.8	47.1	30.6	6.2
f_2 61.2	73.6	94.2	36.8	10.3
f_3 91.8	110.4	141.3	47.1	14.1
f_4 122.4	147.2	188.4	61.2	12.4
f_5 153.0	184.0	235.5	73.6	18.2
f_6 183.6	220.8	282.6	91.8	2.4
f_7 214.2	257.6	329.3	94.2	16.2
f_8 244.8	294.4		110.4	12.0
f_9 275.4	331.2		122.4	18.9
f_{10} 306.0			141.3	5.9
			147.2	5.8
			153.0	30.6
			183.6	0.4
			184.0	4.4
			188.4	25.8
			214.2	6.6
			220.8	14.7
			235.5	9.3
			244.8	12.8
			257.6	17.9
			275.4	7.2
			282.6	11.8
			294.4	11.6
			306.0	

Mean axial mode spacing = 12.0.
Standard deviation = 7.2.

TABLE 21-1 Axial Modes of Audio/Video Room

Proposed Treatment

The proposed treatment of the audio/video work place example is shown in Fig. 21-1. This is a fold-out plan in which the four walls, hinged along the edges of the floor, are laid out flat. It is assumed that the room is of frame construction, with a wooden floor, ½-in drywall on all walls, and carpet covering the floor. The drop ceiling panels are glass fiber. The materials needed to supply absorption are described in Table 21-2. Moreover, the table shows the calculations needed to determine the absorption contributed by each element, at six standard frequencies, using absorption coefficients for the different materials found in the appendix.

Specialized Treatment

Some of the treatment materials shown in Table 21-2 are less common. These elements include polycylindrical diffusing/absorbing units, a bass trap under the polycylindrical modules, and diffuser modules in the ceiling and on one wall.

Polycylindrical modules are inexpensive and reasonably effective. Their simple construction is detailed in Figs. 12-26 through 12-28. As for effectiveness as a diffuser, Fig. 14-21 compares the diffusion of two polys with the diffusion of a flat panel, distributed absorption, and a quadratic residue diffuser. The poly having the smaller chord gives a normal incidence diffusion pattern somewhat similar to the quadratic residue diffuser.

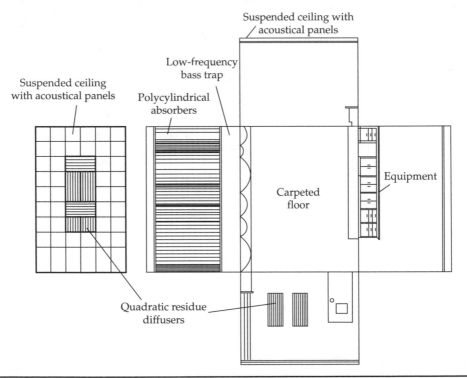

Figure 21-1 Possible treatment of an audio/video room. Elements include suspended ceiling with acoustical tile, carpeted floor, low-frequency Helmholtz resonator bass trap, polycylindrical sound absorbers, and quadratic-residue diffusion panels.

Material	S ft²	125 Hz α	125 Hz Sα	250 Hz α	250 Hz Sα	500 Hz α	500 Hz Sα	1 kHz α	1 kHz Sα	2 kHz α	2 kHz Sα	4 kHz α	4 kHz Sα
Drywall	812	0.10	81.2	0.08	65.0	0.05	40.6	0.03	24.4	0.03	24.4	0.03	24.4
Wood floor	284	0.15	42.6	0.11	31.2	0.10	28.4	0.07	19.9	0.06	17.0	0.07	19.9
Drop ceiling	234	0.69	161.5	0.86	201.2	0.68	159.1	0.87	203.6	0.90	210.6	0.81	189.5
Carpet	284	0.08	27.7	0.24	68.3	0.57	161.9	0.69	196.0	0.71	201.6	0.73	207.3
Polys	148	0.40	59.2	0.55	81.4	0.40	59.2	0.22	32.6	0.20	29.6	0.20	29.6
Bass trap	37	0.65	24.1	0.22	8.1	0.12	4.4	0.10	3.7	0.10	3.7	0.10	3.7
Diffuser	50	0.48	27.8	0.98	49.0	1.2	60.0	1.1	55.0	1.08	54.0	1.15	57.5
Total absorption, sabins			424.1		504.1		513.6		535.2		540.9		531.9
Reverberation time, second			0.39		0.33		0.32		0.31		0.31		0.31

TABLE 21-2 Absorption and Reverberation Calculations for an Audio/Video Room

Instead of polys, this wall could be covered with diffuser modules, which would give improved diffusion plus absorption. Cost should determine which to use. If labor costs to build the polys are too high, diffuser modules may be ideal. The modules are laid into the suspended ceiling framework (50 ft^2) and two panels (16 ft^2) are mounted on the wall near the door to discourage lengthwise flutter. These diffuser modules are also described in Chap. 14.

The bass trap beneath the polys is a simple perforated panel Helmholtz-type absorber tuned to give peak absorption in a low-frequency band. This can be covered with an open weave material for appearance. Full design particulars for the perforated panel bass trap resonator are given in Chap. 12. The bass trap might not be sufficient to control low-frequency modes in the room. If more bass absorption is needed, the two corners opposite the door could be treated in one of the ways described in Fig. 18-4.

Voice-Over Booth

Small closet-sized voice recording spaces can be notorious for their poor acoustics. Boundary surfaces may be treated with acoustic tile or some other material that absorbs high-frequency energy well, but provides very little low-frequency absorption. Thus important voice frequencies may be over absorbed, while low-frequency room modes are untreated. Because of the small dimensions of the booth, these room modes are relatively high in frequency, few and widely spaced. Room-mode analysis should be performed to determine if these will affect frequency response in the speech range.

The booth may be acceptable if room-mode problems are lower than the speech range. If so, mid- and high-frequency absorption may be applied, to the extent needed to control flutter echoes in the booth. Consideration should be given to potential reflections from glass in an observation window, and reflections from a script stand.

Dead versus Live Ambience

In some cases, voice booths are relatively absorbent, and the voice is recorded intentionally dry, and reverberation is added as needed in post production. However, making a small booth too absorbent, for both the high and low frequencies, can produce an eerie effect on the narrator. Such an anechoic space gives essentially no acoustic feedback, which the narrator may need for orientation and voice adjustment. For these reasons and others, an anechoic space is unsuitable for voice-over or announce booths, unless the booth acoustics are simply ignored and only headphones are used.

Another approach to treating a voice-over booth is to obtain a relatively live sound. For example, consider a small (perhaps 6 × 8 ft) voice recording room. The realization of a live sound field is based on a multiplicity of half-round tube traps, as shown in Fig. 21-2. A quarter-round trap is mounted in each wall-wall and ceiling-wall intersection to provide absorption to control normal modes. Half-round traps alternate on the walls and ceiling with hard, reflective drywall strips as wide as the traps (9, 11, or 16 in). The door and window should also be covered in this way to avoid compromise of the sound field. Limited visibility is available through the window glass strips of reflecting surface. The floor remains untreated.

Early Reflections

In the usual small recording room, there are too few early reflections that result in noticeable comb-filter anomalies. In this example, reflections from the script stand or window, which may otherwise cause comb-filter anomalies, are overwhelmed by the

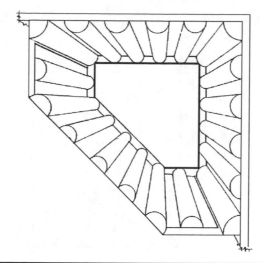

density of diffuse reflections from the many reflecting surfaces of the room. The broadband absorption of the half-round traps combines with the reflections from the hard surfaces between the traps to produce a dense, diffuse group of reflections immediately following the direct signal.

An energy-time-curve (ETC) of the room's response is shown in Fig. 21-3. The horizontal time axis extends to 80 msec in this case. The spike at the left extreme is the direct

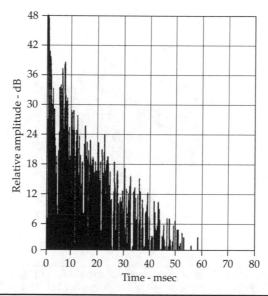

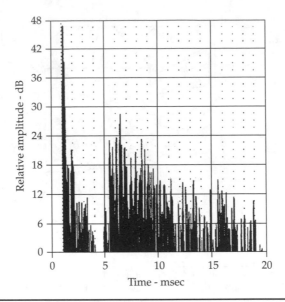

FIGURE 21-4   The early part of the energy-time curve of Fig. 21-3. In this case the expanded time base extends to 20 msec. A well-defined early time gap is shown. (*Acoustic Sciences Corporation*)

signal, followed by a smoothly decaying ambient sound field. This decay is quite fast, corresponding to a reverberation time of 0.08 second (80 msec). The first 20 msec of the same ETC is shown in Fig. 21-4. The initial time-delay gap is distinct and clear, followed by a smooth, dense fill of reflections.

The time-energy-frequency (TEF) "waterfall" record of the room is shown in Fig. 21-5. The vertical amplitude scale is 12 dB/division. The horizontal frequency scale ranges from 100 Hz to 10 kHz. The diagonal scale shows time, ranging from zero

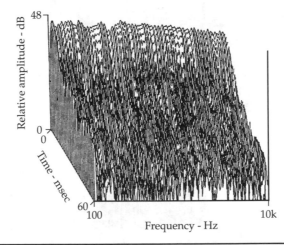

FIGURE 21-5   Time-energy-frequency response of the studio of Fig. 21-3. The vertical scale is energy (12 db/division), the horizontal scale is frequency (100 Hz to 10 kHz), and the diagonal scale is time (zero at the back to 60 msec at the front). The broadband decay of sound is smooth, balanced, and dense. (*Acoustic Sciences Corporation*)

at the rear to 60 msec at the front. This measurement shows no prominent room modal resonances, only a dense series of smooth decays with time throughout the entire 100 Hz to 10 kHz frequency range.

This voice-over studio would yield voice recordings that are accurate and clean-sounding. Furthermore, recordings made at one time match those made at other times, even if different microphone positions are used. Moving the microphone only changes the fine structure of the ambient sound (some thousand reflections per second) and has little effect on the quality of sound.

Live End-Dead End Voice Studio

An alternative to the traditional voice booth uses a live end-dead end approach. This is similar to the live end-dead end (LEDE) design used in some control rooms where monitoring is performed. This live end-dead end voice recording studio requires more space than a typical voice booth, although not much. The object is to obtain a clean direct sound free from early reflections, followed by a normal ambient decay. The microphone is placed in the room's absorbent end so that no reflections reach it, except those delayed and diffused from the room's more distant live end. All walls of the dead end, as well as the floor there, must be absorptive. The observation window is placed in the live end of the room. A small early time gap will exist between the time of arrival of the direct sound and the arrival of the first highly diffused sound from the live end. A natural-sounding voice recording can be obtained with this arrangement. This type of treatment could also be used for isolation booth purposes for musical instrument recording.

CHAPTER 22

Acoustics of Large Halls

In many ways, large performance halls represent the pinnacle of acoustical design. The sheer size of large halls gives them great civic importance, and in most cases, their construction costs dwarf that of any other acoustical space.

Auditoriums, places of worship, drama theaters, and other halls designed for speech can range from intimate spaces to grand edifices. The seating capacity may be less than a hundred, or several thousand. The nature of the application determines the acoustical priorities, and can also create diverse demands. For example, in a place of worship, a sermon may demand clear speech intelligibility; a liturgical service may require that chant be properly heard; another service may incorporate music performances, both from on-stage performers, and singing from the worshippers. Clearly, the type of presentation will profoundly influence the acoustical design of the space. In addition, many halls designed for speech must integrate a sound playback system with the natural acoustics of the space.

Concert halls providing good acoustics for live music performance are revered, while halls with poor acoustics are reviled. The opening of any new concert hall is a great public event, and the evaluation of the quality of its acoustics can make or break the career of an acoustician. Even after extensive computer modeling, it is only on opening night, with a symphony orchestra on stage and a full audience in the house, that the acoustical quality of the hall can be fully ascertained. Different types of music performance spaces, such as concert halls, opera houses, and chamber music halls place particular demands on the design, with different acoustical priorities. For best results, a hall must be tailored for its specific purpose. A well-known axiom is that a multipurpose music hall is a no-purpose hall.

Very generally, we may consider the design of large spaces in two aspects: spaces primarily intended for speech, and those primarily intended for music. Clearly, the former emphasizes speech intelligibility while the latter requires musical sonority.

Essential Design Criteria

In some ways, even the largest hall is no different from the smaller rooms we have considered. In other words, the basic acoustic criteria are the same. A large hall must have a low ambient noise level from internal and external sources; it must provide a reasonable level of acoustic gain; it must provide appropriate reverberation time; it must avoid artifacts such as echoes. Of course, all of these acoustical needs must be reconciled with both aesthetic and practical architectural concerns. Above all, the criteria must be evaluated in terms of the structure's intended purpose. For example, while a low level of noise intrusion in important in any hall, it is particularly important in a place of worship—by definition, a place of refuge from the outside world. From these major design criteria, every small detail of the design is derived. For example, again citing a place of worship, care must be taken to design aisles and doors so that those leaving and entering during services will not disturb others.

Reverberation and Echo Control

As we observed in Chap. 11, reverberation is an important parameter that helps define the sound quality of an acoustic space. This is perhaps especially true in large halls such as concert halls, performance theaters, churches, and auditoriums. Reverberation time RT_{60} is closely linked to the intended purpose for any room, and to room volume. Figure 22-1 shows recommended values for mean reverberation time between the two octave bandwidths 500 and 1,000 Hz when a room is occupied between 80% and 100%. As shown, halls designed for speech have shorter mean reverberation times than halls designed for music performance. Also, the recommended mean reverberation time increases as a function of room volume.

The question of reverberation is more than reverberation time. The frequency response of the reverberant field must also be considered. Figure 22-2 shows the frequency-dependent tolerance ranges of reverberation time referenced to the recommended mean reverberation time described in Fig. 22-1. The response of these tolerances is different for speech (Fig. 22-2A) and music (Fig. 22-2B). In particular, music requires significantly longer reverberation times at low frequencies. This imparts warmth to the sound quality of the reverberant field and is sometimes characterized as the bass ratio, as described later. The converse is true for speech; reverberation time decreases at low frequencies. This improves speech intelligibility.

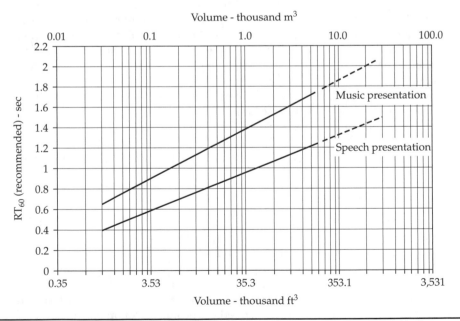

FIGURE 22-1 The recommended mean reverberation time between 500 and 1,000 Hz, for speech and music, with respect to room volume. (*Ahnert and Tennhardt*)

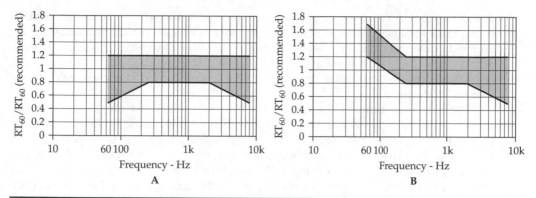

FIGURE 22-2 The frequency-dependent tolerance range of reverberation time, as referenced to recommended reverberation time. (A) Speech. (B) Music. (*Ahnert and Tennhardt*)

Large enclosed spaces are all potentially subject to the problem of discrete echoes. The long path lengths and multiplicity of seating positions near and far from the sound source can easily create echo problems. Architects and acousticians must be alert to surfaces that might produce reflections of sufficient level and delay to be perceived as discrete echoes. This is a gross defect for which there is little tolerance, audible to everyone with normal hearing.

Reverberation time affects the audibility of echoes. As an example, Fig. 22-3 presents acceptable echo levels for speech under reverberant conditions. The heavy broken

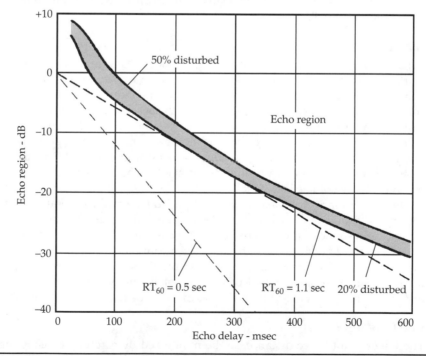

FIGURE 22-3 Acceptable echo levels for speech under reverberant conditions (reverberation time 1.1 seconds). (*Nickson, Muncey, and Dubout*)

line shows the decay rate representing the reverberation time RT_{60} of 1.1 seconds in this example. The shaded area represents combinations of echo level and echo delay, experimentally determined, which result in echoes disturbing to people. The upper edge is for 50%, the lower edge is for 20% of the people disturbed by the echo. Concert hall reverberation time is typically around 1.5 to 2.0 seconds; many churches are closer to 1 second to favor speech.

Other measurements of this type, made in spaces with other reverberation times, show that the shaded area indicating the level/delay region causing troublesome echoes is close to being tangent to the reverberation decay line. This presents the possibility of estimating the echo threat of any large room by plotting the reverberation time. For example, the lighter broken line in Fig. 22-3 is drawn for a reverberation time RT_{60} of 0.5 second. An echo-interference area just above this decay line can be very roughly inferred in this way.

Acoustical consultants and architects design music halls to give lateral reflections of appropriate levels and delays to add a sense of spaciousness to the music. This application emphasizes the importance of informed manipulation of reflections in large spaces to achieve desirable results. Lateral reflections are discussed in more detail later.

Hall Design for Speech

In halls designed for speech applications, many of the same acoustical criteria that are important for any room design will still apply. However, in halls used primarily for speech, some criteria must be modified, and additional important requirements arise. In particular, it is essential that the hall acoustics provide good speech intelligibility in every seat in the hall.

Volume

In halls designed for unamplified speech, it is often necessary to limit the overall room volume. This is because a large volume requires more speech power than a small room. In a very large room, for example, 1,000,000 ft³ in volume, an unamplified voice probably cannot be satisfactorily heard with even the best acoustical design. A smaller room will require less absorption for a given reverberation time. This, in turn, allows more reflectivity which supports greater acoustical gain and a louder speech level. In a relatively large auditorium, a room volume ranging from 100 to 200 ft³ per seat is suggested. This volume minimization is contrary to rooms designed for music, where a relatively large volume is desirable.

In a face-to-face conversation, an unamplified talker may generate a sound-pressure level of only about 65 dBA. This level decreases 6 dB for every doubling of distance. Less significantly, sound is also attenuated as it travels through the hall because of air absorption. To support audible levels, the audience area must be placed as close as possible to the talker. This minimizes sound attenuation, provides a more direct sound path, and also improves visual recognition which improves intelligibility. Very generally, the maximum distance from a talker in an auditorium to the most distant seat should be about 80 ft.

Hall Geometry

The talker-to-audience distance can be minimized by carefully considering the room geometry. In particular, as seating capacity increases, the lateral dimensions of the room must increase, and the side walls should be splayed. A rectangular shoebox-type hall, with the stage across one narrow end, may be excellent for music where an audience can be seated farther away and a greater ratio of reverberant sound is desirable. However, a rectangular geometry is only suitable for a relatively small speech hall. Otherwise, with larger seating capacities, much of the audience is placed far away from the stage, at the far end of the hall.

This limitation can be overcome to some extent by widening the hall. However, for greater seating capacity, the side walls should be splayed from the stage. Figure 22-4 shows a rectangular floor plan, and two examples of halls with splayed side walls. Splayed side walls allow greater seating area that is relatively close to the stage. Care must be taken with wall angles to avoid any flutter echoes, such as the case in Fig. 22-4B for the marked reflection. The splayed walls can usefully reflect sound energy to the rear of the hall (Fig. 22-4B and C). A side-wall splay may occupy the entire length of

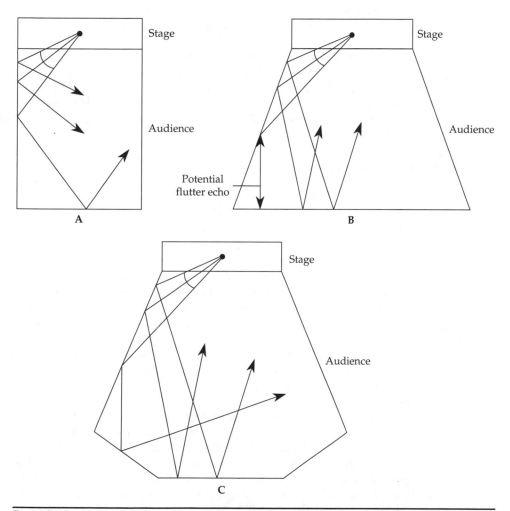

FIGURE 22-4   For greater seating capacity, the side walls can be splayed from the stage.
(A) Rectangular floor plan. (B) Splayed side walls with flat rear wall. Note the potential flutter echo.
(C) Splayed side walls with extended rear walls.

the side walls, or be limited to the rear portions of the side walls. A side-wall splay may range from 30° to 60°; the latter is considered a maximum angle, given the directionality of speech. Very generally, fan-shaped halls are not used for music performance. In addition, the rear walls may be extended outward in the center of the hall (Fig. 22-4C) or the rear wall may be concave with equal radius from the stage, to create a fan-shaped room. Any concave room geometries require careful design to avoid sound focusing.

Absorption Treatment

In small speech halls, the majority of absorption is provided by the audience; therefore, the room surfaces can be relatively reflective. In larger halls, where there is greater room volume per seat, relatively greater room absorption is needed. Beneficially, a reflective front stage area provides strong early reflections (short path length difference) that are integrated (precedence effect) with the direct sound and enhance it. On the contrary, strong late reflections and reverberation, such as from rear walls, would not be integrated and may produce echoes. To accommodate this, the stage area and front of the hall are made reflective, and absorption is placed in the seating area and rear of the hall. To preserve speech intelligibility, although early reflections are useful, the overall reverberation time must be short, perhaps less than 0.5 second. When the front stage area is reflective, an important benefit is increase in overall speech gain. However, the same reflections may produce comb-filter effects; this possible source of timbral change must be considered.

Ceiling, Walls, and Floor

In many large halls, ceiling reflectors, sometimes called clouds, are used to direct sound energy from the stage to the seating area. Both planar and convex reflectors may be used. The size of the reflectors determines the range of frequencies that are reflected; the larger the panel, the lower the cutoff frequency of the reflected sound. Both dimensions of a square reflecting panel should be at least five times the wavelength of the lowest frequency to be reflected. For example, a square panel measuring 5 ft on a side will reflect frequencies of 1 kHz (wavelength is 1 ft) and higher. Panels must also be solid and stiff, and securely mounted to avoid resonances. When ceilings are high, care must be taken to ensure that path-length differences between direct and reflected sound are not too great, and particularly should not exceed 20 msec. In some cases, clouds are made absorptive, to avoid late reflections.

A sloping (raked) floor allows a more direct angle of incidence which in turn allows less absorption. Generally, the slope of an auditorium floor should not be less than 8°. The floor of a lecture-demonstration hall might have a 15° angle of inclination. Staggering of seats is also recommended.

Because of its potential to create undesirable late reflections, the rear wall of a large hall requires special attention. Reflections from the rear wall would create a long path-length difference (the path length of the reflection from the rear wall minus the path length of the direct sound) to a listener at the front of the hall. This can result in audible echoes, particularly because of the otherwise low reverberation level. As noted, a reflective concave rear wall would also undesirably focus sound. For these reasons, the rear wall of a large hall is usually absorptive. In some cases, when added absorption is undesirable because of decreased reverberation time, reflective diffusers can be placed on the rear wall.

Speech Intelligibility

Speech intelligibility is the highest design priority for any hall intended for spoken word. This is the case in many places of worship, auditoriums, and drama theaters. Sound systems are often used to overcome acoustical limitations, and to provide intelligibility in even very large spaces. In a hall where amplification is not used, a room design providing high speech intelligibility begins by recognizing that a normal voice, as noted, will generate a long-term average sound-pressure level of about 65 dBA. Peak sound-pressure levels may be 12 dB higher than the long-term average. Different loudness levels of talkers may range from about 55 to 75 dBA.

Speech Frequencies and Duration

Very generally, the most significant speech frequencies range from 200 Hz to 5 kHz. The great majority of speech power is below 1 kHz, and the maximum speech energy range is 200 to 600 Hz. Speech vowels mainly occupy low frequencies, while consonants occupy higher frequencies. Consonants are most important in intelligibility. The frequencies above 1 kHz, specifically in the 2- to 4-kHz range, are primarily responsible for speech intelligibility. The three bands at 1, 2, and 4 kHz provide 75% of speech intelligibility content.

Generally consonants have a duration of about 65 msec, and vowels a duration of about 100 msec. Syllables may have a duration of 300 to 400 msec, and words a duration of 600 to 900 msec, depending on rate of delivery. Relatively short reverberation time is required so that reverberation from a previous element does not mask a subsequent element. Early reflections (less than 35- to 50-msec delay) tend to be integrated with the direct sound and increase perceived loudness and assist intelligibility. Late reflections (greater than 50-msec delay) particularly degrade speech intelligibility. Also, a high speech-to-noise dynamic range is needed for good speech intelligibility. A slower rate of talker delivery and the ability to articulate the speech also play a large role in intelligibility. For example, in highly reverberant spaces, slowing the delivery rate from five syllables per second to three per second can significantly improve intelligibility.

Subject-Based Measures

Speech intelligibility in a room is often estimated using subject-based measures, that is, by using live experimentation. A talker reads from a list of words and phrases, and listeners in the room writing down what they hear. The list includes examples of important speech sounds. Between 200 and 1,000 words are used per test. For example, Table 22-1 lists some English words used in subject-based intelligibility testing.

The higher the percentage of correctly understood words and phrases, the higher the speech intelligibility. In some cases, listening difficulty is measured. When the level of speech is the same as the noise level, intelligibility can be high, but listeners can still have difficulty in understanding what is being said, and considerable attention is required. When the speech level is raised by 5 or 10 dB over the noise, intelligibility is not greatly improved, but listeners report much less difficulty in hearing.

Analytical Measures

Various analytical measures have been devised to assess speech intelligibility. The articulation index (AI) uses acoustic measurements to estimate speech intelligibility and

aisle	done	jam	ram	tame
barb	dub	law	ring	toil
barge	feed	lawn	rip	ton
bark	feet	lisle	rub	trill
baste	file	live	run	tub
bead	five	loon	sale	vouch
beige	foil	loop	same	vow
boil	fume	mess	shod	whack
choke	fuse	met	shop	wham
chore	get	neat	should	woe
cod	good	need	shrill	woke
coil	guess	oil	sip	would
coon	hews	ouch	skill	yaw
coop	hive	paw	soil	yawn
cop	hod	pawn	soon	yes
couch	hood	pews	soot	yet
could	hop	poke	soup	zing
cow	how	pour	spill	zip
dale	huge	pure	still	
dame	jack	rack	tale	

TABLE 22-1   Examples of Words Used in Subject-Based Intelligibility Testing

conversely, speech privacy. AI uses weighting factors in five octave bands from 250 Hz to 4 kHz (in some cases, $1/_3$-octave bands are used). Each weighting factor accounts for our hearing sensitivity in that band; for example, the weighting factor is highest at 2 kHz, because our hearing sensitivity is greatest there. AI is calculated by multiplying the signal-to-noise (S/N) ratio in each octave band by the weighting factor in each octave band, and summing the result. When the S/N ratio is greater than 30 dB, a value of 30 dB is used. When the S/N ratio is negative, a value of 0 dB is used. In some cases, a correction factor accounting for reverberation is subtracted from the AI value. AI ranges from 0 to 1.0; the higher the value, the better the intelligibility.

Another objective measure used to assess speech intelligibility is known as percentage articulation loss of consonants, known as %Alcons. As its name implies, %Alcons focuses on the perception of spoken consonants. %Alcons can be approximately measured as:

$$\%\text{Alcons} \approx 0.652 \left(\frac{r_{lh}}{r_h} \right)^2 \text{RT}_{60} \tag{22-1}$$

where %Alcons = percentage articulation loss of consonants, percent
r_{lh} = distance from sound source to listener
r_h = reverberation radius, or critical distance for directional sound sources
RT_{60} = reverberation time, sec

Subjective Intelligibility	%Alcons
Ideal	≤3%
Good	3–8%
Satisfactory	8–11%
Poor	>11%
Worthless	>20%*

*Limit value is 15%.

TABLE 22-2 Subjective Weighting for Results of %Alcons Testing

Subjectively, %Alcons scores can be related to speech intelligibility, as shown in Table 22-2. Other methods used to estimate speech intelligibility include the speech transmission index (STI), speech intelligibility index (SII), and rapid acoustics speech transmission index (RASTI). The latter may be correlated to %Alcons by: RASTI = $0.9482 - 0.1845\ln(\%\text{Alcons})$.

According to one criterion, satisfactory speech intelligibility can be achieved by designing for an appropriate reverberation time. In particular, reverberation time at 500 Hz, with the room two-thirds occupied, should be selected so that at the most distant listening position, the ratio of the reflected sound energy to the direct sound energy is no greater than 4. This corresponds to a 6-dB difference between the energy densities, and should provide a low (5%) consonant articulation loss.

Concert Hall Acoustical Design

The design of a large hall for music performance presents special challenges. Perhaps the first complexity rests with the music itself. Clearly, symphonic music, chamber music, and opera each require very different acoustics, as well as size and room functionality. Moreover, different styles of music, such as baroque, classical, and popular have different acoustical requirements. Finally, different music cultures, such as Eastern and Western, require different design criteria. Perhaps the most difficult aspect of hall design is the ambiguity of the goal itself. While it is possible to measure many specific aspects such as reverberation time, there is no measurement, or even specific consensus, on what "good" music acoustics are. The diversity of the requirements, subjectivity of the goal, lack of objective benchmarks, and differences of opinion all conspire to make hall design an art as well as a science.

Reverberation

Very generally, the problem of music hall acoustics may be considered in two parts: early sound and late reverberant sound. Early sound is sometimes considered in terms of early reverberation decay time, intimacy, clarity, and lateral spaciousness. Late reverberant sound can be considered as late reverberation decay time, warmth, loudness, and brilliance.

Reverberation can further be considered in two parts: early and late reverberation. The ear is very sensitive to early reverberation. This is partly because in most music, later reverberation is partly masked by the following music notes. Early reverberation largely defines our subjective impression of the entire reverberation event. The early decay time (EDT) is defined as the time required for sound to decrease 10 dB, multiplied by 6. (Multiplying by 6 allows comparison with late reverberation time RT_{60}.) Unlike dense late reverberation, early reverberation comprises a relatively few primary reflections. These reflections arrive within the Haas fusion zone and are integrated with the direct sound, reinforcing it. This early reverberation can affect the clarity of sound. The greater the energy in the early reverberation, the better the clarity. Late reverberation can affect our perception of the liveness of sound. More late reverberant energy can increase liveness or fullness. As late reverberant energy increases, the clarity relatively decreases.

Clarity

Clarity, measured in decibels, is sometimes defined as the difference between the sound energy in the first 80 msec, and the late reverberation energy arriving after the first 80 msec. This is sometimes referred to as C_{80}. In some cases, a $C_{80}(3)$ value is used, which averages clarity at 500; 1,000; and 2,000 Hz. In large halls with good clarity, the value of $C_{80}(3)$ ranges between −4 and +1 dB.

In some hall designs, to achieve good clarity, as well as good liveness, the reverberation decay is tailored as a two-part slope. The early decay has a steep slope and a short EDT, while the late decay has a shallower slope and a longer RT_{60}. In many halls, the design calls for an EDT of about 10% longer than RT_{60}. For a large concert hall, RT_{60} in mid frequencies is about 2.0 seconds. When there is a single decay slope, then the reverberation time is simply measured as RT_{60}.

Brilliance

Brilliance is another metric used to quantify hall acoustics. Brilliance describes sound that has presence and clearness. Brilliance is achieved with adequate high-frequency energy from reflecting surfaces. On the other hand, a hall with good brilliance should not sound too bright or harsh. Brilliance can be estimated by comparing a high-frequency EDT to an averaged mid-frequency EDT. In particular:

$$\frac{EDT_{2,000}}{EDT_{Mid}} = \frac{EDT_{2,000}}{EDT_{500} + EDT_{1,000}} \tag{22-2}$$

Similarly, $EDT_{4,000}/EDT_{Mid}$ can be calculated. Some sources recommend that $EDT_{2,000}/EDT_{Mid}$ should be at least 0.9, and $EDT_{4,000}/EDT_{Mid}$ should be at least 0.8.

Gain

A good music hall should also provide adequate acoustic gain at all seating positions. Gain (G) can be thought of as the difference in sound-pressure level between the sound-pressure level in the center of the hall from an arbitrary source, minus the sound-pressure level 10 m away from the same source in an anechoic environment. The former gain is a function of direct sound and reflected sound at a particular seating position in the hall, while the latter contains only direct sound at 10 m. Gain thus depends on the volume of the hall and the reverberation time RT_{60} or the early delay time EDT. In particular:

$$G_{Mid} = 10 \log \left(\frac{RT_{Mid}}{V} \right) + 44.4 \tag{22-3}$$

or:

$$G_{Mid} = 10 \ \log\left(\frac{EDT_{Mid}}{V}\right) + 44 \qquad (22\text{-}4)$$

where G_{Mid} = gain as averaged at 500 and 1,000 Hz, dB
   RT_{Mid} = reverberation time averaged at 500 and 1,000 Hz, sec
   EDT_{Mid} = early reflection time averaged at 500 and 1,000 Hz, sec
   V = volume, m^3

In many concert halls with good acoustics, G_{Mid} lies between 4.0 and 5.5 dB. However, in halls with different, more specialized applications, G_{Mid} will vary over a wider range.

Seating Capacity

Given a room volume, the number of seats N may be determined. This relationship partly depends on the type of music that will be performed. Furthermore, the total floor area of the hall (stage and audience) may be estimated from the number of seats. When measured in square meters, the floor area may be estimated as $0.7N$. When measured in square feet, floor area is about $7.5N$. The number of seats may be determined from these equations:

$$N = \frac{0.0057V}{RT_{Mid}} \qquad (22\text{-}5)$$

where N = number of seats
   V = room volume, ft^3 or m^3
   RT_{Mid} = reverberation time averaged at 500 and 1,000 Hz, sec

Note: In metric units, change 0.0057 to 0.2.

Volume

Room volume is profoundly influenced by the intended purpose of the hall. Among other things, volume affects reverberation time and required absorption. Total room volume may be specified; for example, a concert hall may be specified at 900,000 ft^3. In some cases, volume is specified as a minimum volume per listener seat; for example, a concert hall may range from 200 to 400 ft^3 per seat. When a balcony is incorporated into a hall design, volume per seat is generally reduced.

Spaciousness

The characteristics of the early sound field are also important in establishing a sense of spaciousness so the listener feels enveloped by sound in the hall. A sense of spaciousness can be created by early reflections from side walls, often called lateral reflections, which occur within the 80 msec of the arrival of the direct sound. It is important that these reflections arrive at the listener from either side at angles of approximately 20° to 90° relative to the front of the listener. The geometry of rectangular shoebox-type halls lends itself to lateral reflections. In fan-shaped halls, these reflections arrive more from the listener's front; this can decrease the effect. In some designs, a reverse lateral shape can provide good lateral reflections. Spaciousness is also augmented by providing adequate diffusion; the ornate decorations in many older concert halls accomplish this.

Apparent Source Width

Apparent source width (ASW) can be used to describe the perceived breadth of a sound source such as an orchestra, which is wider than the physical source. ASW is improved, that is, the source appears wider when there is a high level of early lateral reflections (before 80 msec) that contribute early spatial impressions. Listener envelopment (LEV) is sometimes used to describe the feeling of being surrounded by and immersed in sound that fills a large space. It is improved by late-arrival lateral reflections (after 80 msec).

Initial Time-Delay Gap

Beranek made an intensive study of concert halls around the world. He noted that those halls rated the highest by qualified listeners had certain technical similarities. Among them was an acoustical metric known as initial time-delay gap (ITDG). This is the time between the arrival of the direct sound at a given seat and the arrival of critically important early reflections. Concert halls that rated high on the quality scale had a well-defined initial time-delay gap of about 20 msec. Halls with an initial time-delay gap flawed by uncontrolled reflections were rated inferior by qualified listeners.

In smaller, more intimate halls, with reflective surfaces closer to the listener, ITDG is small. In larger halls, ITDG is greater for most listeners. However, ITDG depends on seating position; a listener sitting near a side wall, for example, would experience a smaller ITDG. Smaller ITDG values, because they denote a more intimate sound field, are desirable. Very generally, an ITDG value of less than 15 msec is desirable, measured in the center of the hall. Large halls that are relatively narrow, as in a rectangular shoe-box design, can have a relatively small ITDG. In some cases, a small ITDG is achieved by segmenting the audience into smaller areas and placing reflecting walls near them. Alternatively, side balconies, terraces, and other side protrusions can be used to provide early reflections. Since these techniques also provide lateral reflections, they can uniquely impart both a sense of intimacy and spaciousness.

Bass Ratio and Warmth

A sense of acoustic warmth is desirable in most halls. This is often attributed to longer reverberation times at low frequencies. One way to measure this is through the bass ratio (BR). Bass ratio is calculated by summing the reverberation times at 125 and 250 Hz, and dividing by the sum of reverberation times at 500 and 1,000 Hz:

$$BR = \frac{RT_{60/125} + RT_{60/250}}{RT_{60/500} + RT_{60/1,000}} \tag{22-6}$$

An acoustically warm hall with longer bass reverberation times would thus yield a BR greater than 1.0. According to one study, for halls with RT_{60} less than 1.8 seconds, BR should lie between 1.1 and 1.45. For halls with a higher RT_{60}, BR should lie between 1.1 and 1.25. (For speech, a bass ratio between 0.9 and 1.0 is preferred.)

Concert Hall Architectural Design

The architectural design of a concert hall requires close collaboration between the architect and the acoustician. This is particularly true in the sound chamber itself; the stage and seating area of this inner shell must meet very specific acoustical requirements for both performing musicians and the audience while providing amenities and safety for

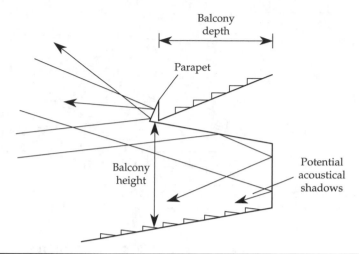

FIGURE 22-5 Ideally, balcony depth should not be more than the height to avoid acoustic shadowing underneath the balcony. The underside of the balcony should be reflective to supplement the direct sound. The front of a balcony parapet should avoid undesirable reflections.

all the occupants. Overarching these practical needs are the more intangible aesthetic requirements; a large concert hall must present a space where the pleasure of hearing music is never compromised.

Balcony

In some halls, a balcony can be used to decrease the distance from the stage to some seating areas, and to provide good sight lines. Care must be taken to avoid acoustic shadowing in the seating areas underneath the balcony, as shown in Fig. 22-5. Very generally, the balcony overhang depth should be less than twice the height of the balcony underside. Ideally, the depth should not be more than the height. Deep balconies can create acoustical shadows in the seats underneath the balcony. In addition, reflecting surfaces on the ceiling and side walls, as well as the underside of the balcony, should be designed to add as much reflected sound as possible to the seating areas on the balcony and under it, to supplement the direct sound from the stage. The front of a balcony parapet should be designed to avoid reflections that could affect sound quality in the seating areas in the front of the hall; this is particularly true when the plan view of the balcony has a concave shape.

Ceiling and Walls

Ceiling height is usually determined by the overall room volume that is required. Very generally, ceiling height should be about one-third to two-thirds of the room width; the lower ratio is used for large rooms, and the higher ratio is used for small rooms. A ceiling that is too high may result in a room volume that is too large, and may also create undesirable late reflections. To avoid potential flutter echo, a smooth ceiling should not be parallel to the floor.

In many halls, the ceiling geometry itself is designed to direct sound to the rear of the hall, or to diffuse it throughout the hall, as shown in Fig. 22-6. A ceiling may have several segments, each sized and angled to reflect sound to different seating areas; for

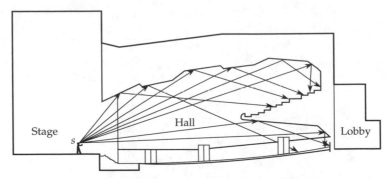

Figure 22-6   The ceiling geometry should direct reflected sound throughout the hall. Several ceiling segments may be sized and angled to reflect sound to particular seating areas in the hall.

example, the ceiling near the stage might reflect sound to the near rows, while the ceiling farther from the stage might reflect sound to the farther rows. In some hall designs, the ceiling can be raised or lowered, often as several independent sections, to vary the acoustics of the hall. For example, when the ceiling is lowered, the hall's acoustics are more intimate.

Concave surfaces such as domes, barreled ceilings, and cylindrical arches should be avoided because of the undesirable sound foci they create. The rear wall must avoid any large, unbroken concave geometry. Side walls must avoid parallelism. This can be avoided by tilting or splaying wall surfaces. These angles can also be advantageously used to direct reflected sound to the audience seating area, and to provide diffusion. Any surface that unavoidably introduces concave geometry or an undesirable angle should be covered with absorptive material.

Raked Floor

In halls designed for either music or speech, a sloping (raked) floor is desirable; this is particularly true for large halls. A sloping floor improves sight lines, and also improves fidelity in the seating area. When sitting on a sloping floor, the listener receives more direct sound than would be available on a flat floor; in either case, the stage should be raised. In some designs, the floor slope is constant with respect to distance from the stage; risers are equal. In other designs, the slope increases with distance from the stage; risers are not equal. In other designs, the floor slope is constant near the stage, then changes to another constant but steeper slope further from the stage. In some halls, a balcony can be used to add seating capacity while decreasing distance from the stage to the audience. It is common for balconies to use relatively steeply sloping floors.

A sloping floor is desirable in halls where audience sound absorption must be minimized. A more direct angle of incidence allows less absorption than a shallow angle that passes sound over a larger audience area. The frequency response of sound passing over an audience can be greatly affected by its angle of incidence. At low angles of incidence, a dip in low-frequency response around 150 Hz, as deep as 10 to 15 dB and extending for two octaves, occurs at the audience head position. Also, this effect is most pronounced for direct and early arrival sound. One solution is to design a low-frequency boost in early-reverberation energy. A steep floor slope is also helpful in this regard, and some evidence shows that strong ceiling reflections are also helpful.

Virtual Image Source Analysis

As we observed in Chaps. 6 and 13, a reflection from a boundary surface such as wall, floor, or ceiling can be considered as a virtual image source. The virtual source is acoustically located behind the reflecting surface, just as viewing an image in a mirror. Moreover, when sound strikes more than one surface, multiple reflections will be created. Thus, images of the images will exist. In a rectangular room, there are six surfaces and the source has an image in all six surfaces, sending energy back into the enclosure, resulting in a highly complex sound field.

Using this modeling technique, we can ignore the boundary surfaces themselves, and consider sound as coming from many virtual sources spaced away from the actual source, arriving at time delays based on their distance from the source. Moreover, the magnitude of the images depends on the absorbency of the reflecting surface, and the number of times the image has been reflected.

This technique can be used to examine the spatial sound field of large spaces, as shown in Fig. 22-7. Four large spaces were tested by sounding an impulse on the stage and measuring the resulting reflecting sound field with a 6-point microphone array. In the figures, the center of each circle represents the virtual source image location, and

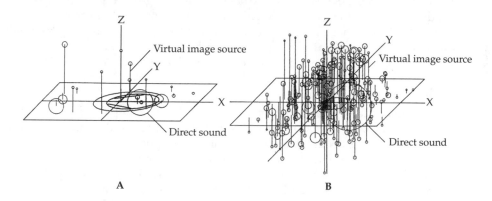

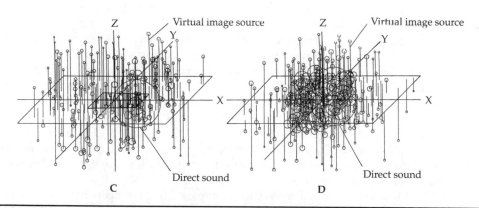

FIGURE 22-7 The virtual image sources created by the reflections from boundary surfaces spatially document the sound fields of four enclosures. (A) Stadium. (B) Opera house. (C) Large concert hall. (D) Small concert hall. (*JVC Corporation*)

the circle size represents the intensity. The distance of the circles from the intersection of the axis represents the time delay. The measuring array is located at the intersection of the axes. The virtual sources created by the reflections from the boundary surfaces spatially document the sound fields of the four enclosures. In the case of the stadium (Fig. 22-7A), there are only a few reflecting surfaces, and the images are widely spaced, revealing probable echoes. In the opera house (Fig. 22-7B), the virtual sources are relatively dense, with a number of strong images overhead. The relative sizes of the large concert hall (Fig. 22-7C) and small concert hall (Fig. 22-7D) are evident from the spatial distribution of the images.

Hall Design Procedure

In practice, an acoustician may begin a hall design by carefully considering the site, and determining ambient noise levels and structure layout. Numerical design may begin by specifying a value for G_{Mid}. For example, a value of 5.0 dB may be assumed. Furthermore, a value of RT_{Mid} or EDT_{Mid} is assumed; this value will depend greatly on the type of music to be performed in the hall. For example, assume a RT_{Mid} value of 2.0 seconds. Given these values, the hall volume may be calculated, as well as the number of seats and total floor area. If these results do not meet the design criteria, the value of G_{Mid} may be adjusted. Using these calculated guidelines, the acoustician and architect may decide on a hall configuration, such as rectangular or fan shaped. Again using the guidelines, elements such as balconies and terraces can be included, taking into account factors such as ITDG. Absorption is added according to the assumed RT and EDT values.

Case Studies

The design of a large concert hall ranks among the most complex of architectural tasks. Hall design is often unique and demonstrates creativity, and daring, on the part of the architect and acoustician. This physical uniqueness guarantees that each concert hall will have a unique sound character that is unlike any other.

Orchestra Hall in Chicago is an example of a traditional rectangular hall, with steeply raked balconies. Figure 22-8A and B show architectural plans for the hall. This hall was constructed in 1904 and acoustical deficiencies were identified shortly after its opening, particularly in the stage area and the shell over the stage. As a result, the hall has been renovated several times. In 1966 much of the plaster ceiling was replaced by a perforated aluminum screen in an effort to increase effective room volume, and thus lengthen reverberation time. However, upholstery was added to the balcony and gallery and the occupied reverberation time did not increase, and the unoccupied reverberation time was severely reduced. In 1981 the hall was again renovated; among other improvements, absorption was reduced in the main floor seating area as well as surfaces surrounding the stage, upper hall surfaces were hardened, hall volume was opened, diffusing plaster was placed on rear wall surfaces, and a new pipe organ was installed. Figure 22-8C shows unoccupied reverberation time before and after the 1981 renovation; the increase in low-frequency reverberation can be seen. Subsequently, another renovation was completed in 1997.

The Philharmonie in Berlin is an example of large concert hall using a vineyard configuration. In this approach, the large hall is segmented into a number of smaller multilevel audience areas separated by low walls. This contributes lateral reflections to provide intimacy within the larger space. This hall, completed in 1963, places the audience

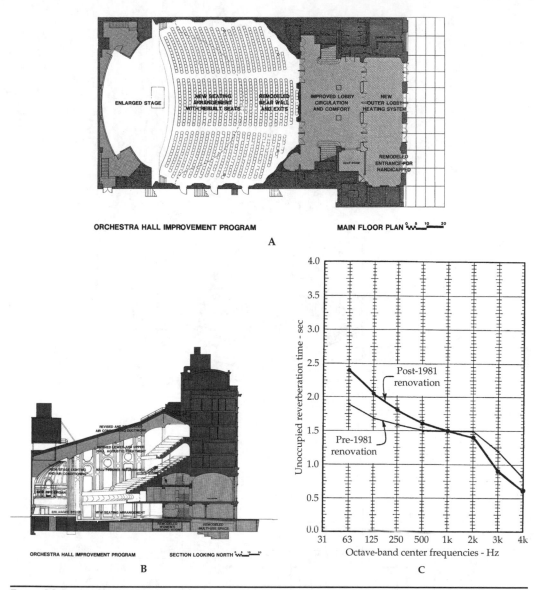

FIGURE 22-8 Architectural plans and reverberation-time measurements for Orchestra Hall in Chicago. (A) Ground plan. (B) Longitudinal section. (C) Unoccupied reverberation time before and after the 1981 renovation. (*Reprinted with permission from the Acoustical Society of America, Halls for Music Performance: Two Decades of Experience, 1962–1982, Richard H. Talaske, Ewart A. Wetherill, and William J. Cavanaugh, eds., 1982*)

around the orchestra and is notable for a tentlike ceiling with reflecting clouds below it. Figure 22-9A and B show architectural plans of the Philharmonie , with sketches of the ground plan and longitudinal section. Figure 22-9C plots reverberation time of the hall in three states: unoccupied; occupied by orchestral musicians as in a recording session; and fully occupied by orchestral musicians, chorus, and audience. In all three cases, the rise in low-frequency bass is evident.

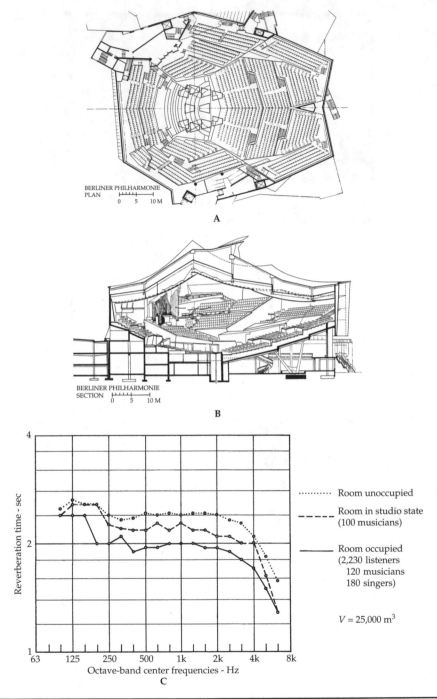

Figure 22-9 Architectural plans and reverberation-time measurements of the Philharmonie in Berlin. (A) Ground plan. (B) Longitudinal section. (C) Reverberation time. (*Reprinted with permission from the Acoustical Society of America, Halls for Music Performance: Two Decades of Experience, 1962–1982, Richard H. Talaske, Ewart A. Wetherill, and William J. Cavanaugh, eds., 1982*)

CHAPTER 23

Acoustic Distortion*

There have been extraordinary advances in the quality of the hardware in our audio systems, but less progress has been made in improving the quality of the acoustical pathway through which all sound must travel from the loudspeaker to our ears. Once the sound has reached our ears, there are still many psychoacoustical factors that determine how we perceive the sound. Much psychoacoustical research is now being vigorously pursued to understand how the mind perceives sound. Between the hardware and the perception of sound there is still that analog sound-transmission path in which many distortions arise. This chapter is devoted to the problem of acoustic distortion.

Acoustic Distortion and the Perception of Sound

There are three psychoacoustic perceptions that are affected by acoustic distortions: frequency response or timbre, imaging, and spatial impression. A uniform frequency response may be a goal, but satisfactory timbre, or overall recognition and appreciation of the harmonic content, is what we are striving for. As we listen to musicians, we form a mental image of those producing the music. The image can be very vivid when things are adjusted properly. We picture the size and shape of the sonic source as well as its height, depth, and width. This image formation is related to specific early lateral reflections of the sound from the side wall.

Sources of Acoustic Distortion

Acoustic distortion in the medium itself is unfortunately easy to hear. The hearing ability of the listener and the distortion of the amplifiers and loudspeakers are important contributors, but this discussion focuses on room-based acoustic distortion. There are many factors in adjusting the acoustics of a room, such as selection of room proportions and the treatment of room surfaces, as described in previous chapters. The four important sources of acoustic distortion covered in this chapter are (1) room modes, (2) speaker-boundary interference response, (3) comb filtering, and (4) diffusion.

Coupling of Room Modes

There are many resonances in most rooms (see Chap. 13). Specifically, the mathematics of the situation requires three modes for a rectangular space: axial, tangential, and oblique. The axial modes are the result of normal (right-angle) reflections between the

---

*Contributed by Peter D'Antonio, RPG Diffusor Systems, Inc., Upper Marlboro, Maryland 20774.

end surfaces of the room, the side-wall surfaces, and the floor and ceiling. Any sound in this room would excite these three resonances or modes. To complicate matters, each fundamental resonance has a series of harmonics, which are also resonances in every sense of the word. These modes are the acoustics of this space and are the source of acoustic distortions as the individual modes interact with each other. A frequency-response measurement of the room would be dictated by these axial, tangential, and oblique modes. Axial modes may be more prominent, but the entire response represents the vector sum of all the other modes. These modes vary from zero to maximum sound pressure and have a major effect on the production or reproduction of sound in this room. Every room has great fluctuations of sound pressure from one point to another, and these are a source of acoustic distortion.

Sound waves are longitudinal waves; that is, they actually oscillate (expand and contract) in the direction of propagation. As the sound waves expand and contract, they cause high-pressure and low-pressure regions. The instantaneous pressure on opposite sides of a pressure minimum has opposite polarity. The pressure on one side is increasing, while the pressure on the other side is decreasing. The position of a loudspeaker and the listener's ear, with respect to these pressure variations, will determine how they couple to the room. We can take advantage of knowing the location of these positive and negative regions to position multiple in-phase subwoofers to cancel certain room modes below the subwoofer crossover frequency. Also, placing a dynamic subwoofer at a null position will not excite this mode.

Speaker-Boundary Interference Response

The next type of acoustic distortion is due to the coherent interference between the direct sound of a loudspeaker and the reflections from the room, in particular the corner immediately surrounding it. This distortion occurs across the entire frequency spectrum but is more significant at low frequencies. It is called the speaker-boundary interference response (SBIR); its effects are discussed in Chap. 25. The room's boundaries surrounding the loudspeaker mirror the loudspeaker, forming virtual images. When these virtual loudspeakers (reflections) combine with the direct sound, they either enhance or cancel it to varying degrees, depending on the amplitude and phase relationship between the reflection and the direct sound at the listening position.

Consider a loudspeaker located 3 ft from each room surface (see Fig. 25-4). The four virtual images on opposite sides of the main room boundaries are responsible for first-order reflections. A virtual image is located at an equivalent distance on the opposite side of a room boundary. The distance from a virtual source to the listener is equal to the reflected path from source to listener. In addition to the four virtual images shown, there are seven more. Three virtual images and one real image exist in the speaker plane and four virtual images of these above the ceiling and floor planes. Imagine that the walls are removed and eleven additional physical speakers are located at the virtual image positions. The resultant sound at a listening position would be equivalent to the sound heard from one source and eleven other loudspeakers.

The effect of the coherent interference between the direct sound and these virtual images is a very uneven frequency response (see Fig. 25-3). The SBIR is averaged over all listening positions with the loudspeaker located 4 ft from one, two, and three walls surrounding the loudspeaker. It can be seen that as each wall is added, the low-frequency response increases by 6 dB and a notch, at roughly 100 Hz, gets deeper. It is important to note that once this notch is created, due to poor placement, it is virtually impossible to eliminate it without moving the listener and loudspeaker, since it is not good practice to electronically compensate for deep notches. Thus the boundary reflections either enhance or cancel the direct sound, depending on the phase relationship between the direct sound and the reflection at the listening position. At the lowest frequencies, the direct sound and reflections are in phase and they add. As the frequency increases, the phase of the reflected sound lags the direct sound. At a certain frequency, the reflection is out of phase with the direct sound and a cancellation occurs. The extent of the null will be determined by the relative amplitudes of the direct sound and reflection. At low frequencies there is typically very little absorption efficiency on the boundary surfaces, and the notches can be between 6 and 25 dB. One important conclusion that can be drawn from this is never place a woofer equidistant from the floor and two surrounding walls.

The low-frequency rise (see Fig. 25-3) illustrates why one can add more bass by moving a loudspeaker into the corner of a room. Actually, there are two choices. Move the loudspeaker either as close to the corner as possible or as far away from the corner as is physically practical. As you move the speaker closer to the corner, the first cancellation notch moves to higher frequencies, where it may be attenuated with porous absorption. (This can be seen in Fig. 25-3 for the last condition in which the loudspeaker is positioned 1 ft from the floor, rear, and side walls.) In addition, the loudspeaker directivity pattern diminishes the backward radiation, thus reducing the amplitude of the reflection relative to the direct sound. This principle is the basis for flush mounting loudspeakers in a corner soffit. The bad news is that with the loudspeaker in the wall-ceiling dihedral corner or wall-wall-ceiling trihedral corner, the loudspeaker very efficiently couples with the room modes. If the dimensional ratios are poor, leading to overlapping or very widely spaced modal frequencies, there will be significant modal emphasis. Many loudspeaker manufacturers provide a roll-off equalization to compensate for the added emphasis of flush mounting the loudspeaker.

We can also move the loudspeaker farther away from the adjacent corner. In this case, the first cancellation notch moves to a lower frequency, hopefully below the lower cutoff frequency of the loudspeaker or the hearing response of the listener. To obtain a 20-Hz first cancellation notch, one needs to position the loudspeaker 14 ft from the rear wall; this is not a very practical location! Clearly, some happy medium must be found, and with the myriad of alternatives available, the most effective approach is to start with a multidimensional computer loudspeaker-listener placement optimization program (described in Chap. 25) and adjust to taste.

Comb Filtering

Another form of acoustic distortion introduced by room reflections is comb filtering. It is due to coherent constructive and destructive interference between the direct sound and a reflected sound. In critical listening rooms, we are primarily concerned with the interaction between the direct sound and the first-order (single-bounce) reflections. Reflections cause time delays, because the reflected path length between the listener and source is longer than the direct sound path.

In this way, when the direct sound is combined with the reflected sound, we experience notches and peaks referred to as comb filtering. The reflections enhance or cancel the direct sound to varying degrees, depending on the path-length difference between the reflection and the direct sound at the listening position. An example of comb filtering between the direct sound and a reflection delayed by 1 msec is shown in Fig. 23-1.

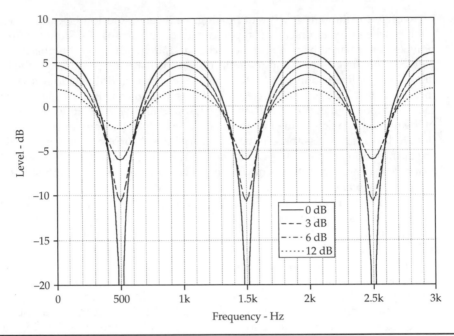

FIGURE 23-1 Comb filtering with one reflection delayed by 1 msec with attenuation of 3, 6, and 12 dB relative to the direct sound.

Four conditions are illustrated. The 0-dB curve refers to the theoretical situation in which the reflection is at the same level as the direct sound. The remaining three interference curves indicate situations in which the reflection is attenuated by 3, 6, and 12 dB. In Fig. 23-2, the locations of the first five interference nulls are indicated as a function of total delay. A delay of 1 msec (1.13 ft) produces a first null at 500 Hz with subsequent notches 1,000 Hz apart. The constructive interference peaks lie midway between successive nulls. When a reflection is at the same level as the direct sound, the nulls theoretically extend to infinity.

The location of the first notch is given by the speed of sound divided by twice the total path-length difference. The spacing between subsequent notches is twice this frequency. The comb filtering due to a series of regularly spaced reflections separated by 1 msec (flutter echo) is shown in Fig. 23-3. Note how the peaks are much sharper than the single reflection.

The audible effect of comb filtering is easy to experience using a delay line. If a signal is combined with a delayed version, you will experience various effects referred to as chorusing or flanging, depending on the length of the delay and the variation of the delay with time. Shorter delays have wider bandwidth notches and thus remove more power than longer delays. This is why microsecond and millisecond delays are so audible.

The effect of a reflection producing comb filtering is illustrated in Figs. 23-4 and 23-5. Figure 23-4 illustrates the time and the frequency response of the right loudspeaker only. The upper curve shows the arrival time, and the lower curve shows the free-field response of the loudspeaker. Figure 23-5 shows the effect of adding a side-wall reflection to the sound of the right loudspeaker. The upper curve shows the arrival time of both the direct sound and the reflected sound. The lower curve shows the severe comb

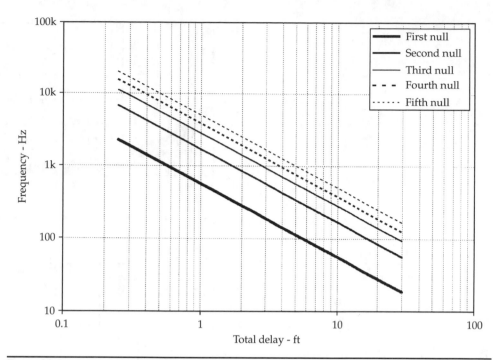

FIGURE 23-2   Comb filter destructive interference null frequencies for a single reflection delayed 1 msec from the direct sound. Constructive interference peaks occur midway between adjacent nulls.

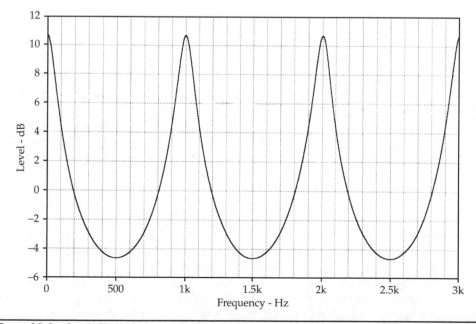

FIGURE 23-3   Comb filter due to equispaced flutter echoes 1-msec apart.

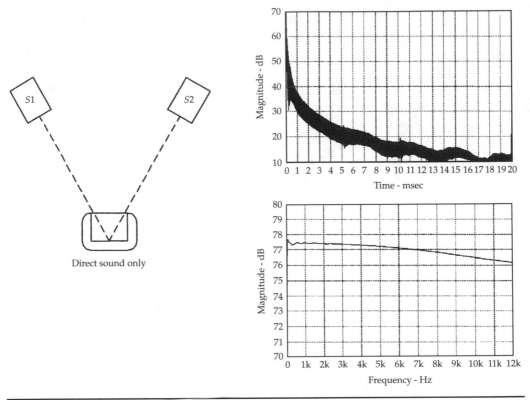

FIGURE **23-4**   Time and frequency response of the direct sound from a loudspeaker.

filtering that a single reflection introduces. If a loudspeaker has a free-field response like this lower curve, it would probably be rejected by most listeners. Yet many rooms are designed without reflection control. In reality our ear-brain combination is more adept at interpreting the direct sound and multiple reflections than the FFT (fast Fourier transform) analyzer, so that the perceived effect may be somewhat less severe. The auditory system can also adapt to different early reflection conditions.

Comb filtering is controlled by attenuating or diffusing the room reflections or by controlling the loudspeaker's directivity to minimize boundary reflections. If the loudspeaker has constant directivity as a function of frequency, any added reflection control must be broadband, so as not to equalize the reflected sound from the room boundaries. Since the directivity of conventional loudspeakers increases with frequency, low-frequency reflection control is important. For this reason, one would not expect to control low-frequency comb filtering with a thin, porous-absorptive foam or fabric-wrapped panel. In addition, when sound diffusors are used they also should scatter sound uniformly over a broad bandwidth. As a rule of thumb, the thickness of absorptive and diffusive surfaces should be roughly 4 in or more.

Comb filtering can be controlled by using absorption, which removes sound energy from the room, or diffusion, which distributes the reflection over time without absorption. Both approaches are valid and produce different psychoacoustical reactions. Using absorption to reduce the effect of a specular reflection will produce pinpoint spatial

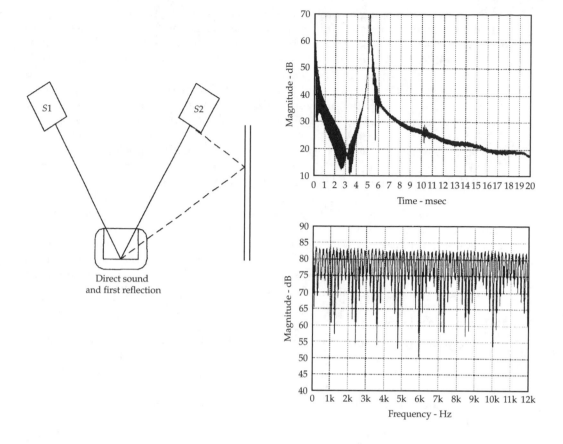

FIGURE 23-5   Time and frequency response of the direct sound combined with a side-wall reflection. The comb filtering is apparent.

phantom images. On the other hand, diffusion will produce sonic images with more spaciousness (width, depth, and height). The degree of this effect can be controlled. Good room design requires an appropriate combination of broad bandwidth absorption and diffusion, with diffusion playing a key role in providing envelopment.

Diffusion

In a performance facility, the acoustical environment contributes to the overall impression of the sound. The role of diffuse reflections in concert hall acoustics has been of interest for many years, this interest being partly fueled by the fact that some of the oldest halls with the best reputations are highly diffuse. In halls like the Grosser Musikvereinssaal in Vienna, the architectural style of the period led to the highly diffuse surface finishings. Due to rising building costs, increased seating requirements, and changes in interior design fashions, flat plaster, concrete, drywall, and cinder block surfaces have become all too commonplace, replacing ornate designs. Acoustic requirements have, however, led to highly modulated surface designs in some more modern halls with a primary aim of creating a highly diffuse sound field.

Small critical listening environments are spaces such as recording and broadcast studios, residential listening rooms, and music practice rooms. These spaces are usually designed to be neutral, adding little of the signature of the room to pre-encoded reproduced sound. A method to achieve this for two-channel audio was described by Davis and Davis in their live end-dead end control room design concept. In 1984, D'Antonio et al. extended the ideas of Davis and Davis by introducing practical ways to provide a temporal and spatial reflection-free zone (RFZ) surrounding the listening position using flush-mounted loudspeakers and a diffuse sound field using reflection phase grating (RPG) diffusors on the rear wall.

The RFZ is created by splaying massive, speaker boundary surfaces, which may also contain porous absorption to further minimize the speaker-boundary interference. The RFZ also creates an initial time delay before the onset of the diffuse energy from the rear wall. The rear wall diffusors create a passive surround sound and a diffuse sound field for the room. They also widen the "sweet spot" and provide low-frequency absorption below the diffusion design frequency. Over the past decade, this design has proved to be a useful approach to designing two-channel listening facilities and has been used in over a thousand projects. The diffusors in effect provide passive surround sound. With the evolution of multichannel audio surround formats and active surrounds, a new design of this type of listening room is warranted.

Diffusion Measurement

With the knowledge of subjective research since the 1960s, the amount and location of diffusive surfaces in critical listening rooms is well understood. Also, measurement standards describing the uniformity of sound diffusion, that is, the diffusion coefficient, clearly rank the effectiveness of potential diffusive surfaces. Diffusion is valuable in promoting lateral reflections to produce a sense of envelopment or spatial impression in rooms.

The AES-4id-2001 document was published in 2001. It outlines a measurement method to determine how uniformly a surface scatters sound. The method measures the directional $1/3$-octave polar responses of the scattering sample and compares it to that of a flat reflective sample of similar size. Directional measurements are obtained for angles of incidence of $0°$, $\pm30°$, and $\pm60°$. The average incidence diffusion coefficient is obtained from the average of the five directional coefficients.

Figure 23-6 shows the proposed presentation format, at an angle of incidence of $-60°$ with respect to the surface normal, for the modulated and optimized diffusor (Modffusor). Three fifth-scale modulated samples are shown in Fig. 23-6A. The diffusion coefficient of the diffusor d_{-60} is compared to that of a flat reflector in Fig. 23-6B. The sample diffusion coefficient is normalized $d_{-60,n}$ by subtracting the diffusion coefficient of the reflector and dividing by 1 minus the diffusion coefficient of the reflector, as shown in Fig. 23-6C. This normalization removes the scattering due to edge diffraction, so that only surface diffusion is shown. It can be seen that the onset of diffusion is roughly 200 Hz. The 12 remaining $1/3$-octave polar responses compare the spatial performance of the diffusor and the reflector. It can be seen that at 4 kHz, the diffusor (light line) is uniformly scattering, while the reflector (heavy line) is directed into the specular $60°$ direction. The AES method is standardized as ISO-17497-2.

Research by Haan and Fricke shows that the surface diffusivity index (SDI), which is a qualitative characterization of surface diffusion, correlates very highly with an acoustic quality index (AQI) in the most highly acclaimed concert halls in the world.

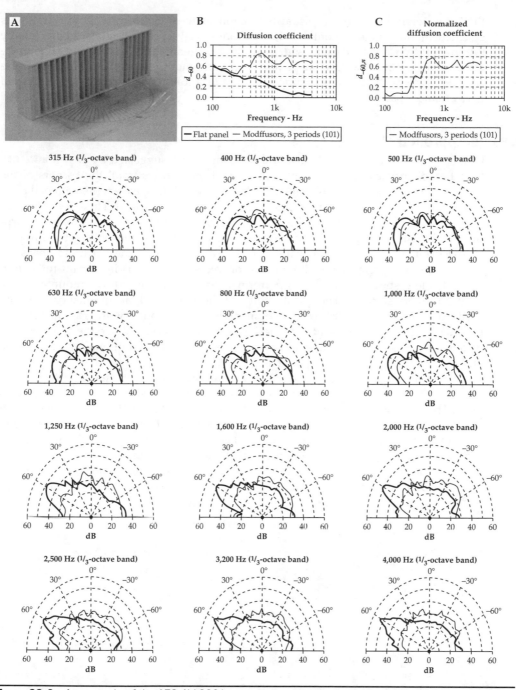

FIGURE 23-6 An example of the AES-4id-2001 measurement method used to determine how uniformly a surface scatters sound. An angle of incidence of −60° with respect to the surface normal is used. (A) Three fifth-scale modulated samples. (B) Comparison between the diffusion coefficients of the sample and the reflector. (C) Normalized diffusion coefficient of the sample. The remaining 12 third-octave polar plots compare the spatial response of the sample (light line) and reflector (heavy line). (*D'Antonio, RPG Diffusor Systems, Inc.*)

SDI was determined from a visual inspection of the surface roughness. While this research requires further confirmation, it bodes well for installations where a large number of diffusors have been used. In addition to geometrical shapes and relief ornamentation, number-theoretic, shape optimized, and modulated diffusors, which represent the state of the art, are finding widespread application in critical listening and performance spaces.

Design Approach

One approach to designing a surround listening room, suggested by D'Antonio, utilizes broadband absorption down to the modal frequencies in all the corners of the room and across the front wall behind the left/center/right speakers (using metal plate resonators which are effective down to 40 Hz with a 4-in thickness). The side-wall space between the front speakers and the listeners can be controlled with broadband absorption or diffusion. Broad bandwidth one-dimensional modulated optimized diffusors are used on the side and rear walls to disperse sound from the surround speakers, which are either preferably wall mounted or free standing. It is important to keep in mind that loudspeakers have better performance when they are surface mounted or far from boundary surfaces. These wall-mounted diffusors work in conjunction with the surround speakers and enhance envelopment. Diffusing clouds, with or without broad bandwidth absorption down to the modal frequencies placed above the listeners, can be effectively used to provide enveloping lateral reflections, additional modal control, and a convenient surface for lighting and HVAC.

The room height can be divided roughly into thirds with diffusive treatment in the central section, to the extent that the ear is covered in both seated and standing positions. The lower and upper areas can remain untreated. There is the possibility of using distributed absorption on the upper third to control flutter echo, if it is a problem. D'Antonio also strongly suggests using wall-ceiling soffits and wall-wall intersections to provide low-frequency absorption. Many dedicated listening rooms utilize massive construction for sound isolation; thus low-frequency control is required. The use of optimal dimensional ratios and multiple in-phase subwoofers properly placed can offer significant advantages.

CHAPTER 24

Room Acoustics
Measurement Software*

Measurements have always been critical to science as they allow us to put assumptions, theories, and equations to the test of reality. As a result, scientists have always searched for ways to improve the availability, accuracy, and precision of measurements. The science of room acoustics is no exception.

Audible sound encompasses the frequency range from 20 to 20,000 Hz. This range of frequencies includes acoustic waves with wavelengths ranging from 56 ft to $\frac{1}{2}$ in. The dimensions of rooms and loudspeakers fall within this range, making them subject to various acoustic effects due to diffraction, resonances, boundary effects, and reflections. Given the complex nature of acoustics, these effects are often difficult to model with a useful level of precision. Therefore, measurements offer a reliable way to accurately quantify room acoustics. In this chapter, we examine how computer software and digital signal processing can provide efficient acoustical measurement tools.

Acoustic Measurements

Around the turn of the twentieth century, Wallace Clement Sabine discovered the fundamental relationship between the size of a room, its sound absorption characteristics, and its reverberation time. From that knowledge, he devised a formula for calculating the reverberation time of a room. Sabine's tools for this research included a portable wind chest, some organ pipes, a stopwatch, seat cushions borrowed from a local theater, and his ears. With these tools, he played sound into a hall, measured the decay rate of its reverberation with a stopwatch and his ears, and modified the decay rate with the seat cushions. While these experiments now seem primitive, Sabine's experiments launched the poorly understood subject of room acoustics into a respected field of study. People began to discover and quantify a variety of acoustical parameters and implement increasingly sophisticated measurement techniques.

Today, personal computers can provide the hardware needed for acoustic measurement. This makes PC-based measurement systems less expensive than most dedicated measurement systems. All that is needed is a test microphone and microphone preamp

---

*Contributed by Geoff Goacher, Acoustical Research Associates, Irvine, California, and Doug Plumb, AcoustiSoft, Peterborough, Ontario, Canada.

to record signals into the computer and software to perform the analysis functions. With a laptop computer, the user has an added advantage that the measurement system is portable, making it easy to take measurements in different locations.

Another significant feature of software-based measurement systems is that the software can account for the imperfections in the hardware (such as sound card frequency response anomalies) and can compensate for them. The time-delay spectrometry (TDS) and maximum length sequence (MLS) test signals (which we will discuss later) used with many measurement programs are insensitive to the nonlinearities present in low-cost computer sound cards.

This is not to say that software-based measurement programs are always better than dedicated measurement systems or that they will replace them altogether. However, software-based systems are extremely cost effective and can provide excellent measurement flexibility and accuracy.

Acoustical measurement software programs can measure a variety of acoustical parameters, many of which have been discussed in previous chapters. With the software that is commonly available, you can commonly measure parameters such as:

- Room reverberation time
- Pseudo-anechoic loudspeaker frequency response in normal rooms
- Reverberant loudspeaker frequency response
- Energy-time curves (specific reflection levels)
- Room resonances
- Impulse response
- Loudspeaker delay times

These measurements can be invaluable in analyzing, troubleshooting, and improving room acoustics and sound reproduction. To provide a better understanding of the sophistication of some of these programs, let's consider the basic principles of how some of the more advanced ones work.

Basic Analysis Instruments

Sound-level meters are probably the most basic sound measurement devices and are useful for tasks in which sound-pressure levels at a particular location must be measured. However, a sound-level meter does not give an indication of the sound quality which is typically sought in most audio applications. Measuring sound quality, at a minimum, also requires the measurement of frequency response. There are a number of instruments and methods for measuring frequency response. Another important requirement for sound-quality measurements is time response. This is frequently used for measuring reverberation time, speech intelligibility, and other time-domain phenomena. Again, there are a variety of instruments used for measuring this factor of sound.

One problem that has traditionally plagued the measurement of these room acoustics parameters is the presence of acoustical background noise. Background noise typically resembles pink noise, which has a power spectrum of −3 dB/octave across the frequency spectrum. Background noise is typically more prevalent at lower frequencies because it travels through walls and structures easily.

It was found that this noise could only be overcome in room acoustics measurements by exciting the room with stimulus energy so great that the background noise becomes insignificant. In the past, this was accomplished by using gunshots as a test signal for impulse response and reverberation time measurements. This method has some technical flaws and provides only a rudimentary look at how sound behaves in rooms. Ultimately, it was concluded that to effectively study room acoustics and the correlation between certain acoustical parameters and "good sound," an instrument with both time and frequency measurement capabilities and significant noise immunity was required.

Time-Delay Spectrometry Techniques

In response to this need for better measurement techniques, in the late 1960s Richard C. Heyser refined a measurement method known as time-delay spectrometry (TDS). By the early 1980s, a TDS measurement system was implemented commercially by Techron in their TEF Analyzer. This digital instrument was implemented on the platform of an industrial portable PC. TEF used time-delay spectrometry to derive both time and frequency response information from measurements.

The basic operating principle of TDS resides in a variable-frequency sweep excitation signal and a receiver with sweep tuning synchronized with that signal. A critical third element is an offset facility that can introduce a time delay between the swept excitation signal and the receiver. The latter is not needed in electrical circuits in which signals travel near the speed of light, but it is vital in acoustical systems in which sound travels at only 1,130 ft/sec.

The coordination of the outgoing sine-wave sweep and the sharply tuned tracking receiver is shown in Fig. 24-1. In this illustration, the horizontal axis is time and the vertical axis is frequency. The signal begins at A (t_1, f_1) and sweeps linearly to B (t_3, f_2). After a time delay (t_d) the receiver sweep begins at C (t_2, f_1) and tracks linearly at a rate identical to that of the signal to D (t_4, f_2). The receiver, being delayed by the time t_d, is now perfectly tuned to receive the swept-signal sine wave after it has traveled in air for the time t_d. At any instant during the sweep, the receiver is offset f_0 Hz from the signal.

The advantages of this arrangement of swept exciting signal and offset-swept tuned receiver are manifold. For example, it takes a certain amount of time for the signal to reach a wall, and it takes a certain amount of time for the reflection to travel back to the measuring microphone. By offsetting the receiver an amount equal to the sum of the two, the receiver examines only that specific reflected component. The offset may be viewed as a frequency offset. When the time offset in the previous example is set to accept the reflection from the wall, the unwanted reflections arrive at the microphone when the analyzing receiver is tuned to some other frequency; hence they are rejected. In this way the receiver is detuned for noise, reverberation, and all unwanted reflections in the measurement. The frequency response of only that particular desired reflection or sound is displayed for study.

Also note that the full signal energy is applied over the testing time throughout the swept spectrum. This is in contrast to applying a brute-force signal to a system, which often results in driving elements into nonlinear regions.

TEF analysis was a breakthrough in measurement tools for evaluating the effects of room acoustics on perceived sound quality. One of its primary benefits is that the original TDS signal from a test can be stored and then repeatedly modulated with delayed

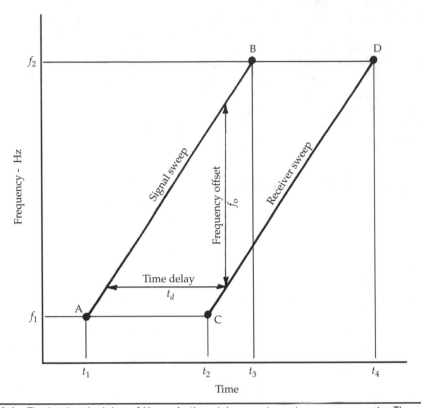

Figure 24-1 The basic principles of Heyser's time-delay spectrometry measurements. The outgoing signal is swept linearly from A to B. After a delay appropriate to select out the reflection desired, the receiver is swept from C to D. Only energy from the desired reflection is accepted; the receiver is detuned to all others.

versions of the sine and cosine chirp to get multiple snapshots of the time versus frequency content. These frequency-response slices allow for generation of three-dimensional waterfall graphs of energy versus time versus frequency.

Another feature of TEF measurements is that it allows for pseudo-anechoic measurements in ordinary rooms by only measuring the initial sound plus the first few (or no) reflections. This technique can be used to take loudspeaker measurements and ignore the room effects by removing reflections from the measurement. The effect of reflections can therefore be studied.

This ability to remove undesired information from the time-response measurements is often referred to as gating in sound measurement. Gating permits audio analyzers to remove reflections from a measurement that occur beyond a certain point in time after the room has initially been stimulated, and hence the term, pseudo-anechoic measurements.

One disadvantage of gating is the loss of resolution that occurs from truncating the time response. For example, room reflections often occur within 5 msec of the direct sound in a typical room with 10-ft ceilings. In a pseudo-anechoic measurement made in such a room, all data that are taken 5 msec after the initial stimulus would have to be removed, limiting the resolution of the measurement. For example, a 5-msec gate limits the lowest frequency that can be measured to $\frac{1}{5}$ msec = 200 Hz. Frequency-response curves are smoothed as a result and frequency effects that appear with resolution finer than 200 Hz appear with reduced detail.

TDS swept tone measurements have two major advantages. First, the post-processing causes unwanted harmonics to be rejected from the measurement, making the measurement accuracy less dependent on system linearity. Devices that add a large distortion component can be reliably measured for frequency response. Second, the swept tone can be used over long durations of time, injecting much energy into the room. This has the effect of increasing measurement signal-to-noise ratio.

The major disadvantage of TDS is that a new measurement must be taken each time a change in the time resolution is desired. If it was desired to take the above measurement again with a time window of say 10 msec, another measurement would have to be taken using a different sweep rate.

Despite the tremendous advantages that were brought about by TDS measurement capabilities, acousticians began searching for a measurement system having the advantages of TDS with the additional capability that a single measurement could later be reprocessed with a different time window. Therefore, only one physical measurement would be required for a complete view of the response. This led to the development of MLS measurement techniques.

Maximum-Length Sequence Techniques

The maximum-length sequence (MLS) measurement technique was refined in the 1980s and has been found by many to be a superior measurement method for room acoustics. MLS measurements use a pseudo-random binary sequence to excite a system and/or room with a test signal resembling wideband white noise. The noise-rejection capabilities of MLS are such that the noise excitation can be played at a low level while maintaining a good signal-to-noise ratio and measurement accuracy. An MLS analyzer can also play a test signal for a long period of time, injecting enough energy into a room to reduce the effects of acoustical background noise to an acceptable level. Like TDS, MLS measurements also have very good distortion immunity.

The binary sequence used to produce the noise signal is precisely known and generated from a logical recursive relationship. Recordings of the test signal played through a system are used to generate an impulse response of the system using a method known as the fast Hadamard transformation.

The ideal impulse can never be realized in practice, but the situation can be approximated perfectly up to a set frequency limit. This limit is due to the aperture effect. An example would be a system excited by a pulse with time duration of 1/1,000 second. The aperture effect will reduce the frequency content above 1,000 Hz when the frequency response is calculated from the recorded data.

Sampling rates of 44.1 kHz and higher in digital measurements prevent the aperture effect from affecting measurements in the audible frequency range. Shannon's sampling theorem proves that all of the information that is contained below 22.05 kHz can be recovered completely. This ideal is realized in practice with PC-based measurement tools and is well approximated with real-time digital components such as CD players and computer sound cards. There is no reason or need to increase this sampling rate when taking measurements for the audible frequency range.

The fact that the impulse response of the system can be determined when using MLS-type stimulus represents a significant advantage of MLS systems over TDS systems. The impulse response can be used to calculate all remaining linear parameters of a system. These include the various forms of frequency-response measurement as well as time measurement. Knowledge of the impulse response of a system permits the construction of a square-wave response, triangular-wave response, or any other form of time response required. All intelligibility calculations can easily be done with only knowledge of the impulse response. When the complete time response of a system is known, the frequency response can be calculated directly using a Fourier transform. Frequency response, by definition, is the Fourier transform of a system time response obtained by impulse excitation.

Numerous companies have taken advantage of the measurement capabilities of MLS techniques and have created software to form the basis of electroacoustic measurement systems. There are several acoustic measurement programs on the market and more will surely become available in the future. For the sake of demonstration, one of these programs and its measurement capabilities are described below.

AcoustiSoft ETF Program

AcoustiSoft introduced their ETF Loudspeaker and Room Acoustics Analysis Program in 1996. The program was one of the first software-based measurement programs for use on a personal computer. The program provides MLS-based measurements and contains calibration features for high accuracy. This presentation describes the fifth version of the program.

The operator connects a test microphone and microphone preamplifier to the computer's sound card line input. A test signal is then played from the sound card's line output through the system (loudspeaker and/or room combination) and is simultaneously recorded through the sound card's line input and stored as a .wav file on the computer for analysis. The software automatically processes the recording and then generates an impulse response of the system being measured, which is stored in the computer for post-processing. Other quantities can be analyzed in post-processing after the measurement takes place. Figure 24-2 illustrates the process in which the test signal is recorded and converted into an impulse response.

As shown in the figure, the ETF system is a two-channel analyzer. The left channel of the input signal is a reference channel and the right channel of the input signal is the test channel. In this way, the program can subtract any response anomalies caused by the sound card so that what remains is actually caused by the system under test.

The measured impulse response of a system can be revealing. Figure 24-3 is an impulse response of a speaker that was measured at a distance of 1 m, on axis with the speaker. As shown in the figure, placing the microphone close to and directly in front of the loudspeaker diminishes the effect of room reflections on the measurement. As will be shown in frequency-response measurements later, this type of impulse response will also be equated with the cleanest frequency response.

Figure 24-4 shows an impulse response of a loudspeaker that was also measured at a distance of 1 m from the loudspeaker, but with a single reflecting surface located nearby. The single wall reflection shows up clearly. These types of reflections also generate a comb-filtering effect in the frequency response. Another interesting feature about reflections in the impulse-response graphs is that reflections containing high-frequency

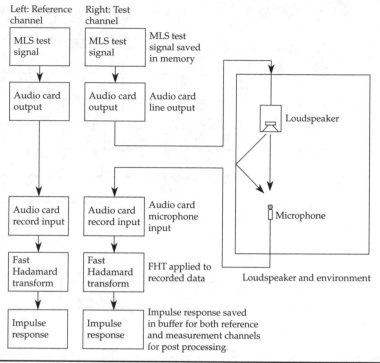

FIGURE 24-2 Block diagram of the ETF software's impulse response measurement methodology. (*Plumb, AcoustiSoft*)

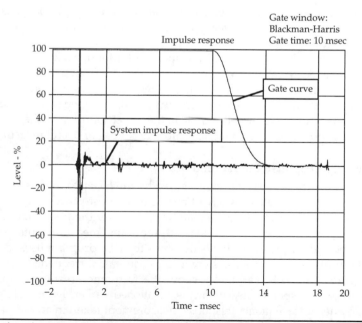

FIGURE 24-3 Impulse response of a loudspeaker measured at a 1-m distance, with minimal reflections. (*Plumb, AcoustiSoft*)

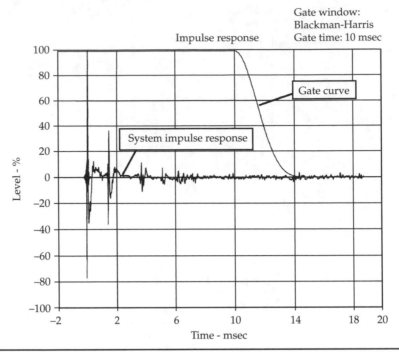

FIGURE 24-4 Impulse response of a loudspeaker measured at a 1-m distance, with a single reflecting surface nearby. (*Plumb, AcoustiSoft*)

content are visible in the impulse-response measurement. Reflections containing only mid-frequency and low-frequency content can be difficult or impossible to see in an impulse response.

Figure 24-5 shows the impulse response of a loudspeaker in a room with the test microphone located at approximately a 3-m distance from the loudspeaker. It shows the effect of multiple reflections in an impulse-response measurement. This type of impulse response generally contains a much more sporadic frequency response due to the interference effects of the room reflections on the direct sound from the loudspeakers. These impulse responses can be converted into other types of useful measurements.

Frequency-Response Measurements

As noted, the impulse response measured by the ETF program is saved in buffers and can be used to compute a variety of results using post-processing. Frequency response is one of these options. Since the gate times used for the frequency-response measurements can be changed as desired in post-processing, measurements usually do not have to be repeated. Figure 24-6 shows the steps the ETF program uses to convert the measured impulse response of a system into its equivalent frequency response.

Figures 24-7, 24-8, and 24-9 show the frequency-response equivalents for the three impulse-response measurements mentioned earlier. Figure 24-7 is a sample of a pseudo-anechoic frequency-response measurement taken in an ordinary room at a distance of 1 m from the loudspeaker. The test setup is the same one used to generate the impulse-response measurement shown in Fig. 24-3. The ability of the gating to cut off

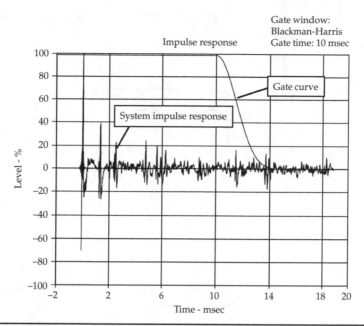

FIGURE 24-5 Impulse response of a loudspeaker measured at a 3-m distance, with multiple reflections. (*Plumb, AcoustiSoft*)

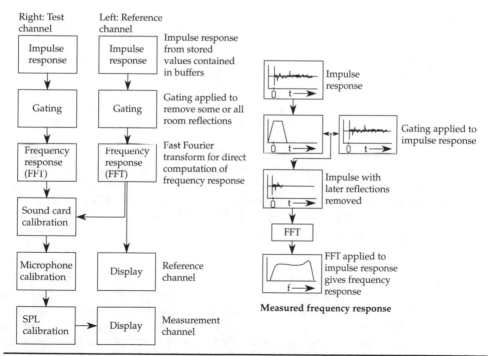

FIGURE 24-6 Block diagram of the ETF software's frequency response computation measurement methodology. (*Plumb, AcoustiSoft*)

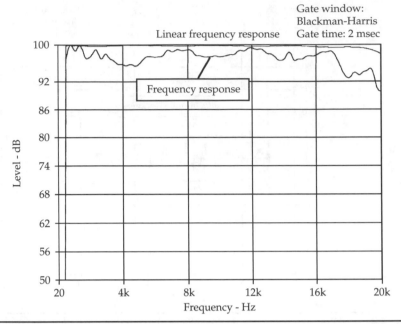

Gate window:
Blackman-Harris
Linear frequency response   Gate time: 2 msec

FIGURE 24-7 Frequency response of a loudspeaker measured at a 1-m distance, with minimal reflections. The configuration of Fig. 24-3 is used. (*Plumb, AcoustiSoft*)

room reflections allows for the measurement of the direct response of the loudspeaker, minus the effect of the room. This is similar to what is possible to measure in an anechoic chamber with no significant reflections; therefore it is a pseudo-anechoic measurement.

In this pseudo-anechoic frequency-response measurement, the impulse response was gated to eliminate information that was recorded after 2 msec of the direct sound so that an accurate picture of the loudspeaker's response could be seen without the effect of room reflections. This frequency response is typical of near-field listening in acoustically treated rooms. Pseudo-anechoic measurements are helpful for speaker designer and are also helpful for finding the actual response of a loudspeaker's direct sound so that it can be differentiated from the response caused by the room. This information can be used to equalize undesirable response problems that are caused by the direct sound of loudspeakers. A linear frequency-response plot is used here to make it easier to diagnose the presence of comb filtering. None is seen in the graph.

Figure 24-8 is a measurement of a loudspeaker at a 1-m distance. However, this measurement has a single reflection that is fairly close in both time and level to that of the direct sound. This single reflection comb filters the response. This results in the equally spaced peaks and dips shown in the figure. This frequency response is typical of near-field listening in acoustically untreated rooms with nearby reflecting surfaces.

Figure 24-9 shows the frequency response at about a 3-m distance from the loudspeaker. This measurement has many reflections in it, and a longer gate time has been selected, which is similar to that of the integration time of the human ear-brain system. This measurement shows the effect of multiple room reflections in the frequency-response calculation. This would be a typical frequency response for far-field listening

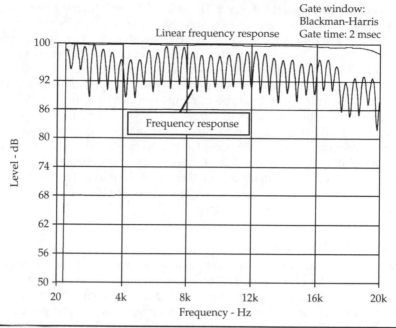

FIGURE 24-8 Frequency response of a loudspeaker measured at a 1-m distance, with a single reflecting surface nearby. The configuration of Fig. 24-4 is used. (*Plumb, AcoustiSoft*)

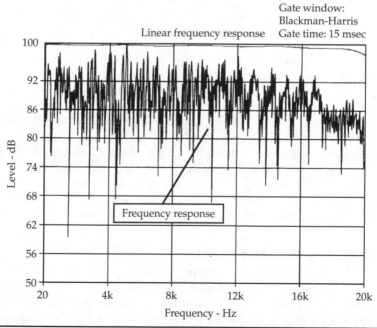

FIGURE 24-9 Frequency response of a loudspeaker measured at a 3-m distance, with multiple reflections. The configuration of Fig. 24-5 is used. (*Plumb, AcoustiSoft*)

in an acoustically untreated room. The effects of comb filtering are present, but much less idealized than they are in the preceding response graph. Multiple reflections causing these sharp dips and peaks can have a negative effect on stereo imaging. Adding absorbers to surfaces causing reflections will tend to smooth this curve.

Compared to the other two measurements, the in-room response graph shown in Fig. 24-9 is of little value in an environment with multiple reflections. A more meaningful response graph for this type of condition can be achieved with further post-processing. Results would yield a fractional octave response (such as $\frac{1}{3}$ octave) that more accurately reflects how we would perceive this frequency response in such a listening environment. This is discussed as part of fractional-octave measurements later.

Pinpointing the level and delay times of room reflections, with respect to the direct sound, is best evaluated using a filtered energy-time curve. The program also generates these energy-time curves for analysis of individual reflection delay times and levels. This is discussed in greater detail later.

Resonance Measurements

The ETF program can also help pinpoint resonances in frequency-response measurements. This is done by examining the later portions of the impulse response for ringing. This ringing results from sharp resonance in a low-frequency room response or in loudspeaker cabinets and drivers. The ringing appears as a sharp peak in the frequency response and can be more easily seen with delayed response graphs, otherwise called time slices. Figure 24-10 shows the steps the program uses to convert the measured impulse response of a system into a time slice frequency-response graph showing resonances.

Figure 24-11 shows a low-frequency room response measurement with time slices. This type of graph is particularly useful for evaluating the effect of room resonances on the system's low-frequency response. The frequency response $t = 0$ (top) curve shows the overall response but does not allow the viewer to easily differentiate between boundary effects and resonances. Boundary effects are the effect of reflections from surfaces near the microphone or loudspeaker. Resonances are frequencies for which the room energetically supports vibrations when excited by a sound source. Room resonances are easy to detect in a graph such as this because the peak structures will have the same shape in each time slice if they are truly resonant-boundary effects will not.

The later slices of the response of Fig. 24-11 show resonances at roughly 35, 65, and 95 Hz. Noise appears to have caused the sharp peaks at 120 Hz because the delayed slices all show the peak at the same level (roughly 68 dB in the graph). It is therefore not decaying, as a resonance would. Measuring resonances in this manner is helpful for optimizing the placement of loudspeaker and listeners in listening rooms. It is also helpful for finding the specific frequency and bandwidth of troublesome resonances to assist tuning of a bass trap.

Fractional-Octave Measurements

The ETF program can average the frequency-response curves into fractional octaves such as $\frac{1}{3}$ octave, $\frac{1}{6}$ octave, and so on. These fractional-octave measurements can also be gated so the curve contains only certain portions of the measurement information. This type of gating and frequency averaging allows for measurements that are highly correlated with our perceptions of sound in various circumstances. Figure 24-12 illustrates the process by which the program creates these fractional-octave displays.

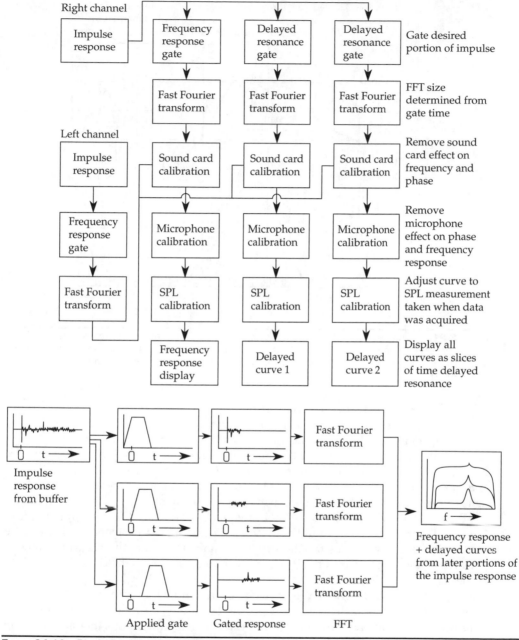

FIGURE 24-10 Block diagram of the ETF software's time slice frequency response computation methodology. (*Plumb, AcoustiSoft*)

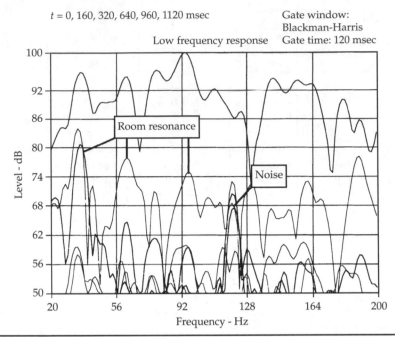

$t = 0, 160, 320, 640, 960, 1120$ msec

Low frequency response

Gate window: Blackman-Harris
Gate time: 120 msec

Figure 24-11   Low-frequency response measurement, showing frequency response and different time intervals or "time slices." Resonances can be identified as peaks that appear in every time slice, with varying level. Peaks that do not vary in level can be assumed to be noise. (*Plumb, AcoustiSoft*)

The $\frac{1}{3}$-octave frequency response is one of the best indicators of subjective frequency balance as it approximates the critical bandwidth accuracy of our hearing. By converting a high-resolution frequency-response measurement into $\frac{1}{3}$ octave, it is easy to see that the sharp dips and peaks previously found in the frequency response are greatly diminished. Only the larger variations in frequency response remain. At mid and high frequencies, reflections are therefore more the cause of stereo imaging problems rather than major subjective frequency balance changes.

Figure 24-13 shows a $\frac{1}{3}$-octave room frequency response. The gate time is set to 20 msec. This measurement would be a good indicator of how a listener would perceive the tonality of the system as the 20-msec gate eliminates what our ears would mostly perceive as room sound, and the $\frac{1}{3}$-octave averaging is close to the critical bandwidth of the ear.

In some instances, higher levels of frequency resolutions are preferred in fractional-octave measurements such as $\frac{1}{10}$ or $\frac{1}{12}$ octave. For these cases, higher-resolution fractional-octave displays are possible. The information gained from this measurement can be used for troubleshooting loudspeaker designs. With an equalizer patched into a system, this measurement can also be used to aid in the flattening of a system's response.

Energy-Time Curve Measurements

The ETF program can also generate energy-time curves which are useful for determining the time and level of reflections versus that of the direct sound. Energy-time curves show the level as a logarithmic quantity, making lower-level behavior more visible.

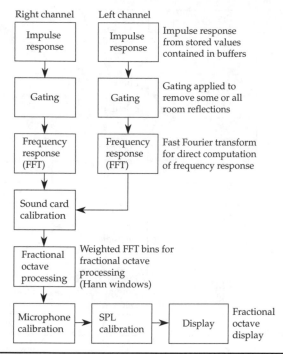

FIGURE 24-12 Block diagram of the ETF software's fractional octave display computation methodology. (*Plumb, AcoustiSoft*)

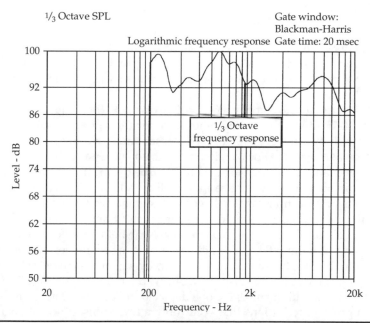

FIGURE 24-13 One-third octave frequency response measurement, with a 20-msec gate. (*Plumb, AcoustiSoft*)

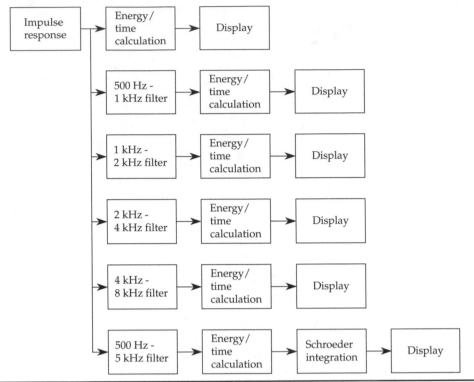

FIGURE 24-14   Block diagram of the ETF software's energy-time curve computation methodology. (*Plumb, AcoustiSoft*)

The energy-time curves can be viewed as full-range levels, or they can be broken into various octave bands for checking the levels of specific ranges of frequencies. Figure 24-14 shows the process by which the program generates energy-time curves from the measured impulse response.

It is instructive to compare energy-time curves to their equivalent impulse response. Figure 24-15 shows the energy-time curve for the same measurement as the impulse response shown in Fig. 24-3. This is a measurement of a loudspeaker taken at 1-m distance, on axis with the speaker. Room reflected energy is clearly present in the energy-time curves and can be seen to be about 20 dB below the initial sound. These energy-time curves can be compared to the results of Olive and Toole's research on the psychoacoustic effects of reflections, which have been summarized in Chaps. 4 and 18. Based on their findings, a listener would not hear much (if any) of the room reflections when positioned this close to the loudspeaker. Near-field monitors take advantage of the principle that room reflections in these conditions become less significant.

Figure 24-16 shows an energy-time curve for a loudspeaker that was also measured at a 1-m distance to the loudspeaker, but with a reflecting surface located nearby. This is the same measurement as the one depicted in Fig. 24-4. The energy-time curve contains high-level reflections that occur only a few milliseconds after the arrival of the direct sound. Also, there are some lower-level reflections visible in this energy-time curve that are not visible in the impulse-response measurement shown in Fig. 24-4.

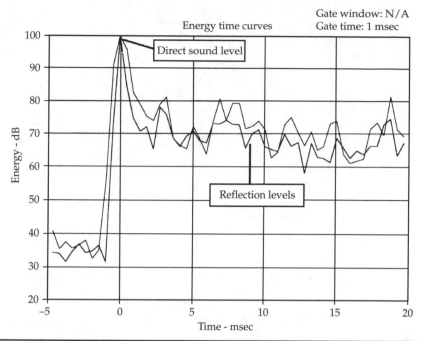

FIGURE 24-15   Energy-time curve of a loudspeaker measured at a 1-m distance, with minimal reflections. The configuration of Fig. 24-3 is used. (*Plumb, AcoustiSoft*)

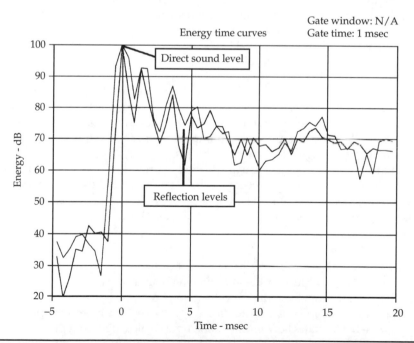

FIGURE 24-16   Energy-time curve of a loudspeaker measured at 1-m distance, with a single reflecting surface nearby. The configuration of Fig. 24-4 is used. (*Plumb, AcoustiSoft*)

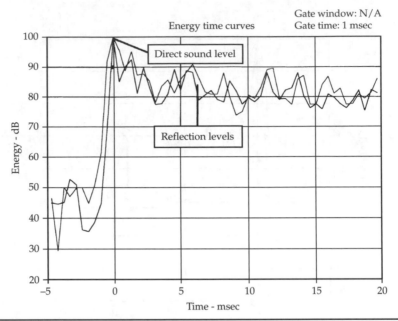

FIGURE 24-17 Energy time curve of a loudspeaker measured at a 3-m distance, with multiple reflections. The configuration of Fig. 24-5 is used. (*Plumb, AcoustiSoft*)

Figure 24-17 is the equivalent energy-time curve for the impulse response shown in Fig. 24-5. This is a measurement of a loudspeaker with the test microphone at approximately a 3-m distance from the loudspeaker. This figure shows a typical response measurement for a room with untreated acoustics. Ideal room treatment for precise sound reproduction would reduce all early reflections occurring within approximately 15 to 20 msec of the direct sound to a level that is 15 to 20 dB below the level of the direct sound. This has the effect of smoothing the phase and frequency response of the perceived sound and it also enables the listener to more clearly hear the direct sound and the acoustics of the space without masking by listening-room reflections. The removal of early reflections also provides an improvement to stereo imagery by improving the coherence of the signal at the listener's ears.

Later reflected sound is usually desired to provide ambience. The longer time delay and mixing of the later reflections causes them to be decorrelated from the original sound and heard as ambience. Loudspeakers in a surround sound system can also be used to provide ambience in an acoustically dead room.

Reverberation Time

The ETF program can calculate the reverberation time of sound in rooms from the measured impulse response. Much of the information about a room's ambience, mentioned above, can be inferred from the room's reverberation time. As shown in Chap. 11, different types of rooms and room sizes require different reverberation times for optimum capabilities. Figure 24-18 illustrates how reverberation time is calculated from the impulse response.

Figure 24-19 shows a sample RT_{60} measurement taken with the program. In the figure, the RT_{60} values average 0.5 second across most of the audible band, rising to

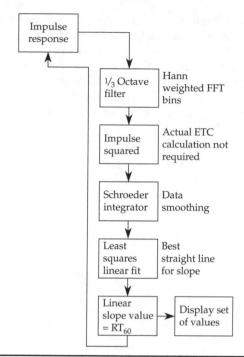

FIGURE 24-18   Block diagram of the ETF software's post-processing for reverberation time computations. (*Plumb, AcoustiSoft*)

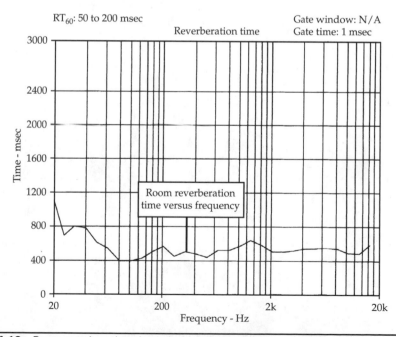

FIGURE 24-19   Room reverberation chart showing measured RT$_{60}$ values at various frequencies. (*Plumb, AcoustiSoft*)

around 1 second at the lowest frequencies. The information gained from a measurement such as this can be used to adjust reverberation-time frequency response until the measurement, and the sound quality, is satisfactory.

The development of advanced acoustic measurement techniques like MLS and the capability of implementing them through software on a PC make it possible to perform highly accurate measurements. The types of measurements these programs provide are an invaluable aid to any sound contractor, acoustical consultant, audiophile, recording engineer, or speaker builder.

CHAPTER 25

Room Optimizer*

The sound that we hear in a critical listening room is determined by the complex interaction among the quality of the electronics, the quality and the placement of the loudspeakers, the hearing ability and placement of the listener, the room dimensions (or geometry if noncuboid), and the acoustical condition of the room's boundary surfaces and contents. All too often these factors are ignored, and emphasis is placed solely on the quality of the loudspeakers. However, the tonal balance and timbre of a given loudspeaker can vary significantly, depending on the placement of the listener and loudspeaker and the room acoustic conditions. The acoustic distortion introduced by the room can be so influential that it dominates the overall sonic impression. Two causes of this acoustic distortion are the acoustical coupling between the loudspeakers and listener with the room's modal pressure variations or room modes and the coherent interaction between the direct sound and the early reflections from the room's boundaries.

Critical listeners have invested considerable time in trial-and-error attempts to minimize these effects. However, no automated method to search for the optimum locations has been proposed. With the advent of 5.1-channel home theater and multichannel music, physical trial-and-error approaches become even less feasible. The task of optimally locating five or more loudspeakers and multiple subwoofers presents a significant challenge. In addition to optimizing the low-frequency response via optimum listener and loudspeaker placement, one must also address imaging and the influence of acoustical surface treatment on the size and location of sonic images, as well as the sense of envelopment or spaciousness experienced in the listening room.

Therefore, to address these acoustical issues we describe an automatic computerized simulation program that suggests optimum locations for loudspeakers, listener, and acoustical surface treatment. In addition, the program can also help with new room design by optimizing the room dimensions. For this discussion, we will focus on the optimum placement of loudspeakers and listener in a room.

Modal Response

All mechanical systems have natural resonances. In rooms, sound waves coherently interfere as they reflect back and forth between hard walls. This interference results in resonances at frequencies determined by the geometry of the room. In lossless cuboid rooms, where the normal component of the particle velocity is zero at the surface, the

---

*Contributed by Peter D'Antonio, RPG Diffusor Systems, Inc., Upper Marlboro, Maryland.

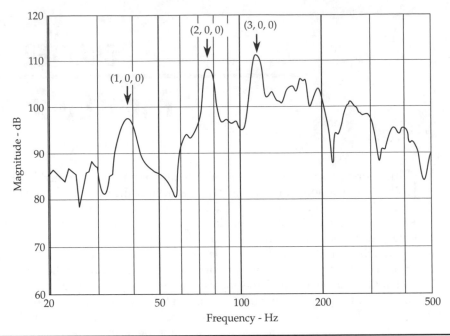

FIGURE 25-1   Measured modal frequency response in a room measuring 2.29 × 4.57 × 2.74 m. The (1, 0, 0), (2, 0, 0), and (3, 0, 0) modes are identified.

modal frequencies are associated with the eigenvalues of the wave equation. These modal frequencies are distributed among axial modes involving two opposing surfaces, tangential modes involving four surfaces, and oblique modes involving all surfaces. For an axial mode between two opposite boundaries, this frequency is equal to the speed of sound divided by twice the room dimension in that direction.

For example, with a speed of sound of 344 m/sec (1,130 ft/sec), a 4.57-m (15-ft) wall-to-wall dimension results in a first-order fundamental room mode of 37.7 Hz. As an example, Fig. 25-1 shows the measured modal frequency response of a room with a 4.57-m dimension. The loudspeaker was located in a corner and the microphone was placed against a wall perpendicular to the 4.57-m dimension, in order to record all axial modes. The first-order (1, 0, 0), second-order (2, 0, 0), and third-order (3, 0, 0) modes are identified in Fig. 25-1 at 37.7, 75.4, and 113 Hz, respectively.

In addition to the modal frequency distribution, the coupling between the loudspeakers and listener with the modal pressure is also important. The loudspeaker placement will accentuate or diminish coupling with the room modes. Similarly, a listener will hear different bass response, depending on where he or she is seated. Figure 25-2 illustrates how the sound pressure is distributed along a room dimension. The room dimension is shown as a fraction ranging from 0 to 1.0. A value of 0.5 would be in the center of the room and 1.0 would be against a wall. Figure 25-2 reveals that the fundamental first-order mode has no energy in the center of the room. Physically, this means that a listener seated in the center of the room would not hear this frequency. The second-order mode, however, is at a maximum. It can be inferred that in the center of the room, all odd-order modal frequencies are absent and all even-order harmonics are at a maximum. Therefore, when we listen to music in a room, the music will be modified by the

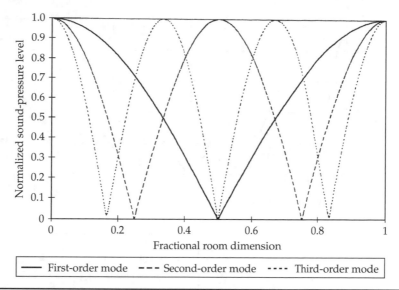

FIGURE 25-2 Normalized energy distribution of the first three modes in a room.

room's modal response, and this acoustic distortion will depend on where the speakers and the listener are located and how they couple with the room. Ideal room dimensions have been variously suggested.

While the distribution of the modal frequencies is important, it is not as important as the placement of the loudspeakers, subwoofers, and the listeners with respect to the boundary surfaces to minimize acoustic distortion introduced by the room. Thus to minimize the modal coloration, we must optimize both the room dimensions and the locations for loudspeakers and listeners.

Speaker-Boundary Interference Response

In addition to modal pressure variations, the interaction of the direct sound from the loudspeakers with reflections from the walls can result in dips and peaks in the spectra due to interference effects. We refer to this as the speaker-boundary interference response (SBIR). This issue has been examined by Allison, Waterhouse, and Waterhouse and Cook. The interference occurs over the entire frequency range, with predominant effects at low frequency. The typical effect is a low-frequency emphasis followed by a notch.

To illustrate the effect in Fig. 25-3, the speaker-boundary interference is averaged over listening positions with the loudspeaker located 1.22 m (4 ft) from one, two, and three walls surrounding the loudspeaker. As each wall is added, the low-frequency response increases by 6 dB and the notch at roughly 100 Hz deepens. This demonstrates the fact that each time the solid angle into which a speaker can radiate is reduced by a factor of 2 (by adding a boundary surface, for example) the sound pressure at low frequencies is increased by a factor of 2 (6 dB). Thus by placing a loudspeaker on the floor near a corner (three boundaries), the full solid angle of 4π steradians is reduced to $\pi/2$, and the total low-frequency gain is increased by roughly 18 dB. Figure 25-3 also illustrates how the notch increases in frequency as the speaker spacing is decreased to 0.31 m (1 ft)

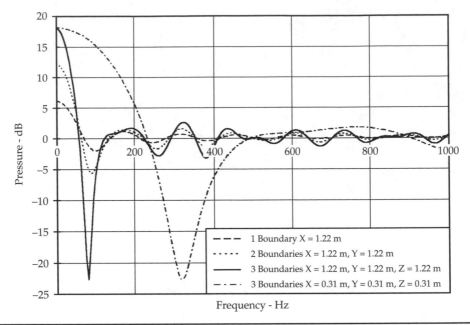

Figure 25-3   Averaged speaker-boundary interference response for several boundary conditions.

from each wall. By moving the loudspeakers and listeners to optimum positions in the room, the coloration produced by the room transfer function can be greatly reduced.

Optimization

Previous chapters describe the complex interaction among the listening room, and the location of the listener and loudspeakers. Guidelines and procedures are already available that address these issues. Modal frequencies for cuboid rooms and their pressure distribution are well known. These can be used to aid listener and loudspeaker placement and room design. Positioning loudspeakers different distances from the nearest floor and walls can reduce the speaker-boundary interference. Simple computer programs that simulate the effect of loudspeaker and listener placement are also available. While these procedures are useful, they can never properly account for the complex sound field, which occurs in real listening rooms. Optimum placement of the loudspeakers and listener must be made taking all of these factors into consideration simultaneously, since the speaker-boundary interference and modal excitation are independent effects. That is, listener and loudspeaker locations that minimize the speaker-boundary interference do not necessarily lead to minimum modal excitation, and vice versa. For this reason an iterative image method was developed to optimize the placement of listener and loudspeakers by monitoring the combined standard deviation of the speaker-boundary interference and modal response spectra.

There has been a great increase in knowledge concerning computer models to predict the acoustics of enclosed spaces. Algorithms which determine the best listener and loudspeaker positions within a space can use complex calculation procedures based on accurate predictions of the sound field received by the listener. These have many advantages over simpler placement theories. For example, they take into account many more reflections from all surfaces in the room. This enables the examination of the subtle effects of many surfaces working in union. By combining the room prediction models with optimization routines, the computer can determine the best positions for the loudspeakers and listener.

In this chapter a program is described which combines an image source model to calculate the room transfer function with a simplex routine to carry out the optimization process. An appropriate cost function to characterize the quality of the spectra received by the listener has been developed. This parameter is based on the extensive subjective evaluations and listening tests of Toole and his colleagues. They have confirmed that loudspeakers that have flat on-axis frequency responses are preferred in standardized listening tests. In addition, similarly good off-axis response is also required, since the listener is hearing the combination of direct and reflected sound from the room's boundary surfaces. At low frequencies speakers are essentially omnidirectional. Since most rooms are not anechoic at low frequencies, we have chosen the flatness of the perceived spectra as a way to evaluate listener and loudspeaker placement and room dimensions. The cost parameter penalizes positions with uneven spectral responses. The optimization program we describe concentrates on lower frequencies (less than 300 Hz), where the individual room modes and the primary speaker-boundary interference are most influential and problematic. The goal of the program is to enable relative nonexperts to determine the best positions for listener, loudspeakers, and acoustical surface treatment in a cuboid room.

Theory of Operation

Considerable research has been carried out in recent decades concerning the predicted room acoustic responses using geometric models. Stephenson details some of the model types available. The fastest and probably simplest prediction model for a cuboid room is based on the image source method. The image solution of a rectangular enclosure rapidly approaches an exact solution of the wave equation as the walls of the room become rigid. While ray tracing allows the analysis of a nonrectangular room, it is not an exact solution of the wave equation. The image model provides a good time-domain transient description of the room response and is appropriate for listener and loudspeaker predictions because of its speed. While the small rooms considered here do not require extensive time to determine their impulse response, the program described will use an iterative optimization process, which will necessitate many hundreds of impulse-response calculations.

The image method includes only those images contributing to the impulse-response and provides appropriate weighting of the modal frequencies. On the other hand, the alternate normal-mode solution of the enclosure, as noted by Morse and Lam, would require calculation of all modes within the frequency range of interest, plus corrections for those outside this range. Allen has derived the exact relationship between the normal-mode and the image solutions for a lossless room. Since the impulse response can be equivalently viewed as a sum of normal modes, there must be repetitive patterns in the impulse response that form early in the response after a transient period. This has been demonstrated by Kovitz, using the algorithm of Burrus and Parks. Kovitz has shown that the full impulse response can be described by an IIR filter that is derived from the early-time FIR impulse response. The equivalence between the impulse response and the modal frequency summation method will be demonstrated later in the chapter using the program.

Prediction of Room Response

The image source model algorithm constructs all possible image sources for the listener and loudspeaker pairs. As we observed in Chap. 13, reflections from room boundaries can be modeled by replacing the surface with an image source at the appropriate position behind the boundary. Single and multiple reflections can be accounted for (see Fig. 13-3). Figure 25-4 shows a source at coordinates (3, 3, 3) and its virtual image sources, which occur at an equivalent perpendicular distance on the opposite side of each boundary. The room is 10 units in height.

The sound at the listener is calculated by propagating the image source wave to the receiver, using the standard equation of a point source, and attenuating the wave according to the absorption coefficient of the boundary it reflects from. The power of computing enables many orders of reflections to be accounted for, not just the first order indicated in Fig. 25-4. The prediction of the pressure received by the listener then

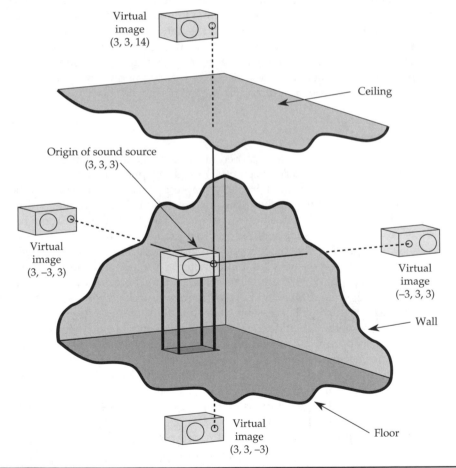

Figure 25-4   The image source model algorithm constructs all possible image sources for the listener and loudspeaker pairs. For example, a source located at coordinates (3, 3, 3) will generate virtual image sources at an equivalent perpendicular distance on the opposite side of each boundary.

reduces to a set of infinite series, which can be encoded and evaluated relatively easily, as demonstrated by Toole. The image source model as described is restricted to cuboid rooms. The model is also limited in that it does not account for any phase change on reflection. This is a common trait among geometric computer models and is also an implicit assumption in other procedures for calculating the position of loudspeaker and listeners. The lack of modeling of the phase change has arisen because it is not a simple process to treat the wall as an extended absorber. This means that the model is most accurate for rooms with relatively hard surfaces where the phase change effects are at their smallest. No account of diffusion caused by surface scattering is made. Fortunately, this is less crucial at the low frequencies being considered here.

The image source model produces an impulse response for the room, where the direct and reflected sounds are clearly distinguishable. An example is shown in Fig. 25-5. A Fourier transform of this gives the spectrum received by a listener if the sound source was producing a continuous tone. This long-term spectrum is shown in Fig. 25-6 and is similar to the modal response of the room (the modal response calculates the sound field at the listener by adding together the effects of individual modes in the frequency domain). In fact, the responses are equivalent, provided that the image source model takes into account an infinite number of sources and the modal calculation includes an infinite number of possible modes.

As an example we examine a room measuring $3 \times 3 \times 3$ m ($10 \times 10 \times 10$ ft). The modes below 300 Hz, calculated with an exact frequency overlap algorithm, are shown in Table 25-1. Figure 25-7 compares this frequency-based calculation with the image model using 30 orders and a wall absorption of 0.12. The time- and frequency-based

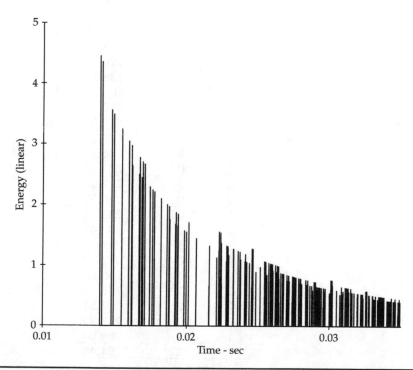

FIGURE 25-5 Typical energy impulse generated by image source model.

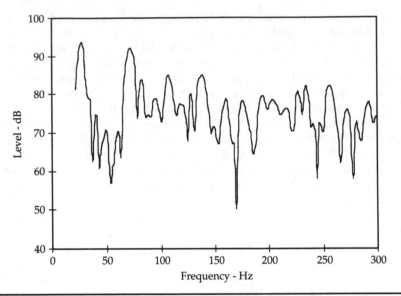

Figure 25-6 Example of long-term spectrum.

n_x	n_y	n_z	f (Hz)
0	0	0	0
0	0	1	56.67
0	1	1	80.14
1	1	1	98.15
2	0	0	113.33
2	1	0	126.71
2	1	1	138.8
0	2	2	160.28
2	1	2	170
3	1	0	179.2
3	1	1	187.94
2	2	2	196.3
3	2	0	204.32
3	2	1	212.03
4	0	0	226.67
4	1	0	233.64
4	1	1	240.42
3	3	1	247
4	2	0	253.42
4	2	1	259.68
3	3	2	265.79
4	2	2	277.61
5	0	0	283.33
5	1	0	288.94
5	1	1	294.45

Table 25-1 Modal Frequencies for a Room
Measuring 3 × 3 × 3 m

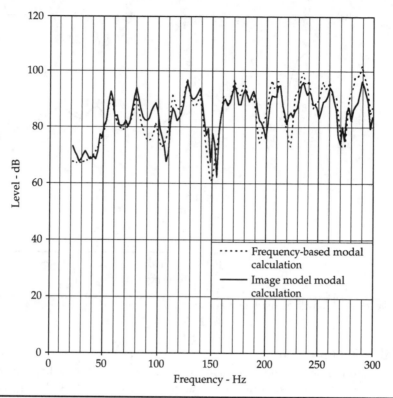

FIGURE 25-7   Comparison of image source calculation (solid line) and modal response (dotted line).

calculations give very similar answers. The discrepancies that do occur could be due to taking insufficient reflection orders and the fact that the impulse response is windowed with a cosine squared term before transforming and that the frequency-based calculation doesn't consider any modes above 300 Hz.

Music is naturally, however, a transient signal, and the ear can distinguish the effects of early arriving reflections. Furthermore, with many audio signals such as music, only the first few reflections are heard before the next musical note arrives and masks the later reflections. Consequently it is also necessary to investigate the response received by the listener for just the first few reflections—this will be referred to as the short-term spectrum. The short-term spectrum is a Fourier transform of the first 64 msec of the impulse response after the direct sound has arrived. The impulse response is windowed using a quarter period cosine squared window. The window starts at 32 msec after the direct sound and gradually weights the impulse response to zero at 64 msec. These times are motivated by the integration time of the ear, which is typically taken to be between 35 and 50 msec. The windowing is necessary to prevent the sudden cutoff of the impulse producing spurious effects in the spectrum. An example of the short-term spectrum is shown in Fig. 25-8. The two spectra shown in Figs. 25-6 and 25-8 are taken to represent the audible characteristics of the propagation from source to receiver, and it is the characteristics of the low-frequency spectra that are optimized to find the best locations for listener and loudspeakers within the room.

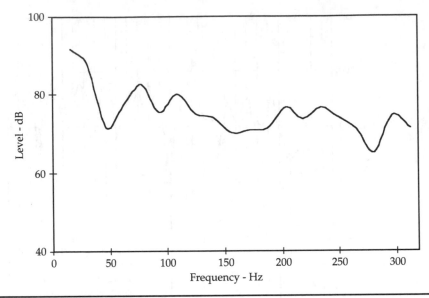

FIGURE **25-8**   Example of short-term spectrum.

Optimizing Procedure

The computer optimizing process finds the best position for the listener and loudspeakers by an iterative process illustrated in Fig. 25-9. The short- and long-term spectra are predicted by the image source method. From these spectra a cost parameter (described below) is derived which characterizes the quality of the sound produced.

Then new listener and loudspeaker positions are repeatedly tried until a minimum in the cost parameter is found, indicating that the best positions are found. The movement of the listener and the loudspeaker positions is carried out using a search engine following a standard minimization procedure described by Press et al.

Cost Parameter

It is assumed that the best position within the room is represented by the position where the short- and long-term spectra have the flattest frequency response. This then motivates the production of a cost parameter that measures how much the true frequency spectrum deviates from a flat response. Previous work by the authors has shown that a standard deviation function is a good measure for characterizing how even the pressure scattered from diffusers is.

In reality, some smoothing over a few adjacent frequency bins is carried out, typically over 1 to 3 bins. This is done to simulate the effect of spatial averaging, which would naturally happen in actual listening rooms. Otherwise there is a risk that the optimization routine will find a solution which is overly sensitive to the exact solution position.

Figures 25-10 and 25-11 show an example of short-term speaker-boundary interference response and long-term modal response for an intermediate (nonoptimized) arrangement of listener and loudspeakers of a stereo pair configuration. The error parameters given in the captions indicate the ability of the standard deviation to measure the quality of the spectra.

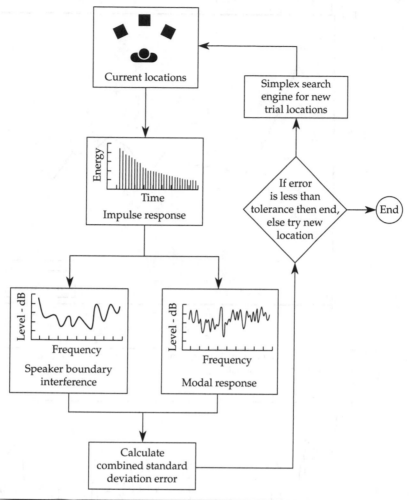

FIGURE 25-9 Iterative cycle to determine optimum listener and loudspeaker locations.

Optimization Procedure

The optimization procedure uses a standard simplex routine. The simplex has a series of nodes, which are different points in the error space (representing different listener and loudspeaker positions). These nodes move around the space until either (1) the difference in the error parameter between the worst and best nodes is less than some tolerance variable or (2) a maximum number of iterations is exceeded (set at 500). Most of the time the program stops because of (1). The simplex routine has the advantage of being a robust system, which does not require first derivatives for calculations. Unfortunately, derivatives of the spectra with respect to the listener and loudspeaker positions are not immediately available from a numerical method such as the image source model. The penalty for using a function-only optimization routine is that the number of iterations used to find a solution is longer and so the procedure takes longer.

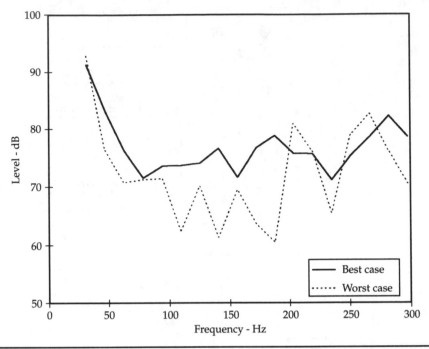

Figure 25-10 Short-term spectrum. The standard deviations are 4.67 and 8.13 dB for the best solution and the worst case, respectively.

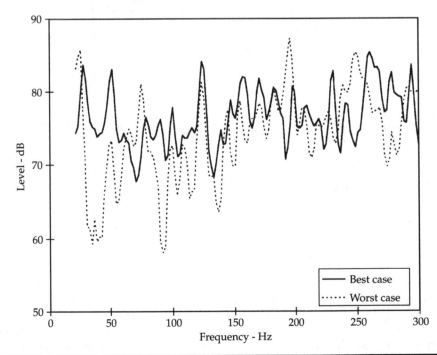

Figure 25-11 Long-term spectrum. The standard deviations are 3.81 and 6.28 dB for the best solution and the worst case, respectively.

Room dimensions (width, length, height)
Rectangular volume defining the x, y, and z limits for the listener position
Rectangular volumes defining x, y, and z limits for the independent loudspeaker(s) position(s)
Number of independent and dependent loudspeakers
Displacement and symmetry constraints relating the dependent loudspeakers to the independent loudspeakers, e.g., simple mirror image symmetry or x, y, z displacement
Stereo constraints
Minimum stereo pair separation
The weighting parameter w that determines the balance between the errors of the short- and long-term spectra
Frequency range of interest (default setting 20–300 Hz)
Number of solutions required

TABLE 25-2   Parameters Presently under User Control

The user of the program inputs various parameters to define the optimization procedure. Table 25-2 lists the parameters that are under the user's control. These can be input via standard Windows dialog boxes. The optimization routines cannot be used completely without some interpretation of the results from the user. For example, there is a tendency for the routines to want to place the source and receivers on the room boundaries as this will minimize the interference effects. Obviously, it is not always possible to build the loudspeakers into the walls and so this may not be a useful solution. Furthermore, there is a risk that the optimization routines will place the loudspeakers close enough to the surfaces to reduce the interference within the frequency band selected (e.g., 20 to 300 Hz), ignoring the fact that there may be audible interference effects just outside this frequency range. Also, the best solution found for the bass response will not always be optimized for stereo imaging, physical listener and loudspeaker placement, and other factors.

Therefore, the user is given the opportunity to limit the search range for the listener and loudspeakers. The limits of the listener and independent loudspeakers are defined in terms of rectangular volumes, determined by the minimum and maximum coordinates. The program allows the user to find the most appropriate solution within those imposed limits. The listener and loudspeakers can vary within the rectangular volume limits for an optimum practical solution. These limit constraints are applied to the simplex routine by brute force. For example, if the simplex routine asks for a prediction for a point outside the listener's rectangle, the program forces the coordinates of the point onto the nearest edge of the listener's constraint boundaries.

The loudspeakers can all be treated as independently varying. However, in most listening situations certain loudspeaker positions are determined by others. For example, in a simple stereo pair, both loudspeakers are related by mirror symmetry about the plane passing through the center of the room. As the number of loudspeakers increases in the 5.1-channel home theater and multichannel music surround formats, we can make the program more efficient by taking advantage of positional relationships

| Minimum stereo pair separation (0.6 m) |
| Absorption coefficient of the surface (0.12) |
| Maximum order of reflections traced in image source model (15) |
| Number of frequency bins to smooth short- and long-term spectra over (1 and 3, respectively) |

TABLE 25-3   Parameters Currently Not under User Control

between the speakers. It is usual to search the room and find several minima. The use of geometric constraints increases the chance of finding the global minimum.

To accomplish this, a system of independent and dependent loudspeakers is adopted with each dependent loudspeaker position being determined by an independent loudspeaker. For the stereo pair example, the left front loudspeaker can be considered the independent loudspeaker and the right front loudspeaker can be defined as the dependent loudspeaker, with its position determined by a simple mirror image of the independent loudspeaker's position about the center of the room. The program allows mirror symmetry operations with respect to the x, y, z planes passing through the listener's position. Mirror symmetry about planes passing through the variable listener position allows constraints to be imposed on the rear surround speakers. For example, in a 5.1-channel multichannel music format with five matching speakers equispaced from the listener, we can set up constraint relationships with one independent speaker (left front) and four dependent speakers (center, right, front, left surround, and right surround).

To include the lessons we have learned about good stereo imaging, the program makes use of a stereo constraint. The stereo constraint refers to the normal angular constraints between a stereo pair and the listener, which are applied to give a good stereo image. In the program this constraint is applied by ensuring that the ratio of the distances between the listener to the center point of the speaker plane, and the distance between the stereo pair is within a specified range. The default is between 0.88 (equilateral triangle) and 1.33. Applying such a nonlinear constraint to the simplex routine can only be achieved by brute force. If a position which violates the constraint is required, the simplex routine moves both the listener and the loudspeaker positions to the nearest points in the room which comply with the constraint. The simplex routine can accommodate such abuses, but there is a risk that this will slow the procedure's finding of the optimum position. When optimizing independent loudspeakers, like subwoofers, for example, the constraint need not be applied. Table 25-3 lists the parameters that are not currently controlled by the user.

Results of Operation

To demonstrate the operation of the optimization program, several typical loudspeaker playback configurations are tested: stereo pair, stereo pair with two vertically displaced woofers, 5.1-channel with dipole surrounds, 5.1-channel with matching satellites, and single subwoofer.

Stereo Pair

The program was required to find the best solution for a stereo pair. The geometry and solution for the optimization are shown in Table 25-4. Figures 25-10 and 25-11 show

	x (m)	*y* (m)	*z* (m)
Room dimensions	7	4.5	2.8
Listener limits	2.0–6.0	2.25 (fixed)	1.14 (fixed)
Independent loudspeaker 1 (front left)	0.5–3	0.5–1.34	0.35–0.8
Dependent loudspeaker 1 (front right)	Mirror of left front at room center *y* = 2.25		
Stereo constraint	0.88–1.33		
Best error parameter	2.24		
Worst error parameter	3.84		
Positions of listener and loudspeakers for best solution:			
Listener position	3.05	2.25	1.14
Independent (left front)	1.28	1.34	0.69
Dependent (right front)	1.28	3.16	0.69

TABLE 25-4   Geometry and Solution for a Stereo Pair Configuration

example intermediate spectra for this configuration. It is apparent how poor the spectral responses can be, and also how they can be improved by the use of a positioning, as outlined here. It is often found that the improvement for the short-term spectrum is more dramatic than for the long-term spectrum. The complexity and number of reflections in the long-term spectrum means that there is less of a chance that there are positions in the room where large improvements of the standard deviation can be found. Even with this complexity, however, useful improvements are found by the optimization routines for the long-term spectrum. The short-term spectrum is much more sensitive to the positions of listener and loudspeaker.

Stereo Pair with Two Woofers per Loudspeaker

We next determine the optimum arrangement of a stereo pair with two woofers vertically displaced by 0.343 m in each loudspeaker. The geometry and best solution are shown in Table 25-5. The optimization assumes constant-impedance loudspeaker floor mounting. The speaker-boundary interference response is shown in Fig. 25-12 and the modal response is shown in Fig. 25-13. The program makes use of mirror symmetry and displacement relationships to reduce the number of independent speakers that have to be optimized. In this case, even though there are four woofers, we actually only need to optimize one, which we will call the lower left front. The upper left front is constrained to follow the lower left-front coordinates, while being displaced vertically in dimension *z* by 0.343 m. The lower and upper right-front woofers are also dependent on the position of the lower left, because of the mirror symmetry of the stereo pair about the center plane of the room located at *y* position of 2.134 m. Thus in this optimization there is one independent loudspeaker, the lower left front, and three dependent loudspeakers, the upper left front, the lower right front, and the upper right front. Of particular note is the avoidance of the roughly 25-dB notch at about 180 Hz in the speaker-boundary interference of Fig. 25-11. The standard deviations for the best and worst solutions were 2.17 and 4.08 dB.

	x (m)	*y* (m)	*z* (m)
Room dimensions	5.791	4.267	3.048
Listener limits	2.591–3.962	2.134 (fixed)	1.14 (fixed)
Independent loudspeaker 1 limits (lower left front)	0.61–1.829	0.457–1.067	0.381 (fixed)
Dependent loudspeaker 1 (upper left front)	Constrained to lower left front	Constrained to lower left front	Displaced 0.343
Dependent loudspeaker 2 (lower right front)	Mirror of lower left front at room center *y* = 2.134		
Dependent loudspeaker 3 (upper right front)	Mirror of lower left front at room center and displaced in *z* by 0.343		
Stereo constraint	0.88–1.33		
Best error parameter	2.1774		
Worst error parameter	4.0839		
Positions of listener and loudspeakers for the best solution:			
Listener position	3.149	2.134	1.14
Independent lower left front	0.947	0.912	0.381
Dependent upper left front	0.947	0.912	0.724
Dependent lower right front	0.947	3.355	0.381
Dependent upper right front	0.947	3.355	0.724

TABLE 25-5 Geometry and Solution for a Stereo Pair Configuration with Two Woofers per Loudspeaker

5.1-Channel Home Theater with Dipole Surrounds

We next consider the 5.1-channel home theater format, with five satellite speakers and a low-frequency effects (subwoofer) channel. The five satellite speakers are placed as a left/center/right (L/C/R) group in front, and two surround speakers are in the rear. In this example, we use a pair of dipoles for the surround channels, as specified by THX. We introduce a new type of constraint in this optimization. The center channel constraint ensures that the center loudspeaker remains on the centerline of the room at a loudspeaker-listener distance equal to that of the left-front listener distance. In this way, all of the arrival times from the front speakers are maintained equal. If this is not desired, the constraint can simply not be applied. In this optimization, the front loudspeakers are allowed to range in *x*, *y*, and also *z*. The *z* search can be used to determine an appropriate elevation above the floor. In some instances this will be useful, while in others the location of the midrange and tweeter may take precedence for good imaging. The dipoles are omnidirectional below 300 Hz, so we will consider them as a point source, for the optimization. The geometry and solution for this configuration are listed in Table 25-6.

Another new constraint relation is added to this optimization to maintain that the dipole surrounds follow the listener's *x* coordinate, so that the listener will remain in the null. Thus, as the listener moves forward and backward during the optimization,

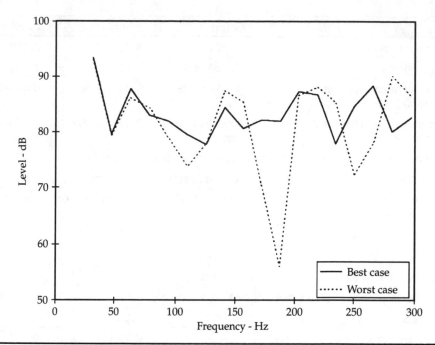

Figure 25-12 Comparison of the speaker-boundary interference response for two listener and loudspeaker arrangements of a stereo pair playback system with two woofers per loudspeaker.

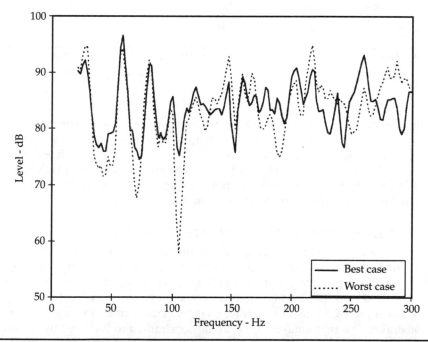

Figure 25-13 Comparison of the modal response for two listener and loudspeaker arrangements of a stereo pair playback system with two woofers per loudspeaker.

	x (m)	y (m)	z (m)
Room dimensions	5.791	4.267	3.048
Listener limits	2.286–3.962	2.134 (fixed)	1.14 (fixed)
Independent loudspeaker 1 limits (left front)	0.61–1.829	0.457–1.067	0.305–0.914
Independent loudspeaker 2 limits (left surround)	Constrained to listener	0.076–0.152	2.134–2.743
Dependent loudspeaker 1 (right front)	Mirror of left front at room center y = 2.134		
Dependent loudspeaker 2 (center)	Constrained to left-front listener distance		
Dependent loudspeaker 3 (right surround)	Mirror of left surround at room center y = 2.134		
Stereo constraint	0.88–1.33		
Best error parameter	1.9476		
Worst error parameter	3.7991		
Positions of listener and loudspeakers for best solution:			
Listener position	3.961	2.134	1.14
Independent left front	1.438	0.723	0.461
Dependent center	1.071	2.134	0.461
Dependent right front	1.438	3.544	0.461
Dependent left dipole surround	3.961	0.123	2.301
Dependent right dipole surround	3.961	4.144	2.301

TABLE 25-6   Geometry and Solution for a 5.1-Channel THX Configuration with Dipole Surrounds

the x coordinate of the dipoles follows this value. The y coordinate, or spacing from the side walls, is optimized over a limited range, and the z coordinate can assume any value within its range limits. A comparison of the speaker-boundary interference response and modal response of the best and worst solutions found for the THX configuration in a 5.791 × 4.267 × 3.048 m room are shown in Figs. 25-14 and 25-15. The standard deviations for the best and worst solutions were 1.95 and 3.80 dB.

5.1-Channel Home Theater with Matched Satellites

As an example of another 5.1-channel configuration, consider an arrangement using five matching loudspeakers (in contrast to the previous example using dipole surrounds). If the physical constraints of the room permit, all five loudspeakers would be equidistant from the listener. To develop the constraints needed for this optimization, we make use of the previous center channel constraint as well as a new rear channel constraint. The rear loudspeakers can be constrained to the front by the use of mirror planes about the listener. This is a dynamic constraint that follows the listener. The geometry and solution for this configuration is given in Table 25-7.

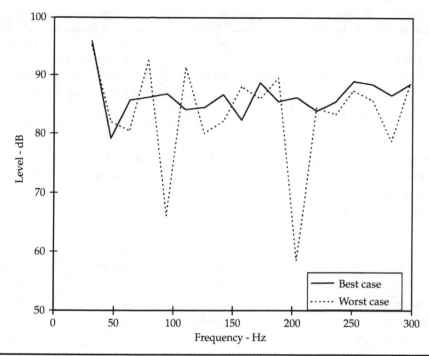

FIGURE 25-14 Comparison of the speaker-boundary interference response for two listener and loudspeaker arrangements of a 5.1-channel playback configuration with dipole surrounds.

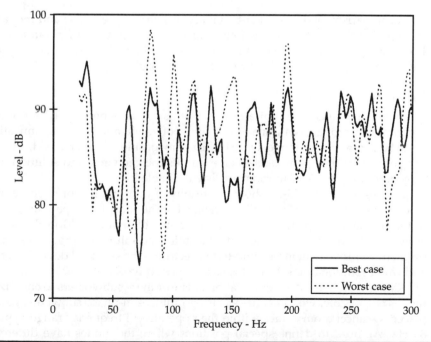

FIGURE 25-15 Comparison of the modal response for two listener and loudspeaker arrangements of a 5.1-channel playback configuration with dipole surrounds.

	x (m)	y (m)	z (m)
Room dimensions	5.791	4.267	3.048
Listener limits	2.591–3.2	2.134 (fixed)	1.14 (fixed)
Independent loudspeaker 1 (left front)	0.914–1.829	0.457–1.067	0.305–0.914
Dependent loudspeaker 1 (right front)	Mirrors left front at room center $y = 2.134$		
Dependent loudspeaker 2 (center)	Distance to listener constrained to equal left-front listener distance and constrained to lie on the $y = 2.134$ line		
Dependent loudspeaker 3 (left rear)	Mirrors left front across x about the listener		
Dependent loudspeaker 4 (right rear)	Mirrors left front across x and y about the listener		
Best error parameter	2.0399		
Worst error parameter	4.0633		
Positions of listener and loudspeakers for best solution:			
Listener position	3.099	2.134	1.14
Independent loudspeaker 1 (left front)	1.05	0.625	0.361
Dependent loudspeaker 1 (right front)	1.05	3.642	0.361
Dependent loudspeaker 2 (center)	0.554	2.134	0.361
Dependent loudspeaker 3 (left rear)	5.148	0.625	0.361
Dependent loudspeaker 4 (right rear)	5.148	3.642	0.361

TABLE 25-7   Geometry and Solution for a 5.1-Channel Configuration with Matched Satellites

A comparison of the speaker-boundary interference response and modal response of the best and worst solutions found for this 5.1-channel configuration in a 5.791 × 4.267 × 3.048 m room are shown in Figs. 25-16 and 25-17. The standard deviations for the best and the worst solutions were 2.04 and 4.06 dB.

Subwoofer

Powered subwoofers are typically used with surround-sound playback formats. The program can provide optimization of subwoofers via a separate optimization over a 20- to 80-Hz frequency range. Once the listening position is determined, any number of subwoofers can be optimized. As an example, the geometry and solution for a single subwoofer in a 10 × 6 × 3 m room are listed in Table 25-8.

Figures 25-18 and 25-19 show the results of an optimization using a single subwoofer operating in the 20- to 80-Hz range. For the short-term spectrum, the variation in the frequency response reduces from a range of about 30 dB in the worst case to around 10 dB for the solution found. The optimized solution provides a somewhat less dramatic improvement in the long-term spectrum. The standard deviations for the best and worst solutions were 2.7 and 5.6 dB, respectively.

Rather than optimizing the location of one or more subwoofers to energize as many modes as possible, we can consider using multiple in-phase subwoofers strategically placed to cancel room modes within their operational frequency range, typically 20 to 80 Hz. We need to define special positions where the modes have different phases. These special positions can be seen in Fig. 25-2 and occur on either side of a null. The null for the first-order mode, for example, can be seen at half of the distance between two opposing room boundaries. Consider what happens when one subwoofer is placed

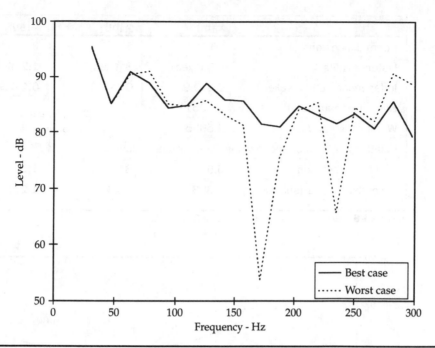

FIGURE 25-16 Comparison of the speaker-boundary interference response for two listener and loudspeaker arrangements of a 5.1-channel playback configuration using five equidistant matching loudspeakers.

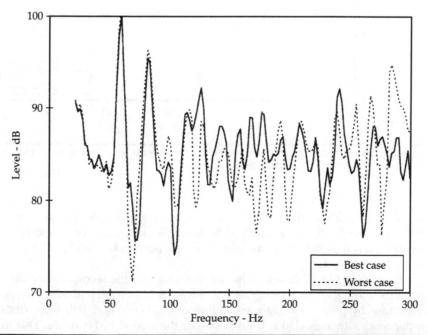

FIGURE 25-17 Comparison of the modal response for the two listener and loudspeaker arrangements of a 5.1-channel playback configuration using five equidistant matching loudspeakers.

	x (m)	y (m)	z (m)
Room dimensions	10	6	3
Listener limits	5.5 (fixed)	3 (fixed)	1.2 (fixed)
Independent loudspeaker 1	5.6–9.9	0.1–2.9	0.1–2.9
Best error parameter	2.7051		
Worst error parameter	5.5535		
Positions of listener and loudspeakers for best solution:			
Listener position	5.5	3	1.2
Loudspeaker 1 (subwoofer)	7.398	0.231	1.227

TABLE 25-8   Geometry and Solution for a Subwoofer

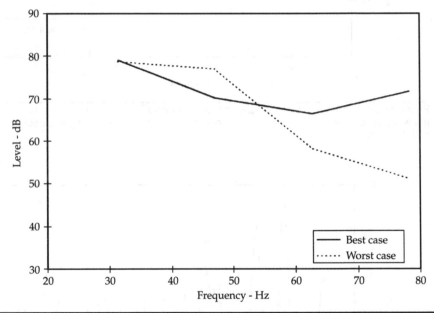

FIGURE 25-18   Comparison of the speaker-boundary interference response for two listener and loudspeaker positions of a subwoofer.

at a quarter (0.25) of the width W of the room, for example, and another is placed at three quarters (0.75) of the width of the room. Because the room's width mode at these locations has the opposite phase, the two in-phase speakers will cancel the mode and it will not be excited.

Examining Fig. 25-2, notice that there is a null for the second-order mode at a quarter of the room's width. Placing a subwoofer at this location will not energize this mode. Therefore, by placing two in-phase subwoofers at $W/4$ and $3W/4$, the first- and third-order modes are canceled and do not excite the second-order mode. This means that a listener can move across the width of the room and experience a flat modal response. If

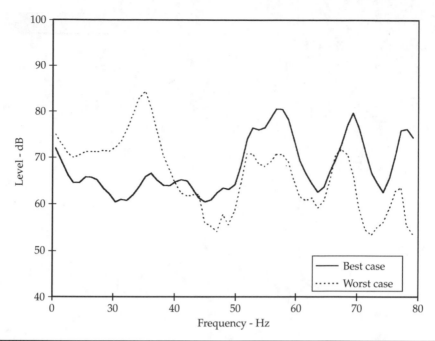

FIGURE 25-19   Comparison of the modal response for two listener and loudspeaker positions of a subwoofer.

we do the same thing with two additional subwoofers in the rear of the room, this will allow a listener to move from front to rear of the room and experience a flat modal response. Unfortunately, these positions are impractical in most listening rooms. However, Welti has determined that by placing four subwoofers in the corners of the room or at the middle of each wall, a similar response can be achieved.

Summary

A software program has been developed that allows automated selection of positions for listeners and loudspeakers within listening rooms. The criterion for optimum listener and loudspeaker positions within the room is the minimum standard deviation of the combined short- and long-term spectra. A cost parameter based on the standard deviation function was developed and used to monitor the quality of the short- and long-term spectra. Predictions of the spectra are carried out using an image source model. The optimization is carried out using a standard simplex routine. Several examples have been presented. All cases demonstrate the ability of the program to find the best positions for listener and loudspeakers within the room.

CHAPTER 26

Room Auralization*

In the past, acousticians used a wide range of design techniques to provide good room acoustics, but acoustical options or the final result could not be tested prior to construction, except by evaluating scale models. Despite these limitations, acousticians have produced some remarkable performance and critical listening rooms.

In contrast to older methods, it would be remarkable if designers could listen to the sound field in a proposed space before the space was built. In this way, we might avoid some acoustical problems and evaluate different design options and surface treatments. This process of acoustical rendering, analogous to visual rendering used by architects, has been called auralization. By definition, auralization is the process of rendering audible, by physical or mathematical modeling, the sound field of a source in a space, in such a way as to simulate the binaural listening experience at a given position in the modeled space. This has been described by Kleiner et al. and Dalenbäck et al. Today, such auralization software provides acousticians with tools to predict and simulate the performance of critical listening rooms.

History of Acoustic Modeling

In the broadest of terms, to characterize a room it would be necessary to find a way to follow the reflection path history of complex sound reflections as they proceed in time through the room.

This reflection path history of sound level versus time as picked up at a location in the room is called an echogram, and the process is called ray tracing. In this approach, we follow each ray through its reflection path history and catalog those rays that pass through a small volume at the listening position. This approach is easy to visualize, and if we use audio slow motion and slow down sound from 1,130 ft/sec to 1 in/sec, we can color-encode each sound ray and record the event.

Consequently, in the late 1960s, acousticians began using ray tracing to determine an echogram and estimate reverberation time. By the 1980s, ray tracing was widely used. In ray tracing, the total energy emitted by a source is distributed according to the radiation characteristics of the source into a specified number of directions. In the simplest form, the energy of each ray is equal to the total energy divided by the number of rays. Depending on the type of surface, each ray from each boundary in its reflection history is either specularly reflected, in which the angle of incidence equals the angle of reflection, or diffusely reflected, in which the direction of the reflected ray is randomized.

*Contributed by Peter D'Antonio, RPG Diffusor Systems, Inc., Upper Marlboro, Maryland.

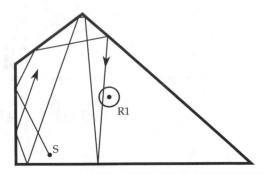

Figure 26-1   Ray tracing example showing emitted ray from source S entering the circular cross-section detection area of receiver cell R1 after three specular reflections.

The reflected energy is diminished by absorption and also by the spherical attenuation due to propagation. (In ray tracing this is achieved automatically because of the fixed receiver size.) The number of rays passing through a receiver cell determines the sound-pressure level.

An example of ray tracing is shown in Fig. 26-1. A ray emitted from source S is shown reflecting from three surfaces before it passes through the circular cross-section detection area of receiver R1. The energy contributions of the various rays to a certain receiver cell are added within prescribed time intervals, resulting in a histogram. Because of time averaging and the strongly random character of the ray arrivals, the histogram will only be an approximation of the true echogram. Ray tracing is straightforward, but its efficiency is achieved at the expense of limited time and spatial resolution in the echogram. The number of rays and the exact angles of emission from the source determine the accuracy of the sampling of room details and the reception of rays within the receiver volume.

In the late 1970s, another approach, called the mirror image source method (MISM) was developed to determine the echogram. In this approach, a virtual image of the actual source is determined by reflecting the source perpendicularly across a room boundary. Thus the image source is located at a distance equal to twice the perpendicular distance d to the reflecting boundary. The distance between S1 and R1 in Fig. 26-2 is equal to the reflection path from S to R1. The reflections of all real and virtual images across the room boundaries create a set of mirror images. The arrival times in the echogram are now simply determined by the distances between these virtual images and the receiver. When applied to a rectangular room, all mirror sources are visible from every position in the room and the calculation is fast. In irregular rooms, however, this is not the case and validity tests are performed. For example, in Fig. 26-2, it can be seen that receiver R1 can be reached by a first-order reflection from surface 1, but receiver R2 cannot. This means R1 is "visible" from S1 and R2 is not. Therefore, validation of each image is required. Since each further reflection order multiplies the number of image sources to be tested (by the number of walls minus one), the number of sources to validate grows very rapidly.

Two approaches were developed in the late 1980s to minimize validation of sources. One was the hybrid method that used specular ray tracing to find potentially valid reflection paths and validated only those, while another used cones or triangular-based pyramidal beams instead of rays. First- and second-order cone tracing examples are

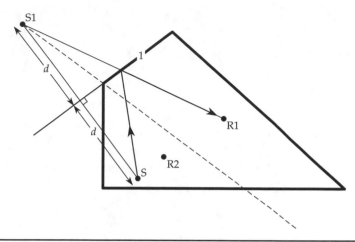

FIGURE 26-2   Mirror image source model (MISM) construction showing real source S, virtual source S1, boundary 1, and receiver R1.

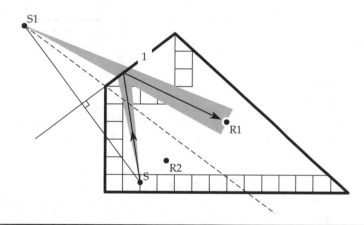

FIGURE 26-3   Equivalence of MISM and cone tracing for first-order reflection from surface 1.

shown in Figs. 26-3 and 26-4. If the receiver point lies within the projection of the beam, then a likely visible image source has been found and validation is not performed. Both of these methods can calculate a more detailed specular echogram than ray tracing and are also faster than MISM at the expense of leaving out reflection paths for the late part in the echogram (when the ray-cone separation is becoming larger than individual walls). However, it was only in ray tracing that diffuse reflection can be included in an efficient way, so the improved algorithms left out a very central acoustical phenomenon. The current state of the art in geometrical room modeling programs all utilize a more complex combination of methods such as MISM for low-order reflections, and randomized cone tracing for higher-order reflections. Randomization is thus reintroduced to handle diffuse reflection.

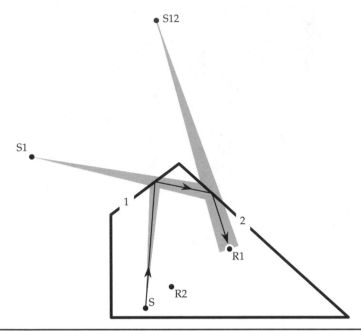

FIGURE 26-4 Second-order reflection of source S from surfaces 1 and 2 to receiver R1 using cone tracing. S1 and S12 are virtual sources.

The Auralization Process

The auralization process begins with predicting octave band echograms based on a 3D CAD model of a room, using geometrical acoustics. Since geometrical acoustics is limited to high frequencies roughly above 125 Hz, there are attempts to model low frequencies using boundary element wave acoustics to extend the range of applicability. Frequency-dependent material properties (absorption and scattering) are assigned to room surfaces, and frequency-dependent source directivities are assigned to sound sources. From these echograms a great number of numerical objective measures, such as reverberation times, early and late energy ratios, and lateral energy fraction can be estimated.

Figure 26-5 illustrates the information needed to generate an echogram. Everything about the room needs to be defined, such as sound sources, the room's geometry and surface treatments, and listeners. Each source, be it a natural source or loudspeaker, is defined by its octave-band directivity at least at 125, 250, and 500 Hz, and 1, 2, and 4 kHz. The directivity balloons (a 3D version of the polar diagrams shown in Fig. 26-5) describe how the sound is directed in each octave band. The surface properties of the room's boundaries are described by their absorption and scattering coefficients.

Scattering Coefficients

A method to determine scattering coefficients has been standardized as ISO 17497-1. The method requires measuring the reverberation time with a suitably sized sample in a reverberation chamber mounted on a circular rotatable turntable in the stationary and rotating condition. In brief, the diffusion coefficient d is a measure of the uniformity of the reflected sound. The purpose of this coefficient is to enable the design of diffusers,

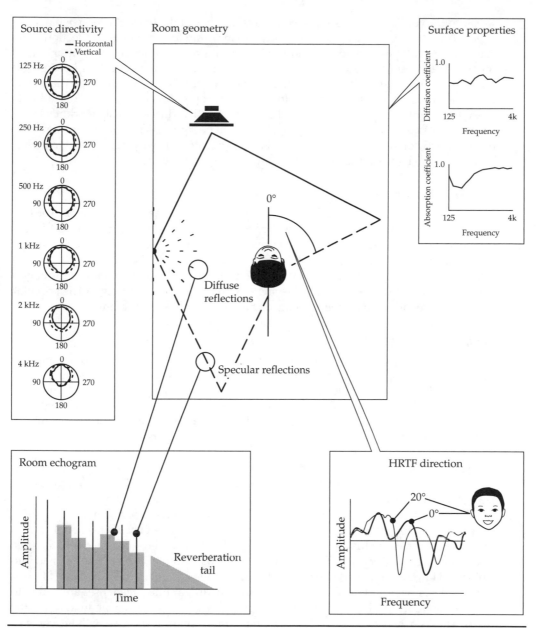

FIGURE 26-5 Simulation of a room echogram using geometrical acoustics and descriptions of the room geometry, the surface properties, the source directivity, and the receiver's HRTFs.

and to also allow acousticians to compare the performance of surfaces for room design and performance specifications. The scattering coefficient s is a ratio of sound energy scattered in a nonspecular manner to the total reflected sound energy. The purpose of this coefficient is to characterize surface scattering for use in geometrical room modeling programs.

Both the coefficients are simplified representations of the true reflection behavior. It is necessary to devise simple metrics, rather than trying to evaluate the reflection characteristics for all possible source and receiver positions, because the amount of data is otherwise too large. The coefficients attempt to represent the reflection by a single parameter, maximizing the information carried by that single number. The difference between diffusion and scattering coefficients is the emphasis on which information is most important to be preserved in the data reduction. For diffuser designers and acoustical consultants, it is the uniformity of all reflected energy which is most important; for room acoustic modeling, it is the amount of energy scattered away from specular angles. The difference between the definitions may appear subtle, but it is significant.

Receiver Characterization

Each receiver is described by an appropriate response. If a binaural analysis is to be carried out, the listener is characterized by the head-related transfer function (HRTF). These frequency responses describe the difference in the response at each ear with and without a listener present. An example of HRTF for sound arriving at 0° and at 20° is shown in Fig. 26-5. Thus in the generation of the echogram, each reflection contains information about its arrival time, its level, and its angle with respect to the receiver. The specular reflections are indicated as lines and the diffuse reflections as histograms.

Echogram Processing

In Fig. 26-6 we illustrate one method for converting the echogram into an impulse response by adding phase information. For each reflection the magnitude is determined by fitting the octave band levels (A through F), determined by a combination of the loudspeaker directivity, the absorption coefficient, and the scattering coefficient. The phase for early specular reflections is often determined by minimum-phase techniques using the Hilbert transform. The impulse response for each reflection is then obtained by an inverse fast Fourier transform (IFFT).

There are several ways in which one can process the predicted echograms for final audition. These include binaural processing for headphones, including headphone equalization or loudspeakers with crosstalk cancellation, mono processing, stereo processing, 5.1-channel processing, or B-format processing (for ambisonic replay). Figure 26-7 illustrates the path for binaural processing. The impulse response of each reflection in turn is modified by the HRTF for the appropriate angle of arrival and transformed into the time domain using an inverse fast Fourier transform. Thus each reflection is transformed into a binaural impulse response, one for the left ear and one for the right. This is done for each of the reflections from 1 to N, and the resultant binaural impulse responses are added to form the total left and right room impulse responses. We are now able to auralize the room. This is achieved by convolving the binaural impulse responses with anechoic or nonanechoic music, as illustrated in Fig. 26-8.

Room Model Data

Figure 26-9 shows a 3D model for a critical listening room created with the CATT-Acoustic program (www.catt.se). 3D models are created either in AutoCAD or with a text program language, which allows the user to use variables, such as length, width, and height. Changing these variables allows one to easily change the model. Objects can also be created with variable orientations, to model the effect of different designs.

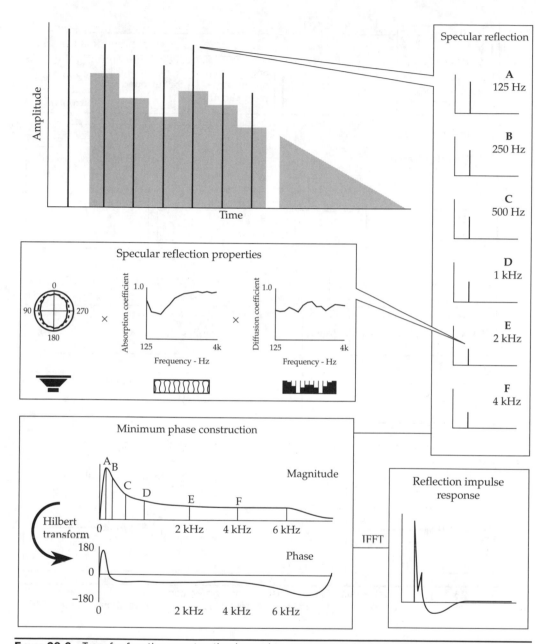

FIGURE 26-6 Transfer function construction is used to convert an echogram reflection into an impulse response.

Figure 26-9 shows the right (A1), center (A0), left (A2), right surround (A3), and left surround (A4) speakers. Also the room is divided vertically into thirds and horizontally in half to allow for different acoustical surface treatment. Adjusting the appropriate variable changes the size of these areas. The corners can be 90° or contain a bass trap, the size of which is variable. The ceiling soffit is also variable. The receiver is indicated by location 01.

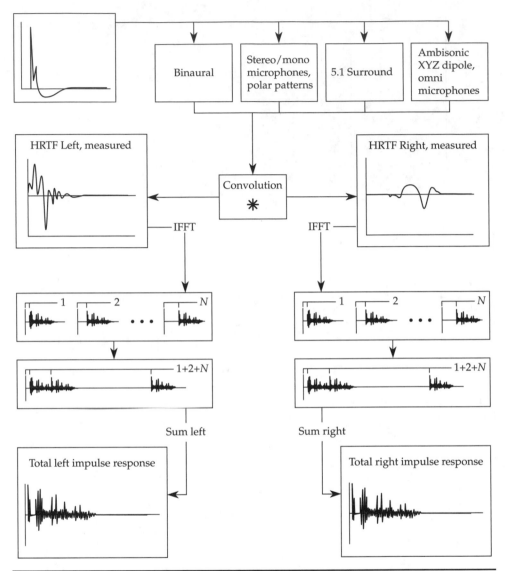

Figure 26-7 For binaural auralization, the postprocessing step converts the room impulse response into a left- and right-ear binaural impulse response through convolution with the HRTFs.

Figure 26-10 illustrates the early part of the echogram showing the specular and diffuse reflections, the emitted angles of the sound from the source, the arrival angles at the receiver, and an isometric view of the room. By moving the cursor to each reflection, the reflection path from source to receiver is visible in the room illustration. In this way one can determine the boundary responsible for a possibly problematic reflection.

Figure 26-11A illustrates the objective parameters that can be determined directly from the echogram after backward integration. The early decay time is the time it takes to reduce the sound-pressure level (SPL) from its steady-state level by 10 dB. T-15 and

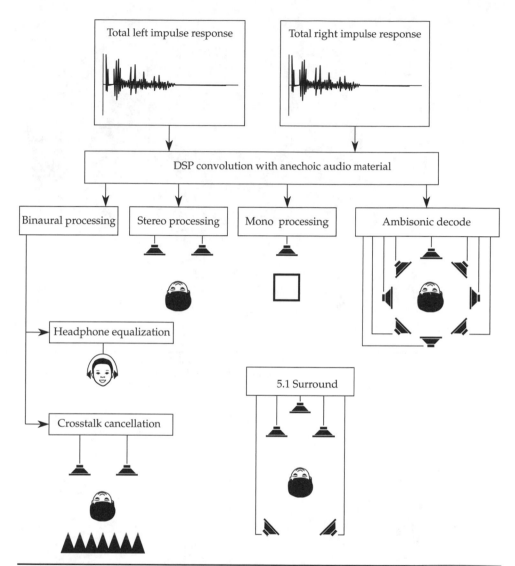

Figure 26-8 Auralization of the room is achieved by convolving the binaural impulse responses with music.

T-30 refer to the time it takes to decrease the SPL from −5 to −20 and from −5 to −35 dB, respectively. There are several objective parameters based on ratios of early- to late-arriving sound. The definition criterion, referred to as D-50 in CATT-Acoustic, is a measure of speech intelligibility. It is a percent ratio of the sound arriving in the first 50 msec following the direct sound to the total sound. The music clarity index, referred to as C-80, is based on a ratio of the first 80 msec of sound to the rest of the sound. Lateral energy fractions, referred to as LEF1 and LEF2, are measures of the impression of spaciousness. They are based on the ratio of laterally arriving reflections between 5 and 80 msec to all of the omnidirectionally arriving reflections after 80 msec.

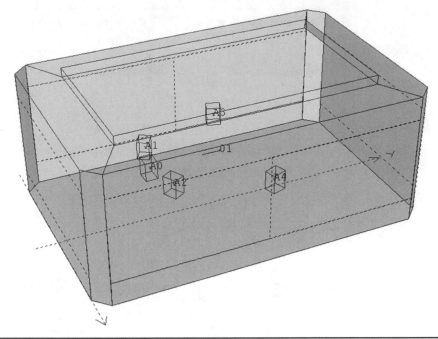

Figure 26-9   3D model of a listening room with five matching speakers in the ITU configuration. The model illustrates how the chosen variables describe the corner detail, soffit, and allocation of acoustical treatment to various sections on the room's boundaries. (*CATT-Acoustic*)

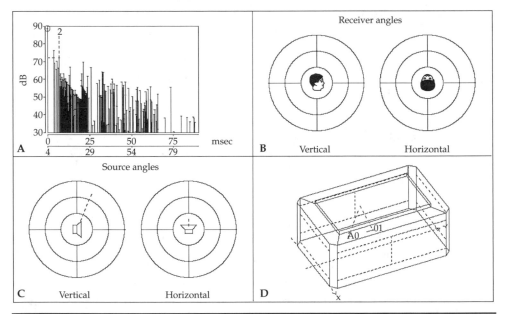

Figure 26-10   Data display from 3D room model. (A) The early part of the echogram. (B and C) The directionality of the receiving and emitted sound rays. (D) Selected reflection from *A*0 to receiver 01 bouncing off the ceiling. (*CATT-Acoustic*)

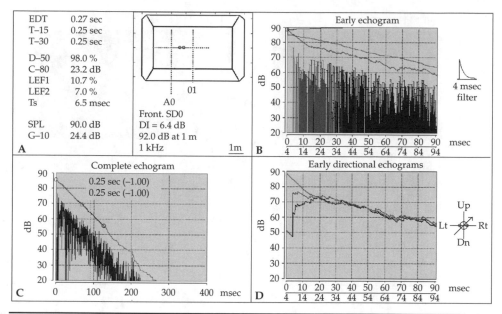

FIGURE 26-11 Data display from 3D room model. (A) List of the objective parameters. (B) Early echogram with backward integration. (C) Complete echogram with T-15 and T-30 from backward integration. (D) Directional echograms (up/down, left/right, and front/back). (*CATT-Acoustic*)

The center-of-gravity time Ts is a measure constructed to describe where the sound is concentrated in the echogram. Ts has a low value if the arriving sound is concentrated in the early part of the echogram and a high value if the early reflections are weak or if the decay is slow. Sound-pressure level is 10 times the 10-based log of the ratio of the pressure squared over a reference pressure squared. G-10 is the sound pressure at a location in a room as compared to what the direct sound from an omnidirectional source would give at a 10-m distance. G-10 is a measure that is used for characterizing how loud the sound will be, and can be compared between halls.

Figures 26-11B, C, and D show additional information: early echogram with backward integration; complete echogram with T-15 and T-30 from the backward integration; directional echograms (up/down, left/right, and front/back).

Room Model Mapping

Another useful function of the geometrical modeling programs is the ability to map the objective parameters, so we can see how a parameter varies over the listening area. (These maps are normally in color, giving a much better visual resolution than the grayscale versions presented here.) Figure 26-12 shows a mapping of RT (which is related to the early decay time), the LEF2, SPL, and D-50 for five matching monopole speakers configured in the International Telecommunications Union (ITU) surround format. Similarly, Fig. 26-13 shows the mapping for five speakers configured in the THX format with dipole surround speakers.

Although these parameters are mostly useful for large-room analysis, they also give some information in the case of a listening room. For example, they give an opportunity to study how the effective listening area is affected by loudspeaker placement and

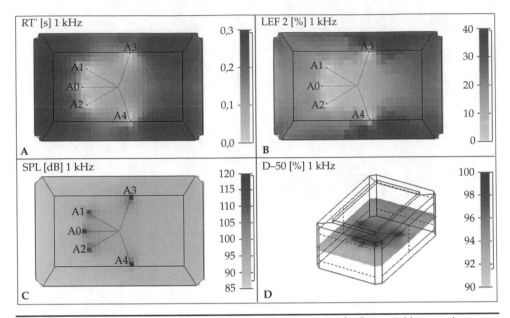

FIGURE 26-12 Parameters are mapped at 1 kHz across the room for five matching speakers in the 5.1-channel ITU surround configuration, left/center/right with monopole surrounds. (A) Perceived reverberation time RT′. (B) Lateral energy fraction LEF2. (C) Sound-pressure level SPL. (D) Energy ratio *D*-50. (*CATT-Acoustic*)

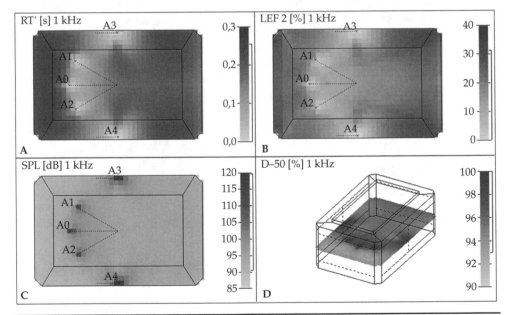

FIGURE 26-13 Parameters are mapped at 1 kHz across the room with a 5.1-channel THX configuration of left/center/right and dipole surrounds. (A) Perceived reverberation time RT′. (B) Lateral energy fraction LEF2. (C) Sound-pressure level SPL. (D) Energy ratio D-50. (*CATT-Acoustic*)

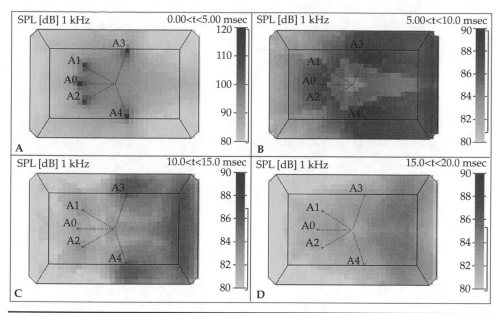

FIGURE 26-14 The sound fields from the five speakers combine over time to energize the room. (A) *t* < 5 msec. (B) 5 < *t* < 10 msec. (C) 10 < *t* < 15 msec. (D) 15 < *t* < 20 msec. (*CATT-Acoustic*)

surface treatment. Note that the RT' value in Fig. 26-12 is very low between the speakers (a distinct sound) while for the THX case in Fig. 26-13, the situation is different since the listener is not reached by much direct sound from the surround dipoles. This is one of the design goals of using dipoles, giving a more diffuse sound from the surround speakers. This shows the ability of the mapping technique to illustrate the behavior of different setups. In current designs, dipoles are no longer used. To guarantee uniform coverage to an extended area of listeners, dynamic speakers are mounted on the side and rear walls.

Another informative mapping plot shows how a given parameter changes with time. In Fig. 26-14, the sound-pressure level is plotted in four time intervals so that we can view how the sounds from the various loudspeakers over time combine to energize the room. If a reflection-free zone is created, it will show up in the upper right map as low sound-pressure level values around the listener position. In the same map, the effect of absorptive surfaces behind the front speakers can be seen as a lower sound-pressure level. By selecting the time windows well, the efficiency of a design can be studied.

Binaural Playback

In addition to analyzing acoustical parameters, we can also listen to what the room might sound like with the specified acoustical design and speaker configuration. To do this we combine the binaural impulse responses of the left and right ear, shown in Fig. 26-15, with music. For binaural auralization, the binaural impulse responses are derived from the room impulse response by convolution of each reflection with HRTFs, which are stored in the program from a mannequin and also from a real person. The next step is to select an anechoic or nonanechoic music sample and convolve it with the binaural

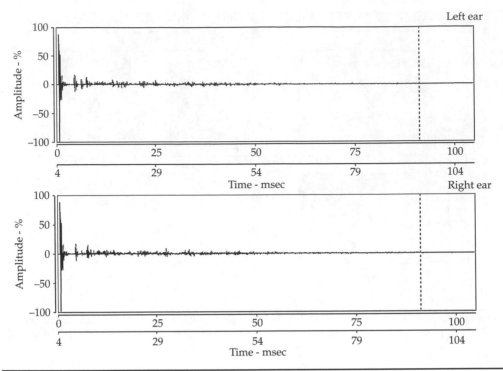

Figure 26-15 Convolution of the echogram reflections with the HRTFs results in a binaural room impulse response shown for the left ear and right ear.

impulse responses. An anechoic music sample can be used to auralize the sound. Since the music is anechoic, it does not contain any room signature. Therefore, convolving with the room response allows us to audition the room at various source and listening positions. A nonanechoic source might be used to determine the extent to which a sound reproduction room colors or influences what is heard by adding its own room signature to the sound. Thus by comparing the nonanechoic music sample with that heard in the room, we can make an evaluation of the coloration introduced.

Figure 26-16 illustrates the left and right ear files that can be listened to by head-phones. As previously indicated, several other auralization formats can be selected during the post-processing step.

Summary

We have described an approach using geometrical acoustics that generates the room impulse response of a virtual room. This impulse response can be post-processed for several types of auralization, including mono, stereo, 5.1-channel, binaural, and ambisonic. We can also auralize real rooms if we measure their impulse response.

Architectural drawings have evolved from plan and section, to 3D, to fully rendered image, to visual walk-throughs. Now we can also auralize the visually rendered environment; auralizations are possible for fixed locations and walk-throughs. Real-time auralizations can also be used for virtual reality applications. It should be reiterated

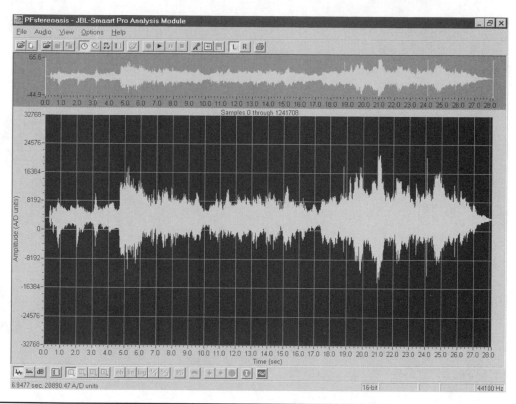

FIGURE 26-16   The convolution of the binaural room impulse responses with music generates the files for the left ear and right ear which can be played back for auralization. (*JBL*)

that geometrical acoustics methods presently used are more accurate at higher frequencies and they cannot accurately represent lower octaves. Wave acoustics can be combined with geometrical acoustics to extend the modeling to lower frequencies. Despite the many approximations involved, auralization is a valid and useful tool, especially in the hands of experienced acousticians.

Bibliography

Chapter 1   Fundamentals of Sound

Backus, J., *The Acoustical Foundations of Music*, Norton, 1969.
Benson, K.B., ed., *Audio Engineering Handbook*, McGraw-Hill, 1988.
Everest, F.A., *Acoustic Techniques for Home and Studio*, 2nd ed., Tab Books, 1984.
Rossing, T.D., R.F. Moore, and P.A. Wheeler, *The Science of Sound*, 3rd ed., Addison Wesley, 2001.
Talbot-Smith, M., ed., *Audio Engineer's Reference Book*, Focal Press, 1999.
Zahm, J.A., *Sound and Music*, A.C. McClurg and Company, 1892.

Chapter 2   Sound Levels and the Decibel

Baranek, L.L., *Acoustics*, McGraw-Hill, 1954.
Brown, P., "Fundamentals of Audio and Acoustics," in *Handbook for Sound Engineers*, ed. G.M. Ballou, 4th ed., Elsevier Focal Press, 2008.
Campbell, M. and C. Greated, *The Musician's Guide to Acoustics*, Oxford University Press, 1994.
Davis, D., and C. Davis, *Sound System Engineering*, Elsevier Focal Press, 1997.
Hall, D.E., *Musical Acoustics*, 2nd ed., Brooks/Cole, 1991.

Chapter 3   Sound in the Free Field

Rossing, T.D., ed., *Springer Handbook of Acoustics*, Springer, 2007.
Whitaker, J. and K.B. Benson, eds., *Standard Handbook of Audio and Radio Engineering*, 2nd ed., McGraw-Hill, 2002.
Woram, J.M., *Sound Recording Handbook*, Howard W. Sams, 1989.

Chapter 4   The Perception of Sound

Ashihara, K., "Hearing Thresholds for Pure Tones above 16 kHz," *JASA*, 122, pp. EL 52–57, 2007.
Blauert, J., *Spatial Hearing: The Psychophysics of Human Sound Localization*, MIT Press, 1996.

Blesser, B., and L.-R. Salter, *Spaces Speak, Are You Listening?*, MIT Press, 2007.

Bloom, P.J., "Creating Source Illusions by Special Manipulation," *JAES*, 25, pp. 560–565, 1977.

Djelani, T. and J. Blauert, "Investigations into the Build-Up and Breakdown of the Precedence Effect," *Acta Acustica*, 87, pp. 253–261, 2001.

Everest, F.A., "The Filters in Our Ears," *Audio*, 70:9, pp. 50–59, 1986.

Fletcher, H., "The Ear as a Measuring Instrument," *JAES*, 17:5, pp. 532–534, 1969.

Fletcher, H. and W.A. Munson, "Loudness, Its Definition, Measurement, and Calculations," *JASA*, 5, pp. 82–108, 1933.

Haas, H., "The Influence of a Single Echo on the Audibility of Speech," *JAES*, 20:2, pp. 146–159, 1972. (English translation from the German by K.P.R. Ehrenberg of Haas' original paper in *Acustica*, 1:2, 1951.

Handel, S., *Listening*, MIT Press, 1993.

Hartmann, W.M., *Signals, Sound, and Sensation*, Springer, 2005.

Hartmann, W.M. and B. Rakerd, "Localization of Sound in Rooms IV: The Franssen Effect," *JASA*, 86, pp. 1366–1373, 1999.

Hawkes, R.J. and H. Douglas, "Subjective Experience in Concert Auditoria," *Acustica*, 28:5, pp. 235–250, 1971.

ISO 226, "Acoustics—Normal Equal-Loudness Contours," 2003.

Litovsky, R.Y., H.S. Colburn, W.A. Yost, and S.J. Guzman, "The Precedence Effect," *JASA*, 106, pp. 1633–1654, 1999.

Lochner, J.P.A. and J.F. Burger, "The Subjective Masking of Short Time Delayed Echoes by Their Primary Sounds and Their Contribution to the Intelligibility of Speech," *Acustica*, 8, pp. 1–10, 1958.

Mehrgardt, S. and V. Mellert, "Transformation Characteristics of the External Human Ear," *JASA*, 61:6, pp. 1567–1576, 1977.

Meyer, E., "Physical Measurements in Rooms and Their Meaning in Terms of Hearing Conditions," *Proc. 2nd Int. Congress on Acoustics*, pp. 59–68, 1956.

Meyer, E. and G.R. Schodder, "On the Influence of Reflected Sounds on Directional Localization and Loudness of Speech," *Nachr. Akad. Wiss., Göttingen, Math., Phys., Klasse IIa*, 6, pp. 31–42, 1952.

Moore, B.C.J., *An Introduction to the Psychology of Hearing*, 5th ed., Academic Press, 2003.

Moore, B.C.J. and B. Glasberg, "Suggested Formulae for Calculating Auditory Filter Bandwidths and Excitation Patterns," *JASA*, 74:3, pp. 750–753, 1983.

Moore, B.C.J. and C-T. Tan, "Development and Validation of a Method for Predicting the Perceived Naturalness of Sounds Subjected to Spectral Distortion," *JAES*, 52, pp. 900–914, 2004.

Perrot, D.R., K. Marlborough, P. Merrill, and T.S. Strybel, "Minimum Audible Angle Thresholds Obtained under Conditions in Which the Precedence Effect Is Assumed to Operate," *JASA*, 85, pp. 282–288, 1989.

Plomb, R. and H.J.M. Steeneken, "Place Dependence of Timbre in Reverberant Sound Fields," *Acustica*, 28:1, pp. 50–59, 1973.

Robinson, D.W. and R.S. Dadson, "A Re-Determination of the Equal-Loudness Relations for Pure Tones," *British J. Appl. Psychology*, 7, pp. 166–181, 1956. (Adopted by the International Standards Organization as ISO-226.)

Schroeder, M.R., D. Gottlob, and K.F. Siebrasse, "Comparative Study of European Concert Halls: Correlation of Subjective Presence with Geometric and Acoustic Parameters," *JASA*, 56:4, pp. 1195–1201, 1974.

Stelmachowicz, P.G., K.A. Beauchaine, A. Kalberer, W.J. Kelly, and W. Jesteadt, "High-Frequency Audiometry: Test Reliability and Procedural Considerations," *JASA*, 85, pp. 879–887, 1989.

Stevens, S.S. and J. Volkman, "The Relation of Pitch to Frequency: A Revised Scale," *Am. J. Psychology*, 53, pp. 329–353, 1940.

Tobias, J.V., ed., *Foundations of Modern Auditory Theory,* Vol. 1, Academic Press, 1970.

Tobias, J.V., ed., *Foundations of Modern Auditory Theory,* Vol. 2, Academic Press, 1972.

Toole, F.E., "Loudness—Applications and Implications for Audio," *dB Magazine*, Part I: 7:5, pp. 7–30, 1973, Part II: 7:6, pp. 25–28, 1973.

Toole, F.E., "Loudspeakers and Rooms for Stereophonic Sound Reproduction," *Proc. AES 8th Intl. Conf.*, Washington, D.C., pp. 71–91, May, 1990.

Warren, R.M., *Auditory Perception: A New Analysis and Synthesis*, Cambridge University Press, 1999.

Zhang, P.X., "Psychoacoustics," in *Handbook for Sound Engineers*, ed. G.M. Ballou, 4th ed., Elsevier Focal Press, 2008.

Zwicker, C.G., G. Flottorp, and S.S. Stevens, "Critical Bandwidths in Loudness Summation," *JASA*, 29:5, pp. 548–557, 1957.

Chapter 5   Signals, Speech, Music, and Noise

Buff, P.C., "Perceiving Audio Noise and Distortion," *Recording Eng./Prod.*, 10:3, p. 84, June, 1979.

Cabot, R.C., "A Comparison of Nonlinear Distortion Measurement Methods," *AES 66th Conv.*, Los Angeles, preprint 1638, 1980.

Clark, D., "High-Resolution Subjective Testing Using a Double-Blind Comparator," *JAES*, 30:5, pp. 330–338, 1982.

Flanagan, J.L., "Voices of Men and Machines, Part 1," *JASA*, 51:5, pp. 1375–1387, 1972.

Fletcher, H., "The Ear as a Measuring Instrument," *JAES*, 17:5, pp. 532–534, 1969.

Hutchins, C.M. and F.L. Fielding, "Acoustical Measurement of Violins," *Physics Today*, pp. 35–41, July, 1968.

Jung, W.G., M.L. Stephens, and C.C. Todd, "An Overview of SID and TIM," *Audio*, Part I, 63:6, pp. 59–72, June, 1979, Part II, 63:7, pp. 38–47, July, 1979.

Kuttruff, H., *Room Acoustics,* Applied Science Publishers, Ltd., 1979.

Peterson, A.P.G. and E.E. Gross, Jr., *Handbook of Noise Measurements*, 7th ed., GenRadio, 1974.

Pohlmann, K.C., *Principles of Digital Audio*, 5th ed., McGraw-Hill, 2005.

Sivian, L.J., H.K. Dunn, and S.D. White, "Absolute Amplitudes and Spectra of Certain Musical Instruments and Orchestras," *JASA*, 2:3, pp. 330–371, 1931.

Chapter 6   Reflection

Knudsen, V.O. and C.M. Harris, *Acoustical Designing in Architecture*, Acoustical Society of America, 1978.

Lochner, J.P.A. and J.F. Burger, "The Subjective Masking of Short Time Delayed Echoes by Their Primary Sounds and Their Contribution to the Intelligibility of Speech," *Acustica*, 8, pp. 1–10, 1958.

Meyer, E. and G.R. Schodder, "On the Influence of Reflected Sound on Directional Localization and Loudness of Speech," *Nachr. Akad. Wiss., Göttingen; Math. Phys. Klasse IIa*, 6, pp. 31–42, 1952.

Olive, S.E. and F.E. Toole, "The Detection of Reflections in Typical Rooms," *JAES*, 37:7/8, pp. 539–553, 1989.

Toole, F.E., "Loudspeakers and Rooms for Stereophonic Sound Reproduction," *Proc. AES. 8th Intl. Conf.*, Washington, D.C., pp. 71–91, 1990.

Chapter 7   Diffraction

Kaufman, R.J., "With a Little Help from My Friends," *Audio*, 76:9, pp. 42–46, 1992.

Kessel, R.T., "Predicting Far-Field Pressures from Near-Field Loudspeaker Measurements," *JAES*, Abstract, 36, preprint 2729, p. 1026, 1988.

Muller, C.G., R. Black, and T.E. Davis, "The Diffraction Produced by Cylinders and Cubical Obstacles and by Circular and Square Plates," *JASA*, 10:1, p. 6, 1938.

Olson, H.F., *Elements of Acoustical Engineering*, Van Nostrand, 1940.

Rettinger, M., *Acoustic Design and Noise Control*, Chemical Publishing Co., 1973.

Vanderkooy, J., "A Simple Theory of Cabinet Edge Diffraction," *JAES*, 39:12, pp. 923–933, 1991.

Wood, A., *Acoustics*, Interscience Publishers, Inc., 1941.

Chapter 8   Refraction

Heaney, K.D., W.A. Kuperman, and B.E. McDonald, "Perth-Bermuda Sound Propagation (1960): Adiabatic Mode Interpretation," *JASA*, 90:5, pp. 2586–2594, 1991.

Shockley, R.C., J. Northrop, P.G. Hansen, and C. Hartdegen, "SOFAR Propagation Paths from Australia to Bermuda," *JASA*, 71:51, 1982.

Spiesberger, J., K. Metzger, and J.A. Ferguson, "Listening for Climatic Temperature Changes in the Northeast Pacific 1983-1989," *JASA*, 92:1, pp. 384–396, July, 1992.

Chapter 9   Diffusion

AES-4id-2001, *AES Information Document for Room Acoustics and Sound Reinforcement Systems—Characterization and Measurement of Surface Scattering Uniformity*, 2001. (See also ISO 17497-2.)

Angus, J.A.S., "Controlling Early Reflections Using Diffusion," *AES 102nd Conv.*, preprint 4405, 1997.

Angus, J.A.S., "The Effects of Specular versus Diffuse Reflections on the Frequency Response at the Listener," *AES 106th Conv.*, preprint 4938, 1999.

Boner, C.P., "Performance of Broadcast Studios Designed with Convex Surfaces of Plywood," *JASA*, 13, pp. 244–247, 1942.

Cox, T. and P. D'Antonio, *Acoustics, Absorbers and Diffusers*, Spon Press, 2004.

D'Antonio, P. and T. Cox, "Two Decades of Diffuser Design and Development, Part 1: Applications and Design, Part 2: Prediction, Measurement and Characterization," *JAES*, 46, pp. 955–976, 1075–1091, 1998.

Nimura, T. and K. Shibayama, "Effect of Splayed Walls of a Room on Steady-State Sound Transmission Characteristics," *JASA*, 29:1, pp. 85–93, 1957.

Randall, K.E. and F.L. Ward, "Diffusion of Sound in Small Rooms," *Proc. Inst. Elect. Engs.*, 107B, pp. 439–450, Sept., 1960.

Somerville, T. and F.L. Ward, "Investigation of Sound Diffusion in Rooms by Means of a Model," *Acustica*, 1:1, pp. 40–48, 1951.

van Nieuwland, J.M. and C. Weber, "Eigenmodes in Non-Rectangular Reverberation Rooms," *Noise Control Eng.*, 13:3, pp. 112–121, Nov./Dec., 1979.

Volkmann, J.E., "Polycylindrical Diffusers in Room Acoustical Design," *JASA*, 13, pp. 234–243, 1942.

Chapter 10   Comb-Filter Effects

Bartlett, B., "A Scientific Explanation of Phasing (Flanging)," *JAES*, 18:6, pp. 674–675, 1970.

Blauert, J., *Spatial Hearing*, MIT Press, 1983.

Burroughs, L, *Microphones: Design and Application*, Sagamore Publishing Co., 1974.

Moore, B.C.J. and B. Glasberg, "Suggested Formulae for Calculating Auditory-Filter Bandwidths," *JASA*, 74:3, pp. 750–753, 1983.

Streicher, R. and F.A. Everest, *The New Stereo Soundbook*, 3rd ed., Audio Engineering Associates, 2006.

Chapter 11   Reverberation

Arau-Puchades, H., "An Improved Reverberation Formula," *Acustica*, 65, pp. 163–180, 1988.

Balachandran, C.G., "Pitch Changes during Reverberation Decay," *J. Sound and Vibration*, 48:4, pp. 559–560, 1976.

Beranek, L.L., "Broadcast Studio design," *J. SMPTE*, 64, pp. 550–559, Oct., 1955.

Beranek, L.L., *Concert and Opera Halls—How They Sound*, Acoustical Society of America, 1996.

Beranek, L.L., *Music, Acoustics, and Architecture*, John Wiley and Sons, 1962.

Dalenback, B.-I., "Reverberation Time, Diffuse Reflection, Sabine, and Computerized Prediction—Part I," http://rpginc.com/research/reverb01.htm.

D'Antonio, P. and D. Eger, "T_{60}—How Do I Measure Thee, Let Me Count the Ways," *AES 81st Conv.*, preprint 2368, 1986.

Everest, F.A., *Acoustic Techniques for Home and Studio*, 2nd ed., Tab Books, 1984.

Jackson, G.M. and H.G. Leventhall, "The Acoustics of Domestic Rooms," *Applied Acoustics*, 5, pp. 265–277, 1972.

Klein, W., "Articulation Loss of Consonants as a Basis for the Design and Judgment of Sound Reinforcement Systems," *JAES*, 19:11, pp. 920–922, 1971.

Kuttruff, H., *Room Acoustics*, 4th ed., Spon Press, 2000.

Long, M., *Architectural Acoustics*, Elsevier Academic Press, 2006.

Mankovsky, V.S., *Acoustics of Studios and Auditoria*, Focal Press, 1971.

Matsudaira, T.K., et al., "Fast Room Acoustic Analyzer (FRA) Using Public Telephone Line and Computer," *JAES*, 25:3, pp. 82–94, 1977.

Olive, S.E. and F.E. Toole, "The Detection of Reflections in Typical Rooms," *JAES*, 37, pp. 539–553, 1989.

Parker, S.P., ed., *Acoustics Source Book*, McGraw-Hill, 1988.

Peutz, V.M.A., "Articulation Loss of Consonants as a Criterion for Speech Transmission in a Room," *JAES*, 19:11, pp. 915–919, 1971.

Schroeder, M.R., "New Method of Measuring Reverberation Time," *JASA*, 37, pp. 409–412, 1965.

Schultz, T.J., "Problems in the Measurement of Reverberation Time," *JAES*, 11:4, pp. 307–317, 1963.

Spring, N.F. and K.E. Randall, "Permissible Bass Rise in Talk Studios," *BBC Engineering*, 83, pp. 29–34, 1970.

Young, R.W., "Sabine Equation and Sound Power Calculations," *JASA*, 31:12, p. 1681, 1959.

Chapter 12   Absorption

Acoustical Ceilings: Use and Practice, Ceiling and Interior Systems Contractors Association, 1978.

Beranek, L.L., *Acoustics*, McGraw-Hill, 1954.

Bradley, J.S., "Sound Absorption of Gypsum Board Cavity Walls," *JAES*, 45, pp. 253–259, 1997.

Brüel, P.V., *Sound Insulation and Room Acoustics,* Chapman and Hall, 1951.

Callaway, D.B. and L.G. Ramer, "The Use of Perforated Facings in Designing Low Frequency Resonant Absorbers," *JASA*, 24:3, pp. 309–312, 1952.

Cox, T. and P. D'Antonio, *Acoustics, Absorbers and Diffusers*, Spon Press, 2004.

Davern, W.A., "Perforated Facings Backed with Porous Materials as Sound Absorbers—An Experimental Study," *Applied Acoustics*, 10, pp. 85–112, 1977.

Evans, E.J. and E.N. Bazley, *Sound Absorbing Materials*, National Physical Laboratories, 1960.

Everest, F.A., "The Acoustic Treatment of Three Small Studios," *JAES*, 15:3, pp. 307–313, 1968.

Gilford, C., *Acoustics for Radio and Television Studios*, Peter Peregrinus, Ltd., 1972.

Harris, C.M., "Acoustical Properties of Carpet," *JASA*, 27:6, pp. 1077–1082, 1955.

Ingard, U. and R.H. Bolt, "Absorption Characteristics of Acoustic Materials with Perforated Facings," *JASA*, 23:5, pp. 533–540, 1951.

Kingsbury, H.F. and W.J. Wallace, "Acoustic Absorption Characteristics of People," *J. Sound and Vibration*, 2:2, 1968.

Mankovsky, V.S., *Acoustics of Studios and Auditoria*, Focal Press, Ltd., 1971.

Noise Control Design Guide, www.owenscorning.com.

Noise Control Manual, Owens-Corning Fiberglas Corp., publication No. 5-BMG-8277-A., 1980.

Rettinger, M., "Bass Traps," *Recording Eng./Prod.*, 11:4, pp. 46–51, Aug., 1980.

Rettinger, M., "Low-Frequency Slot Absorbers," *db Magazine*, 10:6, pp. 40–43, June, 1976.

Rettinger, M., "Low Frequency Sound Absorbers," *db Magazine*, 4:4, pp. 44–46, Apr., 1970.

Sabine, P.E. and L.G. Ramer, "Absorption-Frequency Characteristics of Plywood Panels," *JASA*, 20:3, pp. 267–270, May, 1948.

Schultz, T.J. and B.G. Watters, "Propagation of Sound Across Audience Seating," *JASA*, 36:5, pp. 885–896, May, 1964.

Siekman, W., "Private Communication," Riverbank Acoustical Laboratories, measurements reported to Acoustical Society of America, Apr., 1969.

Szymanski, J., "Acoustical Treatment for Indoor Areas," in *Handbook for Sound Engineers*, ed. G.M. Ballou, 4th ed., Elsevier Focal Press, 2008.

Young, R.W., "On Naming Reverberation Equations," *JASA*, 31:12, p. 1681, Dec., 1959.

Young, R.W., "Sabine Reverberation and Sound Power Calculations," *JASA*, 31:7, pp. 912–921, July, 1959.

Chapter 13   Modal Resonances

Blesser, B. and L.R. Salter, *Spaces Speak, Are You Listening?*, MIT Press, 2007.

Bolt, R.H., "Note on Normal Frequency Statistics for Rectangular Rooms," *JASA*, 18:1, pp. 130–133, July, 1946.

Bolt, R.H., "Perturbation of Sound Waves in Irregular Rooms," *JASA*, 13, pp. 65–73, July, 1942.

Bonello, O.J., "A New Computer-Aided Method for the Complete Acoustical Design of Broadcasting and Recording Studios," *IEEE Intl. Conf. Acoustics and Signal Processing, ICASSP 79*, Washington, pp. 326–329, 1979.

Bonello, O.J., "A New Criterion for the Distribution of Normal Room Modes," *JAES*, 29:9, pp. 597–606, Sept., 1981. (Correction, *JAES*, 29:12, p. 905, 1981).

Brüel, P.V., *Sound Insulation and Room Acoustics*, Chapman and Hall, 1951.

Cox, T., P. D'Antonio, and M.R. Avis, "Room Sizing and Optimization at Low Frequencies," *JAES*, 52, pp. 640–651, 2004.

Everest, F.A., *Acoustic Techniques for Home and Studio*, 2nd ed., Tab Books, 1984.

Fazenda, B.M., M.R. Avis, and W.J. Davies, "Perception of Modal Distribution Metrics in Critical Listening Spaces—Dependence on Room Aspect Ratios," *JAES*, 53:12, pp. 1128–1141, Dec., 2005.

Gilford, C.L.S., "The Acoustic Design of Talk Studios and Listening Rooms," *Proc. Inst. Elect. Engs.*, 106, Part B, 27, pp. 245–258, May, 1959. Reprinted in *JAES*, 27:1/2, pp. 17–31, 1979.

Hunt, F.V., "Investigation of Room Acoustics by Steady-State Transmission Measurements," *JASA*, 10, pp. 216–227, Jan., 1939.

Hunt, F.V., L.L. Beranek, and D.Y. Maa, "Analysis of Sound Decay in Rectangular Rooms," *JASA*, 11, pp. 80–94, July, 1939.

Knudsen, V.O., "Resonances in Small Rooms," *JASA*, pp. 20–37, July, 1932.

Louden, M.M., "Dimension-Ratios of Rectangular Rooms with Good Distribution of Eigentones," *Acustica*, 24, pp. 101–103, 1971.

Mayo, C.G., "Standing-Wave Patterns in Studio Acoustics," *Acustica*, 2:2, pp. 49–64, 1952.

Meyer, E., "Physical Measurements in Rooms and Their Meaning in Terms of Hearing Conditions," *Proc. 2nd Intl. Congress on Acoustics*, pp. 59–68, 1956.

Morse, P.M. and R.H. Bolt, "Sound Waves in Rooms," *Review of Modern Physics*, 16:2, pp. 69–150, Apr., 1944.

Sepmeyer, L.W., "Computed Frequency and Angular Distribution of the Normal Modes of Vibration in Rectangular Rooms," *JASA*, 37:3, pp. 413–423, Mar., 1965.

van Leeuwen, F.J., "The Damping of Eigentones in Small Rooms by Helmholtz Resonators," *European Broadcast Union Review*, A, 62, pp. 155–161, 1960.

Walker, R., "Optimum Dimensional Ratios for Studios, Control Rooms and Listening Rooms," *BBC Research Department*, Report No. BBC RD 1993/8, 1993. http://www.bbc.co.uk/rd/pubs/index.shtml.

Chapter 14   Schroeder Diffusers

Acoustics—Sound-Scattering Properties of Surfaces—Part 1: Measurement of the Random-Incidence Scattering Coefficient in a Reverberation Room, ISO 17497-1:2004.

AES Information Document for Room Acoustics and Sound Reinforcement System—Characterization and Measurement of Surface Scattering Uniformity, AES, 2001.

Beiler, A.H., *Recreations in the Theory of Numbers*, 2nd ed., Dover Publications, Inc., 1966.

Berkout, D.W., van W. Palthe, and D. deVries, "Theory of Optimal Plane Diffusors," *JASA*, 65:5, pp. 1334–1336, May, 1979.

D'Antonio, P., "The Reflection-Phase-Grating Acoustical Diffusor: Diffuse It or Lose It," *dB Magazine*, 19:5, pp. 46–49, Sept./Oct., 1985.

D'Antonio, P. and J.H. Konnert, "Advanced Acoustic Design of Stereo Broadcast and Recording Facilities," *1986 NAB Eng. Conf. Proc.*, pp. 215–223, 1986.

D'Antonio, P. and J.H. Konnert, "Incorporating Reflection-Phase-Grating Diffusors in Worship Spaces," *AES 81st Conv.*, Los Angeles, preprint 2364, Nov., 1986.

D'Antonio, P. and J.H. Konnert, "New Acoustical Materials Improve Broadcast Facility Design," *1987 NAB Eng. Conf. Proc.*, pp. 399–406, 1987.

D'Antonio, P. and J.H. Konnert, "The Acoustical Properties of Sound Diffusing Surfaces: The Time, Frequency, and Directivity Energy Response," *AES 79th Conv.*, New York, preprint 2295, Oct., 1985.

D'Antonio, P. and J.H. Konnert, "The Directional Scattering Coefficient: Experimental Determination," *JAES*, 40:12, pp. 997–1017, Dec., 1992.

D'Antonio, P. and J.H. Konnert, "The Reflection Phase Grating: Design Theory and Application," *JAES*, 32:4, pp. 228–238, Apr., 1984.

D'Antonio, P. and J.H. Konnert, "The QRD Diffractal: A New One- or Two-Dimensional Fractal Sound Diffusor," *JAES*, 40:3, pp. 117–129, Mar., 1992.

D'Antonio, P. and J.H. Konnert, "The RPG Reflection-Phase-Grating Acoustical Diffusor: Applications," *AES 76th Conv.*, New York, preprint 2156, Oct., 1984.

D'Antonio, P. and J.H. Konnert, "The RPG Reflection-Phase-Grating Acoustical Diffusor: Experimental Measurements," *AES 76th Conv.*, New York, preprint 2158, Oct., 1984.

D'Antonio, P. and J.H. Konnert, "The Schroeder Quadratic-Residue Diffusor: Design Theory and Application," *AES 74th Conv.*, New York, preprint 1999, Oct., 1983.

Davenport, H., *The Higher Arithmetic*, Dover Publications, Inc., 1983.

deJong, B.A. and P.M. van den Berg, "Theoretical Design of Optimum Planar Sound Diffusors," *JASA*, 68:4, pp. 1154–1159, Oct., 1980.

Mackenzie, R., *Auditorium Acoustics*, Applied Science Publishers, Ltd., 1975.

Schroeder, M.R., "Binaural Dissimilarity and Optimum Ceilings for Concert Halls: More Lateral Sound Diffusion," *JASA*, 65:4, pp. 958–963, Apr., 1979.

Schroeder, M.R., "Diffuse Sound Reflection by Maximum-Length Sequences," *JASA*, 57:1, pp 149–151, Jan., 1975.

Schroeder, M.R., *Number Theory in Science and Communication*, 2nd ed., Springer, 1986.

Schroeder, M.R., and R.E. Gerlach, "Diffuse Sound Reflection Surfaces," *Proc. 9th Intl. Congress on Acoustics*, Madrid, paper D-8, 1977.

Strube, H.W., "Scattering of a Plane Wave by a Schroeder Diffusor: A Mode-Matching Approach," *JASA*, 67:2, pp. 453–459, Feb., 1980.

Chapter 15   Adjustable Acoustics

Brüel, P.V., *Sound Insulation and Room Acoustics,* Chapman and Hall, 1951.

Snow, W.B., "Recent Application of Acoustical Engineering Principles in Studios and Review Rooms," *J. SMPTE,* 70:1, pp. 33–38, Jan., 1961.

Chapter 16   Control of Interfering Noise

Berger, R. and T. Rose, "Partitions," *Mix magazine,* 9:10, Oct., 1985.

Crocker, M.J., ed., *Handbook of Noise and Vibration Control,* John Wiley and Sons, 2007.

Everest, F.A., "Glass in the Studio," *dB Magazine* Part I, 18:3, pp. 28–33, Apr., 1984, Part II, 18:4, pp. 41–44, May, 1984.

Green, D.W. and C.W. Sherry, "Sound Transmission Loss of Gypsum Wallboard Partitions, Report No. 3, 2x4 Wood Stud Partitions," *JASA,* 71:4, pp. 908–914, 1982.

Harris, C.M., *Handbook of Acoustical Measurements and Noise Control,* American Institute of Physics, 1998.

Jones, D., "Acoustical Noise Control," in *Handbook for Sound Engineers,* ed. G.M. Ballou, 4th ed., Elsevier Focal Press, 2008.

Jones, R.E., "How to Design Walls of Desired STC Ratings," *J. Sound and Vibration,* 12:8, pp. 14–17, 1978.

OSHA Regulation, Federal Register, 39:207, Oct. 24, 1974.

Rettinger, M., *Handbook of Architectural Acoustics and Noise Control,* TAB Books, 1988.

U.S. Department of Housing and Urban Development, *The Noise Guide Book,* HUD-953-CPD, Washington, D.C., 1985.

U.S. Department of Transportation, Federal Highway Administration, *Summary of State Highway Agency Noise Planning Definitions,* Office of Environment and Planning, Washington, D.C., 1991.

Ver, I.L. and L.L. Benanek, eds. *Noise and Vibration Control Engineering: Principles and Applications,* John Wiley and Sons, 2005.

Yerges, L.F., *Sound, Noise & Vibration Control,* Van Nostrand Reinhold, 1978.

Chapter 17   Noise Control in Ventilating Systems

American Society of Heating, Refrigerating and Air-Conditioning Engineers, *ASHRAE Handbook of Fundamentals,* 1993.

Beranek, L.L., "Balanced Noise-Criterion (NCB) Curves," *JASA,* 86:2, pp. 650–664, Aug., 1989.

Broner, N., "Rating and Assessment of Noise," *Australian Institute of Refrigeration, Air Conditioning and Heating Conf.,* www.airah.org.au, 2004.

Doelling, N., "How Effective Are Packaged Attenuators?," *ASHRAE Journal,* 2:2, pp. 46–50, Feb., 1960.

Harris, C.M., ed., *Handbook of Acoustical Measurements and Noise Control,* 3rd ed., McGraw-Hill, 1991.

Knudsen, V.O. and C.M. Harris, *Acoustical Designing in Architecture,* Acoustical Society of America, 1978.

Rettinger, M., *Handbook of Architectural Acoustics and Noise Control,* TAB Books, 1988.

Sanders, G.J., "Silencers: Their Design and Application," *J. Sound and Vibration,* 2:2, pp. 6–13, Feb., 1968.

Soulodre, G.A., "Evaluation of Objective Loudness Meters," *AES 116th Conv.*, preprint 6161, 2004.

Soulodre, G.A. and S.G. Norcross, "Objective Measures of Loudness," *AES 115th Conv.*, preprint 5896, 2003.

Tocci, G.C., "Room Noise Criteria—The State of the Art in the Year 2000," www.cavtocci .com/portfolio/publications/tocci.pdf, 2000.

Chapter 18   Acoustics of Listening Rooms

Allison, R.F. and R. Berkowitz, "The Sound Field in Home Listening Rooms," *JAES*, 20:6, pp. 459–469, July/Aug., 1972.

Benjamin, E. and B. Gannon, "Effect of Room Acoustics on Subwoofer Performance and Level Setting," *AES 109th Conv.*, preprint 5232, 2000.

Celestrinos, A. and S.B. Nielsen, "Low Frequency Sound Field Enhancement System for Rectangular Rooms Using Multiple Low Frequency Loudspeakers," *AES 120th Conv.*, preprint 6688, 2006.

Celestrinos, A. and S.B. Nielsen, "Optimizing Placement and Equalization of Multiple Low Frequency Loudspeakers in Rooms," *AES, 119th Conv.*, preprint 6545, 2005.

Cox, T., P. D'Antonio, and M.R. Avis, "Room Sizing and Optimization at Low Frequencies," *JAES*, 52, pp. 640–651, 2004.

Cremer, L. and H.A. Muller, *Principles and Applications of Room Acoustics*, Vols. 1 and 2, Applied Science Publishers, 1982.

Fiedler, L.D., "Dynamic-Range Requirements for Subjectively Noise-Free Reproduction of Music," *JAES*, 30:7/8, pp. 504–511, 1982.

ITU-R BS.775-2, *Multichannel Stereophonic Sound System With and Without Accompanying Picture*, International Telecommunication Union/ITU Radiocommunication Sector, 2006.

Olive, S.E. and F.E. Toole, "The Detection of Reflections in Typical Rooms," *JAES*, 37:7/8, pp. 539–553, July/Aug., 1989.

Toole, F.E., "Loudspeaker and Rooms for Stereophonic Sound Reproduction," *AES 8th Intl. Conf.*, Washington, D.C., paper 18-001, 1990.

Voetmann, J. and J. Klinkby, "Review of the Low-Frequency Absorber and Its Application to Small Room Acoustics," *AES 94th Conv.*, preprint 3578, 1993.

Chapter 19   Acoustics of Small Recording Studios

Bech, S., "Spatial Aspects of Reproduced Sound in Small Rooms," *JASA*, 103, pp. 434–445, 1998.

Bech, S., "Timbral Aspects of Reproduced Sound in Small Rooms II," *JASA*, 99, pp. 3539–3549, 1996.

Beranek, L.L., *Acoustics*, Acoustical Society of America, 1986.

Beranek, L.L., *Music, Acoustics and Architecture*, John Wiley and Sons, 1962.

Gilford, C.L.S., "The Acoustic Design of Talk Studios and Listening Rooms," *Proc. Inst. Elect. Engs.*, 106, part B, 27, pp. 245–258, May, 1959. Reprinted in *JAES*, 27:1/2, pp. 17–31, 1979.

Jones, D., "Small Room Acoustics," in *Handbook for Sound Engineers*, ed. G.M. Ballou, 4th ed., Elsevier Focal Press, 2008.

Kuhl, W., "Optimal Acoustical Design of Rooms for Performing, Listening, and Recording," *Proc. 2nd. Intl. Congress on Acoustics*, pp. 53–58, 1956.

Kuttruff, H., "Sound Fields in Small Rooms," *AES 15th Intl. Conf.*, paper 15-002, 1998.

Chapter 20   Acoustics of Control Rooms

Augspurger, G.L., "Loudspeakers in Control Rooms and Living Rooms," *AES 8th Intl. Conf.*, Washington, D.C., 1990.

Beranek, L.L., *Music, Acoustics, and Architecture*, John Wiley and Sons, 1962.

Berger, R.E., "Speaker/Boundary Interference Response (SBIR)," Synergetic Audio Concepts Tech Topics, Winter, 11:5 p. 6, 1984.

D'Antonio, P., "Control-Room Design Incorporating RFZ™, LFD™, and RPF™ Diffusors," *dB Magazine*, 20:5, pp. 47–55, Sept./Oct., 1986.

D'Antonio, P. and J.H. Konnert, "New Acoustical Materials and Designs Improve Room Acoustics," *AES 81st Conv.*, preprint 2365, 1986.

D'Antonio, P. and J.H. Konnert, "The RFZ™/RPG™ Approach to Control Room Monitoring," *AES 76th Conv.*, preprint 2157, 1984.

D'Antonio, P. and J.H. Konnert, "The Role of Reflection-Phase-Grating Diffusors in Critical Listening and Performing Environments," *AES 78th Conv.*, Anaheim, CA, preprint 2255, May, 1985.

Davis, D., "The Role of the Initial Time-Delay Gap in the Acoustic Design of Control Rooms for Recording and Reinforcing Systems," *AES 64th Conv.*, preprint 1574, 1979.

Davis, D. and C. Davis, "The LEDE-Concept for the Control of Acoustic and Psychoacoustic Parameters in Recording Control Rooms," *JAES*, 28:9, pp. 585–595, Sept., 1980.

Davis, C. and G.E. Meeks, "History and Development of the LEDE™ Control-Room Concept," *AES 64th Conv.*, preprint 1954, 1982.

Muncy, N.A., "Applying the Reflection-Free Zone RFZ™ Concept in Control-Room Design," *dB Magazine*, 20:4, pp. 35–39, July/Aug., 1986.

Chapter 21   Acoustics of Audio/Video Rooms

Everest, F.A., *Acoustic Techniques for Home and Studio*, 2nd ed., Tab Books, 1984.

Gilford, C., *Acoustics for Radio and Television Studios*, Peter Peregrinus, Ltd., 1972.

Noxon, A.M., "Sound Fusion and the Acoustic Presence Effect," *AES 89th Conv.*, preprint 2998, Sept., 1990.

Olive, S.E. and F.E. Toole, "The Detection of Reflections in Typical Rooms," *JAES*, 37:7/8, pp. 539–553, 1989.

Chapter 22   Acoustics of Large Halls

Ahnert, W. and H.P. Tennhardt, "Acoustics for Auditoriums and Concert Halls," in *Handbook for Sound Engineers*, ed. G.M. Ballou, 4th ed., Elsevier Focal Press, 2008.

American National Standards Institute, *American National Standard Method for the Calculation of the Articulation Index*, ANSI S3.5-1969, p. 7.

Ando, Y., *Concert Hall Acoustics*, Springer-Verlag, 1985.

Ando, Y., "Subjective Preference in Relation to Objective Parameters of Music Sound Fields with a Single Echo," *JASA*, 62:6, pp. 1436–1441, Dec., 1977.

Ando, Y., Sakai, H., and S. Sato, "Formulae Describing Subjective Attributes for Sound Fields Based on a Model of the Auditory-Brain System," *J. Sound and Vibration*, 232:1, pp. 101–127, 2000.

Barron, M., "Measured Early Lateral Energy Fractions in Concert Halls and Opera Houses," *J. Sound and Vibration*, 232:1, pp. 79–100, 2000.

Barron, M. and A.H. Marshall, "Spatial Impression due to Early Lateral Reflections in Concert Halls: The Derivation of a Physical Measure," *J. Sound and Vibration*, 77:2, pp. 211–232, 1981.

Barron, M., "The Subjective Effects of First Reflections in Concert Halls—The Need for Lateral Reflections," *J. Sound and Vibration*, 15:4, pp. 475–494, 1971.

Beranek, L.L., "Audience and Chair Absorption in Large Halls," *JASA*, 45:1, pp. 13–19, 1969.

Beranek, L.L., *Concert Halls and Opera Houses*, 2nd ed., Springer-Verlag, 2004.

Bradley, J.S., "Experience with New Auditorium Acoustic Measurements," *JASA*, 73:6, pp. 2051–2058, 1983.

Bradley, J.S., "Some Further Investigations of the Seat Dip Effect," *JASA*, 90:1, pp. 324–333, 1991.

Bradley, J.S. and G.A. Soulodre, "The Influence of Late Arriving Energy on Spatial Impression," *JASA*, 97:4, pp. 2263–2271, 1995.

Bradley, J.S., R.D. Reich, and S.G. Norcross, "On the Combined Effects of Early- and Late-Arriving Sound on Spatial Impression in Concert Halls," *JASA*, 108:2, pp. 651–661, 2000.

Bradley, J.S., H. Sato, and M. Picard, "On the Importance of Early Reflections for Speech in Rooms," *JASA*, 113:6, pp. 3233–3244, 2003.

Cowan, J., *Architectural Acoustics Design Guide*, McGraw-Hill, 2000.

Davies, W.J., T.J. Cox, and Y.W. Lam, "Subjective Perception of Seat Dip Attenuation," *Acustica*, 82:5, pp. 784–792, 1996.

Egan, M.D., *Architectural Acoustics*, McGraw-Hill, 1988.

Griesinger, D., "General Overview of Spatial Impression, Envelopment, Localization, and Externalization," *AES 15th Intl. Conf.*, paper 15-013, 1998.

Griesinger, D., "Objective Measures of Spaciousness and Envelopment," *AES 16th Intl. Conf.*, paper 16-003, 1999.

Johnson, V.L., *Acoustical Design of Concert Halls and Theaters*, Applied Science Publishers, 1980.

Knudsen, V.O. and C.M. Harris, *Acoustical Designing in Architecture*, Acoustical Society of America, 1978.

Kuttruff, H., *Room Acoustics*, Applied Science Publishers, 1979.

Kwon, Y. and G.W. Siebein, "Chronological Analysis of Architectural and Acoustical Indices in Music Performance Halls," *JASA*, 121:5, pp. 2691–2699, 2007.

Marshall, L., *Architectural Acoustics*, Elsevier Academic Press, 2006.

Mehta, M., J. Johnson, and J. Rocafort, *Architectural Acoustics Principles and Design*, Prentice-Hall, 1999.

Nickson, A.F.B., R.W. Muncey, and P. Dubout, "The Acceptability of Artificial Echoes with Reverberant Speech and Music," *Acustica*, 4, pp. 515–518, 1954.

Sato, H., J.S. Bradley, and M. Masayuki, "Using Listening Difficulty Ratings of Conditions for Speech Communication in Rooms," *JASA*, 117:3, pp. 1157–1167, 2005.

Schroeder, M.R., "Toward Better Acoustics for Concert Halls," *Physics Today*, 33:10, pp. 24–30, 1980.

Schroeder, M.R., D. Gottlob, and K.F. Siebrasse, "Comparative Study of European Concert Halls," *JASA*, 56:4, pp. 1195–1201, 1974.

Schultz, T.J. and B.G. Watters, "Propagation of Sound Across Audience Seating," *JASA*, 36:5, pp. 885–896, 1964.

Soulodre, G.A., N. Popplewell, and J.S. Bradley, "Combined Effects of Early Reflections and Background Noise on Speech Intelligibility," *J. Sound and Vibration*, 135:1, pp. 123–133, 1989.

Talaske, R.H., E.A. Wetherill, and W.J. Cavanaugh, *Halls for Music Performance, Two Decades of Experience: 1962-1982*, American Institute of Physics for the Acoustical Society of America, 1982.

Toole, F.E., *Sound Reproduction*, Elsevier Focal Press, 2008.

Watkins, A.J., "Perceptual Compensation for Effects of Echo and of Reverberation on Speech Identification," *Acta Acustica*, 91, pp. 892–901, 2005.

Chapter 23   Acoustic Distortion

Cox, T.J. and P. D'Antonio, *Acoustic Absorbers and Diffusors: Theory, Design and Application*, Spon Press, 2004.

D'Antonio, P. and J.H. Konnert, "The RFZ™/RPG™ Approach to Control Room Monitoring," *AES 76th Conv.*, preprint 2157 (I-6), Oct., 1984.

Davis, D. and C. Davis, "The LEDE™ Concept for the Control of Acoustic and Psychoacoustic Parameters in Recording Control Rooms," *JAES*, 28:9, pp. 585–595, 1980.

Davis, D. and C. Davis, *Sound System Engineering*, 2nd ed., Howard W. Sams, pp. 168–169, 1987.

Haan, C.N. and F.R. Fricke, "Surface Diffusivity as a Measure of the Acoustic Quality of Concert Halls," *Proc. of Australia and New Zealand Architectural Science Association Conference*, Sydney, 1993.

Haan, C.N. and F.R. Fricke, "The Use of Neural Network Analysis for the Prediction of Acoustic Quality of Concert Halls," *Proc. of WESTPRAC V '94*, Seoul, pp. 543–550, 1994.

Welti, T. and A. Devantier, "Low-Frequency Optimization Using Multiple Subwoofers," *JAES*, 54:5, pp. 347–364, 2006.

Chapter 24   Room Acoustics Measurement Software

Dunn, C. and M.O. Hawksford, "Distortion Immunity of MLS-Derived Impulse Measurements," *JAES*, 41:5, pp. 314–335, May, 1993. (Correction: *JAES*, 42:3 p. 152, 1994).

Haykin, S., *An Introduction to Analogue and Digital Communications*, John Wiley and Sons, 1989.

Heyser, R.C., "Acoustical Measurements by Time Delay Spectrometry," *JAES*, 15:4, pp. 370–382, 1967.

Rife, D.D., "Transfer Function Measurement with Maximum Length Sequences," *JAES*, 37:6, pp. 419–444, 1989.

Toole, F.E., "Loudspeaker Measurements and Their Relationship to Listener Preferences, Part 2," *JAES*, 34:5, pp. 323–348, 1986.

Toole, F.E., "Loudspeakers and Rooms for Stereophonic Sound Reproduction," *AES 8th Intl. Conf.*, Washington, D.C., preprint 1989, 1990.

Vanderkooy, J., "Another Approach to Time Delay Spectrometry," *JAES*, 34:7/8, pp. 523–538, July/Aug., 1986.

Chapter 25   Room Optimizer

AES20-1996: AES Recommended Practice for Professional Audio—Subjective Evaluation of Loudspeakers, JAES, 44:5, pp. 386–401, 1996.

Allen, J.B. and D.A. Berkeley, "Image Method for Efficiently Simulating Small-Room Acoustics," *JASA*, 65:4, pp. 943–950, 1979.

Allison, R.F., "The Influence of Room Boundaries on Loudspeaker Power Output," *JAES*, 22:5, pp. 314–320, May, 1974.

Allison, R.F., "The Sound Field in Home Listening Rooms II," *JAES*, 24:1, pp. 14–19, Jan./Feb., 1976.

Ballagh, K.P., "Optimum Loudspeaker Placement Near Reflecting Planes," *JAES*, 31:12, pp. 931–935, Dec., 1983. (Letters to the Editor, *JAES*, 31:9, p. 677, Sept., 1984.)

Bonello, O.J., "A New Criterion for the Distribution of Normal Room Modes," *JAES*, 29:9, pp. 597–606, Sept., 1981. (Correction, *JAES*, 29:12, p. 905, Dec., 1981.)

Cox, T.J., "Designing Curved Diffusors for Performance Spaces," *JAES*, 44:5, pp. 354–364, May, 1996.

Cox, T.J., "Optimization of Profiled Diffusors," *JASA*, 97:5, pp. 2928–2936, May, 1995.

D'Antonio, P., C. Bilello, and D. Davis, "Optimizing Home Listening Rooms," *AES 85th Conv.*, Los Angeles, preprint 2735, Nov., 1988.

Groh, A.R., "High Fidelity Sound System Equalization by Analysis of Standing Waves," *JAES*, 22:10, pp. 795–799, Dec., 1974.

Kovitz, P., *Extensions to the Image Method Model of Sound Propagation in a Room*, Pennsylvania State University Thesis, Dec., 1994.

Lam, Y.W. and D.C. Hodgson, "The Prediction of the Sound Field due to an Arbitrary Vibrating Body in a Rectangular Enclosure," *JASA*, 88:4, pp. 1993–2000, Oct., 1990.

Louden, M.M., "Dimension-Ratios of Rectangular Rooms with Good Distribution of Eigentones," *Acustica*, 24, pp. 101–104, 1971.

Morse, P.M. and K.U..Ingard, *Theoretical Acoustics*, McGraw-Hill, 1968.

Olive, S.E. "A Method for Training of Listeners and Selecting Program Material for Listening Tests," *AES 97th Conv.*, preprint 3893, Nov., 1994.

Olive, S.E., P. Schuck, S. Sally, and M. Bonneville, "The Effects of Loudspeaker Placement on Listener Preference Ratings," *JAES*, 42:9, pp. 651–669, Sept., 1994.

Press, W.H., B.P. Flannery, S.A. Teukolsky, and W.T. Vetterling, *Numerical Recipes, The Art of Scientific Computing*, 2nd ed., Cambridge University Press, 1992.

Schuck, P.L., S. Olive, J. Ryan, F.E. Toole, S. Sally, M. Bonneville, V. Verreault, and K. Momtohan, "Perception of Reproduced Sound in Rooms: Some Results from the Athena Project," *AES 12th Intl. Conf.*, pp. 49–73, June, 1993.

Stephenson, U., "Comparison of the Mirror Image Source Method and the Sound Particle Simulation Method," *Applied Acoustics*, 29, pp. 35–72, 1990.

Toole, F.E., "Loudspeaker Measurements and Their Relationship to Listener Preferences, Part 1" *JAES*, 34:4, pp. 227–235, Apr., 1986.

Toole, F.E., "Loudspeaker Measurements and Their Relationship to Listener Preferences, Part 2," *JAES*, 34:5, pp. 323–348, May, 1986.

Toole, F.E., "Subjective Evaluation," in *Loudspeaker and Headphone Handbook,* ed. J. Borwick, 2nd ed., Elsevier Focal Press, 1994.

Toole, F.E. and S.E. Olive, "Hearing is Believing vs. Believing Is Hearing: Blind vs. Sighted Listening Tests and Other Interesting Things," *AES 97th Conv.*, preprint 3894, Nov., 1994.

Toole, F.E. and S.E. Olive, "The Modification of Timbre by Resonances: Perception and Measurement," *JAES*, 36:3, pp. 122–142, Mar., 1988.

Waterhouse, R.V., "Output of a Sound Source in a Reverberation Chamber and Other Reflecting Environments," *JASA*, 30:1, pp. 4–13, Mar., 1965.

Waterhouse, R.V., "Output of a Sound Source in a Reverberant Sound Field," *JASA*, 27, pp. 247–258, Mar., 1958.

Waterhouse, R.V. and R.K. Cook, "Interference Patterns in Reverberant Sound Fields II," *JASA*, 37, pp. 424–428, Mar., 1965.

Welti, T. and A. Devantier, "Low-Frequency Optimization Using Multiple Subwoofers," *JAES*, 54:5, pp. 347–364, 2006.

Chapter 26   Room Auralization

Chéenne, D.J., "Acoustical Modeling and Auralization," in *Handbook for Sound Engineers*, ed. G.M. Ballou, 4th ed., Elsevier Focal Press, 2008.

Cox, T.J. and P. D'Antonio, *Acoustic Absorbers and Diffusors: Theory, Design and Application*, Spon Press, 2004.

Kleiner, M., B.-I. Dalenbäck, and P. Svensson, "Auralization—an Overview," *JAES*, 41:11, pp. 861–875, Nov., 1993.

Dalenbäck, B.-I., M. Kleiner, and P. Svensson, "A Macroscopic View of Diffuse Reflection," *JAES*, 42:10, pp. 973–807, Oct., 1994.

Selected Absorption Coefficients

Material	125 Hz	250 Hz	500 Hz	1 kHz	2 kHz	4 kHz
Porous Type						
Drapes: cotton 14 oz/yd^2						
Draped to $^7/_8$ area	0.03	0.12	0.15	0.27	0.37	0.42
Draped to $^3/_4$ area	0.04	0.23	0.40	0.57	0.53	0.40
Draped to $^1/_2$ area	0.07	0.37	0.49	0.81	0.65	0.54
Drapes: medium velour, 14 oz/yd^2						
Draped to $^1/_2$ area	0.07	0.31	0.49	0.75	0.70	0.60
Drapes: heavy velour, 18 oz/yd^2						
Draped to $^1/_2$ area	0.14	0.35	0.55	0.72	0.70	0.65
Carpet: heavy on concrete	0.02	0.06	0.14	0.37	0.60	0.65
Carpet: heavy on 40-oz hair felt	0.08	0.24	0.57	0.69	0.71	0.73
Carpet: heavy with latex backing on foam or 40-oz hair felt	0.08	0.27	0.39	0.34	0.48	0.63
Carpet: indoor/outdoor	0.01	0.05	0.10	0.20	0.45	0.65
Acoustical tile, ave, $^1/_2$-in thick	0.07	0.21	0.66	0.75	0.62	0.49
Acoustical tile, ave, $^3/_4$-in thick	0.09	0.28	0.78	0.84	0.73	0.64
Miscellaneous Building Materials						
Concrete block, coarse	0.36	0.44	0.31	0.29	0.39	0.25
Concrete block, painted	0.10	0.05	0.06	0.07	0.09	0.08
Concrete floor	0.01	0.01	0.015	0.02	0.02	0.02
Floor: linoleum, asphalt-tile, or cork tile on concrete	0.02	0.03	0.03	0.03	0.03	0.02
Floor: wood	0.15	0.11	0.10	0.07	0.06	0.07
Glass: large panes, heavy glass	0.18	0.06	0.04	0.03	0.02	0.02
Glass, ordinary window	0.35	0.25	0.18	0.12	0.07	0.04

Material	125 Hz	250 Hz	500 Hz	1 kHz	2 kHz	4 kHz
Owens-Corning Frescor: painted, $5/8$-in thick, mounting 7	0.69	0.86	0.68	0.87	0.90	0.81
Plaster: gypsum or lime, smooth finish on tile or brick	0.013	0.015	0.02	0.03	0.04	0.05
Plaster: gypsum or lime, smooth finish on lath	0.14	0.10	0.06	0.05	0.04	0.03
Gypsum board: $1/2$-in 2 × 4 studs, 16-in centers	0.29	0.10	0.05	0.04	0.07	0.09
Resonant Absorbers						
Plywood panel: $3/8$-in thick	0.28	0.22	0.17	0.09	0.10	0.11
Polycylindrical						
Chord 45-in, height 16-in, empty	0.41	0.40	0.33	0.25	0.20	0.22
Chord 35-in, height 12-in, empty	0.37	0.35	0.32	0.28	0.22	0.22
Chord 28-in, height 10-in, empty	0.32	0.35	0.3	0.25	0.2	0.23
Chord 28-in, height 10-in, filled	0.35	0.5	0.38	0.3	0.22	0.18
Chord 20-in, height 8-in, empty	0.25	0.3	0.33	0.22	0.2	0.21
Chord 20-in, height 8-in, filled	0.3	0.42	0.35	0.23	0.19	0.2
Perforated Panel						
$5/32$-in thick, 4-in depth, 2-in glass fiber						
Perf: 0.18%	0.4	0.7	0.3	0.12	0.1	0.05
Perf: 0.79%	0.4	0.84	0.4	0.16	0.14	0.12
Perf: 1.4%	0.25	0.96	0.66	0.26	0.16	0.1
Perf: 8.7%	0.27	0.84	0.96	0.36	0.32	0.26
8-in depth, 4-in glass fiber						
Perf: 0.18%	0.8	0.58	0.27	0.14	0.12	0.1
Perf: 0.79%	0.98	0.88	0.52	0.21	0.16	0.14
Perf: 1.4%	0.78	0.98	0.68	0.27	0.16	0.12
Perf: 8.7%	0.78	0.98	0.95	0.53	0.32	0.27
With 7-in airspace with 1-in mineral fiber, 9-10 lb/ft^3 density, $1/4$-in cover						
Wideband, 25% perf or more	0.67	1.09	0.98	0.93	0.98	0.96
Midpeak, 5% perf	0.60	0.98	0.82	0.90	0.49	0.30
Low peak, 0.5% perf	0.74	0.53	0.40	0.30	0.14	0.16
With 2-in airspace filled with mineral fiber, 9-10 lb/ft^3 density						
Perf: 0.5%	0.48	0.78	0.60	0.38	0.32	0.16

Glossary

Abffusor A proprietary panel offering both absorption and diffusion of sound.

absorption In acoustics, when some energy is taken away from sound, for example, the changing of sound energy to heat.

absorption coefficient The fraction of sound energy that is absorbed at any surface. It has a theoretical value between 0 and 1 and varies with the frequency and angle of incidence of the sound.

acoustics The science of sound. It can also refer to the effect a given environment has on sound.

AES Audio Engineering Society.

algorithm Procedure for solving a mathematical problem.

ambience The distinctive acoustical characteristics of a given space.

amplifier, line An amplifier designed to operate at intermediate levels. Its output is usually on the order of one volt.

amplifier, output A power amplifier designed to drive a loudspeaker or other load.

amplitude The instantaneous magnitude of an oscillating quantity such as sound pressure. The peak amplitude is the maximum value.

amplitude distortion A distortion of the wave shape of a signal.

analog An electrical signal whose frequency and level vary continuously in direct relationship to the original electrical or acoustical signal.

anechoic Without reflection or echo.

anechoic chamber A room designed to suppress internal sound reflections. Used for acoustical measurements.

antinodic Point of maximum vibration in a vibrating body.

articulation A quantitative measure of the intelligibility of speech. For example, the percentage of speech items correctly perceived.

artificial reverberation Reverberation generated usually by algorithm to simulate the soundfield of real or imagined acoustic spaces.

ASA   Acoustical Society of America.

ASHRAE   American Society of Heating, Refrigerating, and Air-Conditioning Engineers.

attack   The beginning of a sound, the initial transient of a musical note.

attenuate   To reduce the level of an electrical or acoustical signal.

attenuator   A device, usually a variable resistance, used to control the level of an electrical signal.

audio frequency   An acoustical or electrical signal of a frequency that falls within the audible range of the human ear, usually taken as 20 Hz to 20 kHz.

audio spectrum   *See* audio frequency.

auditory area   The sensory area lying between the threshold of hearing and the threshold of feeling or pain.

auditory cortex   The region of the brain receiving nerve impulses from the ear.

auditory system   The human hearing system made up of the external ear, the middle ear, the inner ear, the nerve pathways, and the brain.

aural   Having to do with the auditory mechanism.

A-weighting   A frequency response adjustment of a sound-level meter that makes its reading conform, very roughly, to human response.

axial mode   The room resonances associated with one pair of parallel walls.

baffle   A movable barrier (also known as a gobo) used in the recording studio to achieve separation of signals from different sources. Also, the surface or board upon which a loudspeaker is mounted.

bandpass filter   A filter that attenuates signals both below and above the desired passband.

bandwidth   The frequency range passed by a given device or structure.

basilar membrane   A membrane inside the cochlea that vibrates in response to sound, exciting the hair cells.

bass   The lower range of audible frequencies.

bass boost   The increase in level of the lower range of frequencies, usually achieved by digital or analog circuits.

beats   Periodic fluctuations that are heard when sounds of slightly different frequencies are superimposed.

binaural   A recording and playback configuration emulating hearing with two ears.

bit   The elemental state of a binary number system.

boomy   Colloquial expression for excessive bass response in a recording, playback, or sound-reinforcing system.

byte   A grouping of bits in a digital system. One byte usually comprises 8 bits of data.

capacitor   An electrical component that passes alternating currents but blocks direct currents. Also called a condenser, it is capable of storing electrical energy.

clipping   An electrical signal is clipped if the signal level exceeds the capabilities of the amplifier. It is a distortion of the signal.

cochlea   The portion of the inner ear that changes the mechanical vibrations of the cochlear fluid into electrical signals. It can be modeled as the frequency-analyzing portion of the auditory system.

coloration   The distortion of an audio signal detectable by the ear.

comb filter   A distortion produced by combining an electrical or acoustical signal with a delayed replica of itself. The result is constructive and destructive interference that results in peaks and nulls introduced into the frequency response. When plotted to a linear frequency scale, the response resembles a comb, hence the name.

compression   Reducing the dynamic range of a signal by digital or analog circuits that reduce the level of loud passages.

condenser   *See* capacitor.

correlogram   A graph showing the correlation of one signal with another.

cortex   *See* auditory cortex.

crest factor   Peak value divided by rms (root-mean-square) value.

critical band   In human hearing, a narrow band over which frequency components within the band will mask a given tone or noise. Critical bandwidth varies with frequency but is usually between ⅙ and ⅓ octave.

crossover frequency   The frequency corresponding to the −3-dB point of a network that divides signal energy. For example, in a loudspeaker with multiple radiators, the frequency where the response from different radiators crosses.

crosstalk   The signal of one channel or circuit interfering with another.

cycles per second   The frequency of an electrical signal or sound wave. Measured in hertz (Hz).

dB   *See* decibel.

dB(A)   A sound-level meter reading with an A-weighting network simulating the human-ear response at a loudness level of 40 phons.

dB(B)   A sound-level meter reading with a B-weighting network simulating the human-ear response at a loudness level of 70 phons.

dB(C)   A sound-level meter reading with a C-weighting network simulating the human-ear response at a loudness level of 100 phons.

decade   Ten times any quantity or frequency range. The range of the human ear is about three decades.

decay rate   A measure of the decay of acoustical signals, expressed as a slope in decibels/second.

decibel   The bel is the logarithm of the ratio of two powers, and decibel is $^1/_{10}$ bel. The human ear responds logarithmically and it is expedient to deal in logarithmic units in audio systems.

delay line   A digital or analog device employed to delay one audio signal with respect to another.

diaphragm   Any surface that vibrates in response to sound or is vibrated to emit sound, such as in microphones and loudspeakers. Also applied to wall and floor surfaces vibrating in response to sound or in transmitting sound.

dielectric   An insulating material. The material between the plates of a capacitor.

diffraction   The distortion (bending) of a wavefront caused by the presence of an obstacle in the sound field. The angle of diffraction is a function of wavelength relative to the size of the obstacle.

diffuser   A device or physical surface causing the diffusion of sound.

diffusor   A proprietary device for the diffusion of sound through reflection phase-grating means.

digital   A numerical representation of an analog signal. Pertaining to the application of digital techniques to audio tasks.

distance double law   In pure spherical divergence of sound from a point source in free space, the sound-pressure level decreases 6 dB for each doubling of the distance. This condition is rarely encountered in practice, but can be used to estimate sound changes with distance.

distortion   Any change in the waveform or harmonic content of an original signal as it passes through a device. The result of nonlinearity within the device.

distortion, harmonic   Changing the harmonic content of a signal by passing it through a nonlinear device.

DSZ   Diffused sound zone.

dynamic range   All audio systems are limited by inherent noise at low levels and by overload distortion at high levels. The usable region between these two extremes is the dynamic range of the system. Expressed in decibels.

dyne   The force that will accelerate a 1-gram mass at the rate of 1 cm/sec. A standard reference level for sound pressure is 0.0002 dyne/cm^2, also expressed as 20 micropascals, or 20 μPa.

ear canal   The external auditory meatus; the canal between the pinna and the eardrum.

eardrum   The tympanic membrane located at the end of the ear canal that is attached to the ossicles of the middle ear.

echo   A delayed return of sound that is perceived by the ear as a discrete repetition.

echogram   A record of the very early reverberatory decay of sound in a room.

EES   Early, early sound. Structureborne sound may reach the microphone in a room before the airborne sound because sound travels faster through the denser materials.

EFC   Energy-frequency curve.

EFTC   Energy-frequency-time curve.

eigen function   A defining constant in the wave equation.

ensemble   Musicians must hear each other to function properly; in other words, ensemble must prevail. Diffusing elements surrounding the stage area contribute greatly to ensemble.

equal loudness contour   A contour representing a constant loudness for all audible frequencies. The contour having a sound-pressure level of 40 dB at 1,000 Hz is arbitrarily defined as the 40-phon contour.

equalization   The process of adjusting the frequency response of a device or system to achieve a flat or other desired response.

equalizer   A device for adjusting the frequency response of a device or system.

ETC   Energy-time curve.

ETF   AcoustiSoft loudspeaker and room acoustic analysis program.

Eustachian tube   The tube running from the middle ear into the pharynx that equalizes middle-ear and atmospheric pressure.

external meatus   The ear canal terminated by the eardrum.

feedback, acoustic   Unwanted interaction between the output and input of an acoustical system, for example, between the loudspeaker and the microphone of a system.

FFT   Fast Fourier transform. An iterative program that computes the Fourier transform in a shorter time.

fidelity   As applied to sound quality, the faithfulness to the original.

filter, band pass   A filter that passes all frequencies between a low-frequency and a high-frequency cutoff point.

filter, high pass   A filter that passes all frequencies above a cutoff frequency.

Filter, FIR   Digital filter of finite impulse response type.

Filter, IIR   Digital filter of infinite impulse response type.

filter, low pass   A filter that passes all frequencies below a cutoff frequency.

flanging   The term applied to the use of comb filters to obtain special sound effects.

flutter   A repetitive echo set up by parallel reflecting surfaces.

Fourier analysis   Application of the Fourier transform to a signal to determine its spectrum.

frequency   The measure of the rapidity of alterations of a periodic signal, expressed in hertz.

frequency response   The changes in the amplitude or sensitivity of a circuit or device with frequency.

FTC   Frequency-time curve.

fundamental   The basic pitch of a musical note.

fusion zone   All reflections arriving at the observer's ear within 20 to 40 msec of the direct sound are integrated, or fused together, with a resulting apparent increase in level. *See* also Haas effect.

gain   The increase in power level of a signal produced by an amplifier.

graphic-level recorder   A device for recording signal level in decibels versus time. The level in decibels versus angle can be recorded also for directivity patterns.

grating, diffraction   An optical grating consists of minute, parallel lines used to break light down into its component colors. The principle is also used to achieve diffraction of acoustical waves.

grating, reflection phase   An acoustical diffraction grating to produce diffusion of sound.

Haas effect   Delayed sounds are integrated by the auditory apparatus if they fall on the ear within 20 to 40 msec of the direct sound. The level of the delayed components contributes to the apparent level of the sound, and sound is localized at the first-arriving source. Also called the precedence effect. *See* also fusion zone.

hair cell   The sensory elements of the cochlea that transduce the mechanical vibrations of the basilar membrane to nerve impulses that are sent to the brain.

harmonic distortion   *See* distortion, harmonic.

harmonics   Integral multiples of the fundamental frequency. The first harmonic is the fundamental, and the second is twice the frequency of the fundamental, etc.

hearing loss   The loss of sensitivity of the auditory system, measured in dB below a standard level. Some hearing loss is age related; some is related to exposure to high-level sound.

Helmholtz resonator   A reactive, tuned, sound absorber. A bottle is such a resonator. It can employ a perforated cover or slats over a cavity.

henry   The unit of inductance.

hertz   The unit of frequency, abbreviated Hz. The same as cycles per second.

high-pass filter   *See* filter, high pass.

HRTF   Head-related transfer function. Describes how sound is changed by reflection and diffraction at the head, torso, and outer ear.

IEEE   Institute of Electrical and Electronic Engineers.

image source   A virtual reflected source located at an image point.

impedance   The opposition to the flow of electric or acoustic energy measured in ohms.

impedance matching   Maximum power is transferred from one circuit to another when the output impedance of the one is matched to the input impedance of the other. Maximum power transfer may be less important in many electronic circuits than low noise or voltage gain.

impulse   A very short, transient, electric or acoustic signal.

in phase   Two periodic waves reaching peaks and going through zero at the same instant are said to be "in phase."

inductance   An electrical characteristic of circuits, especially of coils, that introduces inertial lag because of the presence of a magnetic field. Measured in henrys.

initial time-delay gap   The time gap between the arrival of the direct sound and the first sound reflected from the surfaces of the room.

intensity   Acoustic intensity is sound energy flux per unit area. The average rate of sound energy transmitted through a unit area normal to the direction of sound transmission.

interference   The combining of two or more signals results in an interaction called interference. This may be constructive or destructive. Another use of the term is to refer to undesired signals.

intermodulation distortion   Distortion produced by the interaction of two or more signals. The distortion components are not harmonically related to the original signals.

inverse-square law   *See* spherical divergence.

ITDG   Initial time-delay gap.

JASA   Journal of the Acoustical Society of America.

JAES   Journal of the Audio Engineering Society.

kHz   1,000 Hz.

Korner Killer   A proprietary sound absorbing/diffusing unit for use in corners of rooms.

law of the first wavefront   The first wavefront falling on the ear determines the perceived direction of the sound.

LEDE   *See* live end-dead end.

LFD   Low-frequency diffusion.

level   A sound-pressure level in decibels means that it is calculated with respect to the standard reference level of 20 µPa. The term "level" associates that figure with the appropriate standard reference level.

linear   A device or circuit with a linear characteristic means that a signal passing through it is not distorted.

live end-dead end   An acoustical treatment plan for rooms in which the front end is highly absorbent and the back end is reflective and diffusive.

logarithm   An exponent of 10 in the common logarithms to the base 10. For example, 10 to the exponent 2 is 100; the log of 100 is 2.

loudness   A subjective term for the sensation of the magnitude of sound.

loudspeaker   An electroacoustical transducer that changes electrical energy to acoustical energy.

masking   The amount (or the process) by which the threshold of audibility for one sound (maskee) is raised by the presence of another (masking) sound.

maximum length sequence   *See* sequence, maximum length.

mean free path   For sound waves in an enclosure, it is the average distance traveled between successive reflections.

microphone   An acoustical-to-electrical transducer by which sound waves in air may be converted to electrical signals.

middle ear   The cavity between the eardrum and the cochlea housing the ossicles connecting the eardrum to the oval window of the cochlea.

millisecond   One-thousandth of a second, abbreviated ms or msec.

mixer   A resistive device, sometimes very elaborate, that is used for combining signals from many sources.

MLS   Maximum length sequence.

modal resonance   *See* mode.

mode   A room resonance, a standing wave. Axial modes are associated with pairs of parallel walls. Tangential modes involve four room surfaces and oblique modes all six surfaces. The effect of modes is greatest at low frequencies and for small rooms.

monaural   *See* monophonic.

monitor   Loudspeaker used in the control room of a recording studio.

monophonic   Single-channel sound.

NAB   National Association of Broadcasters.

node   Point of no vibration.

noise   Interference of an electrical or acoustical nature. Random noise is a desirable signal used in acoustical measurements. Pink noise is random noise whose spectrum falls at 3 dB/octave; it is useful for sound analyzers with constant percentage bandwidths.

noise criteria   Standard spectrum curves by which a given measured noise may be described by a single NC number.

nonlinear   A device or circuit is nonlinear if a signal passing through it is distorted.

normal mode   A room resonance. *See* mode.

null   A low or minimum point on a graph. A minimum pressure region in a room.

oblique mode   A room mode produced by reflections from all of the six surfaces of a rectangular room.

octave   The interval between two frequencies having a ratio of 2:1.

ossicles   A linkage of three tiny bones providing the mechanical coupling between the eardrum and the oval window of the cochlea consisting of the hammer, anvil, and stirrup.

out of phase   The offset in time of two related signals.

oval window   A tiny membranous window on the cochlea to which the foot plate of the stirrup ossicle is attached. The sound from the eardrum is transmitted to the fluid of the inner ear through the oval window.

overtone   A component of a complex tone having a frequency higher than the fundamental.

partial   One of a group of frequencies related to the fundamental, which appear in a complex tone. Partials may be harmonically related to the fundamental, or not.

pascal   A unit of pressure, a measure of perpendicular force per unit area, equivalent to $1 \, \text{N/m}^2$.

passive absorber   A sound absorber that dissipates sound energy as heat.

PFC   Phase-frequency curve.

phase   The time relationship between two signals.

phase shift   The time or angular difference between two signals.

phon   The unit of loudness level of a tone, related to the ear's subjective impression of signal strength.

pink noise   A noise signal whose spectrum level decreases at a 3-dB/octave rate. This gives the noise equal energy per octave.

pinna   The exterior ear.

pitch   A subjective term for the perceived frequency of a tone.

place effect   The theory that pitch perception is related to the pattern of excitation on the basilar membrane of the cochlea.

plenum   A building cavity that is absorbent-lined to reduce noise, through which conditioned air is routed.

polar pattern   A graph showing the 360° directional characteristics of a microphone or loudspeaker.

polarity   The relative position of the high (+) and the low (–) signal leads in an audio system.

PRD   Primitive root diffusor.

preamplifier   An amplifier designed to optimize the amplification of weak signals, such as from a microphone.

precedence effect   *See* Haas effect.

pressure zone   As sound waves strike a solid surface, the particle velocity is zero at the surface and the pressure is high, thus creating a high-pressure layer near the surface.

primitive root sequence   *See* sequence, primitive root.

psychoacoustics   The study of the ear-brain auditory system and its perception and reaction to auditory stimuli.

pure tone   A tone with no harmonics, a sine wave. All energy is concentrated at a single frequency.

Q-factor   Quality-factor. A measure of the losses in a resonance system. The sharper the tuning curve, the higher the Q factor.

QRD   Quadratic-residue diffusor.

quadratic residue sequence   *See* sequence, quadratic residue.

random noise   A noise signal, commonly used in measurements, which has constantly shifting amplitude, phase, and frequency and a uniform spectral distribution of energy.

ray   At higher audio frequencies, sound may be considered to travel in straight lines as a beam in a direction normal to the wavefront.

reactance   The opposition to the flow of electricity posed by capacitors and inductors.

reactive absorber   A sound absorber, such as the Helmholtz resonator, which involves the effects of mass and compliance as well as resistance.

reactive silencer   A silencer in air-conditioning systems that uses reflection effects for its action.

reflection   For surfaces large compared to the wavelength of impinging sound, sound is reflected much as light is reflected, with the angle of incidence equaling the angle of reflection.

refraction   The bending of sound waves traveling through layered media with different sound velocities.

resistance   That quality of electrical or acoustical circuits that results in dissipation of energy through heat.

resonance   A resonant system vibrates at maximum amplitude when tuned to its natural frequency.

resonator silencer   An air-conditioning silencer employing tuned stubs and their resonating effect for its action.

response   *See* frequency response.

reverberation   The tailing off of sound in an enclosure because of multiple reflections from the boundaries.

reverberation chamber   A room with reflective boundaries used for measuring sound absorption coefficients.

reverberation time   The time required for the sound in an enclosure to decay 60 dB from an initial steady-state level.

RFZ   Reflection-free zone.

ringing   High-Q electrical circuits and acoustical devices have a tendency to oscillate (or ring) when excited by a suddenly applied signal.

room mode   The normal modes of vibration of an enclosed space. *See* mode.

round window   The tiny membrane of the cochlea that opens into the middle ear that serves as a pressure release for the cochlear fluid.

RPG   Reflection-phase grating. A diffusor of sound energy using the principle of the diffraction grating.

RPG cloud   Reflection-phase gratings arranged for overhead mounting.

RT$_{60}$ *See* reverberation time.

sabin   The unit of sound absorption. One square foot of open window has an absorption of 1 sabin.

Sabine   W.C. Sabine, originator of the Sabine reverberation formula.

SBIR   Speaker boundary interference response.

Schroeder plot   A reverberation decay computed by the mathematical process defined by Manfred Schroeder.

semicircular canals   The three sensory organs for balance that are a part of the cochlear structure.

sequence, maximum length   A measurement method using a pseudo-random binary sequence to excite a system and/or room with a test signal resembling wideband white noise. Also a mathematical sequence used in determining the well depth of diffusers.

sequence, primitive root   A mathematical sequence used in determining the well depth of diffusers.

sequence, quadratic residue   A mathematical sequence used in determining the well depth of diffusers.

signal-to-noise ratio   The difference between the nominal or maximum operating level and the noise floor in decibels, (S/N).

sine wave   A pure wave, periodic wave related to simple harmonic motion.

slap back   A discrete reflection from a nearby surface.

sone   The unit of measurement for subjective loudness.

sound absorption coefficient   The practical unit between 0 and 1 expressing the absorbing efficiency of a material. It is determined experimentally.

sound-power level   A power expressed in decibels above the standard reference level of 1 pW.

sound-pressure level   A sound pressure expressed in dB above the standard sound pressure of 20 μPa.

sound spectrograph   An instrument that displays the time, level, and frequency of a signal.

spectrum   The distribution of the energy of a signal with frequency.

spectrum analyzer   An instrument for measuring and recording the spectrum of a signal.

specularity   A term expressing the efficiency of diffraction-grating types of diffusors.

spherical divergence   Sound diverges spherically from a point source in free space.

splaying   Walls are splayed when they are constructed "off square," that is, a few degrees from the normal rectilinear form.

standing wave   A resonance condition in an enclosed space in which sound waves traveling in one direction interact with those traveling in the opposite direction, resulting in a stable condition. *See* mode.

steady state   A continuous condition devoid of transient effects.

stereo   A sterophonic system with two channels.

subwoofer   A low-frequency loudspeaker typically in conjunction with other high-frequency satellite loudspeakers.

superposition   Many sound waves may traverse the same point in space, the air molecules responding to the vector sum of the demands of the different waves.

T_{60} *See* RT_{60}.

tangential mode   A room mode produced by reflections from four of the six surfaces of a rectangular room.

TDS   *See* time-delay spectrometry.

TEF   See time, energy, frequency.

threshold of feeling (pain)   The sound-pressure level that makes the ears tickle, located about 120 dB above the threshold of hearing.

threshold of hearing   The lowest level sound that can be perceived by the human auditory system. This is close to the standard reference level of sound pressure, 20 µPa.

timbre   The quality of a sound related to its harmonic structure.

time-delay spectrometry   A measurement method for obtaining anechoic results in echoic spaces.

Time, energy, frequency   An analysis method employing TDS.

tone   A tone results in an auditory sensation of pitch.

tone burst   A short signal used in acoustical measurements to make possible differentiating desired signals from spurious reflections.

tone control   An electrical circuit to allow adjustment of frequency response.

transducer   A device for changing electrical signals to acoustical or vice versa, such as a microphone or loudspeaker.

transient   A short-lived aspect of a signal, such as the attack and decay of musical tones.

treble   The higher frequencies of the audible spectrum.

Tube Traps   Proprietary sound-absorbing units.

volume   Colloquial equivalent of sound level.

watt   The unit of electrical or acoustical power.

wave   A regular variation of an electrical signal or acoustical pressure.

wavelength   The distance a sound wave travels in the time it takes to complete one cycle.

weighting   Adjustment of sound-level meter response to achieve a desired measurement.

white noise   Random noise having uniform distribution of energy with frequency.

woofer   A low-frequency loudspeaker used with the main loudspeaker.

Index

Z